51.94

LIQUEFIED PETROLEUM GASES HANDBOOK

LIQUEFIED PETROLEUM GASES HANDBOOK

Based on the 1989 Edition of NFPA 58,
*Standard for the Storage and Handling of
Liquefied Petroleum Gases*

Theodore C. Lemoff, P.E., Editor

Second Edition

National Fire Protection Association
Quincy, Massachusetts

This *Handbook* has not been processed in accordance with NFPA Regulations Governing Committee Projects. Therefore, the commentary in it shall not be considered the official position of NFPA or any of its committees and shall not be considered to be, nor relied upon as, a Formal Interpretation of the meaning or intent of any specific provision or provisions of NFPA 58, *Standard for the Storage and Handling of Liquefied Petroleum Gases.*

Project Manager: Jennifer Evans
Project Editor: Paula Eissmann
Composition: Kathleen Barber
Illustrations: George Nichols
Cover Design: Toby Wolk
Production: Donald McGonagle

NFPA 58HB89
ISBN: 0-87765-360-7
Library of Congress No.: 89-60576
Printed in U.S.A.

First Printing, April 1989

CONTENTS

FIGURE CREDITS

The following institutions and individuals have supplied artwork for the commentary of this *Handbook*.

Union Carbide Corporation, Linde Division: History Photo.

Manchester Tank: Figures 2.1, 2.2, 2.3, 3.20(c).

National Propane Gas Association: Tables 2.1 and 2.2; Figures 2.4 (illustration), 2.7, 2.11, 2.13(a), 2.14, 2.15(d), 2.16(a), 2.16(d), 2.17(a), 2.17(b), 2.17(c), 2.18(b) (bottom), 2.19, 2.20(b), 2.20(c), 2.22(b), 2.26(a), 2.26(b), 2.26(c), 2.26(d), 2.27(a), 3.4(a), 3.5, 3.6, 3.7(a), 3.7(b), 3.8, 3.9, 3.10(a-g), 3.11(b), 3.12, 3.13, 3.14, 3.15, 3.16, 3.17, 3.20(a), 3.20(b), 3.21, 3.22, 4.1, 4.2, 4.3, 4.5, 4.7, 4.11, 6.2, 6.4, 6.5, 6.6 (left), 6.8, 6.9 (bottom), 6.10, C.1.

American Welding and Tank Company: Figure 2.5.

Fisher Controls Company: Figures 2.6, 2.9, 2.10(c), 2.12 (right), 2.13(b), 2.15(b), 2.15(c), 2.16(b), 2.16(c), 2.18(a) (bottom), 2.18(b) (top), 2.27(b), 3.11(a), 6.6 (right).

Trinity Industries, Inc.: Figure 2.8.

RegO Company: Figures 2.10(a), 2.10(b), 2.10(d), 2.12 (left), 2.15(a), 2.18(a) (top), 3.10(h).

Dover Corporation, Blackmer Pump Division: Figures 2.20(a), 2.21(a).

Smith Precision Products Co.: Figure 2.20(d).

Corken International Corporation: Figure 2.21(b).

Pete Freeman: Figure 2.22(a).

International Gas Systems: Figures 2.23, 3.24, 3.28.

H. Emerson Thomas: Figures 2.24, 3.4(b), 3.4(c), 3.25, 3.26, 3.29, 4.8.

Liquid Controls, Inc.: Figure 2.25.

Wilbur L. Walls: Figures 2.4 (labels), 3.2, 3.3.

Pioneer/Eclipse Corp.: Figure 3.23.

Eastern Propane: Figures 3.19(a), 3.19(b), 6.1.

Alternate Energy Installations, Inc.: Figures 3.27, 4.9.

Suburban Gas, Inc.: Figures 3.5(a).

Schultz Gas Service, Inc.: Figures 4.4, 4.6, 6.3, 6.7, 6.9 (top).

Ferrellgas: Figure 4.10.

H & H Equipment Co., Inc.: Figure 5.1.

PREFACE TO THE FIRST EDITION

Since liquefied petroleum gas (LP-Gas, LPG) was introduced in North America as a commercial fuel gas energy source shortly before World War I, the hazards associated with its use have proven confusing to many — especially to consumers, insurers, and public safety regulators not familiar with its technology.

Unlike manufactured or natural gas, which is stored, transported, and consumed in the gaseous state,[1] and fuel oil or gasoline, which is stored and transported up to the point of burning in the liquid state, LP-Gas is stored, transported, and consumed in both the gaseous and liquid states. Over the entire range of temperatures encountered on the inhabited surface of the earth, the most common LP-Gas (propane) can exist in the liquid state only if it is under a pressure exceeding that of the earth's atmosphere. Generally, other LP-Gases, e.g., butane, are used only where climatic temperatures result in pressure existing in their containers.

The pressure in an LP-Gas container is continually changing with the changes in temperature of the liquid. The temperature fluctuates with changes in air temperature and solar radiation, and the rate at which vapor is withdrawn from the container. However, the pressure is normally quite substantial; often over 100 psi (0.7 MPa) in the case of propane. Gasoline and fuel oil are seldom under pressure. Manufactured and natural gas are under pressure but, in the consumers' environment, this pressure is usually quite low.[2]

When manufactured or natural gas escapes from storage, transportation, or consuming equipment, it behaves solely as a gas. When fuel oil or gasoline escapes, it behaves solely as a liquid. When propane escapes, it will ultimately behave as a gas. However, if it escapes in liquid form there will be a brief period of time during which it will be rapidly vaporizing to the gaseous state. Such circumstances result in a hazard situation different from that existing with natural gas, fuel oil, or gasoline.

The properties of LP-Gas require that a high degree of technical sophistication be applied in the design and use of the equipment in which LP-Gas is stored and handled. Additionally, fire prevention and protection safeguards must be well understood. As an energy source which can be packaged in small, portable containers, LP-Gas is continually being used in new ways, and as additional uses by the general public are developed, they must be covered in the standards.

From their inception, the NFPA Technical Committees responsible for NFPA 58, *Standard for the Storage and Handling of Liquefied Petroleum Gases,* have relied a great deal upon the technical expertise of the LP-Gas industry. The technical content of the standard, therefore, essentially reflects the fact that the industry

---

[1]Natural gas is occasionally stored in the liquid state but not on the premises of the ultimate consumer.

[2]The use of natural gas as a vehicle fuel is a recent exception. However, such an application [in which the vehicle containers are under pressures of 2000 - 3000 psi (13.7 - 20.6 MPa)] is not yet found in privately owned personal vehicles.

believes it necessary to provide the needed degree of safety. This *Handbook* inevitably reflects the backgrounds and views of its contributors, therefore the commentary on the provisions in NFPA 58 is quite technical and written largely in the language of the LP-Gas industry. Furthermore, as a standard written for use as a public safety regulatory document, the extent to which commentary can explain why a provision is needed or just how it can be accomplished is restricted.

Considering the above, and the fact that NFPA 58 has been frequently revised over the nearly 60 years of its existence, it is easy to see why many users have difficulty in understanding and applying its provisions; even members of the LP-Gas industry itself are not isolated from this problem. Contributing to this problem is the fact that the 1983 edition of NFPA 58 was the first to be adopted using the Technical Committee Report (TCR)/Technical Committee Documentation (TCD) procedure, wherein the reasons for amendments are explained in a formal manner. Prior to 1983, reasons for changes to the content are found only in the agendas and minutes of Committee meetings and as the Committee has had a formal secretary only since 1954, this record is far from complete.

Wilbur L. Walls, P.E.
Editor, First Edition

PREFACE TO THE SECOND EDITION

The purpose of this second edition is to provide additional information on the provisions in the standard in order to aid the reader in understanding and applying the provisions. The *Handbook* commentary will assist the first-time user of the standard by providing helpful background information and will aid the experienced user in understanding the Technical Committee's intent in making the changes in this 1989 edition. Also, the new commentary clarifies some provisions that have been frequently misinterpreted in previous years.

A handbook of this type is never complete. In order to stay current it must change as technology changes. Users may find that problems they encounter may not be addressed in the commentary, or that the commentary does not cover certain aspects of a real problem. The editor invites suggestions from any reader for improvements in the next edition.

Theodore C. Lemoff, P.E.
Editor, Second Edition

NFPA standards concerned with gases date from 1900—only four years after the establishment of NFPA itself. The first standards were concerned with acetylene (actually used as a household cooking, lighting, and heating fuel in those days) and manufactured ("city") gas derived from coal and oil.

An Early Magazine Advertisement for Acetylene Lighting.

These standards were developed by the NFPA Technical Committee on Gases. By 1924, the use of liquefied petroleum gas (LP-Gas), primarily as a household cooking and heating fuel in rural areas, had become common and the need for a national firesafety standard was recognized. These systems used compressed gas cylinders for storage of the LP-Gas. Because of the cylindrical shape, LP-Gas was widely referred to as simply "bottled gas" — an identification rather lacking in specificity (acetylene was also a "bottled gas") but one still used today.

In 1927, the Committee on Gases secured NFPA approval of the first NFPA standard on LP-Gas, a four-page document whose title, "Regulations for the Installation and Operation of Compressed Gas Systems Other than Acetylene for Lighting and Heating," comprised a fair proportion of the text of the standard itself. The standard covered only systems in which the LP-Gas was stored in cylinders (bottles) fabricated to regulations of the U.S. Interstate Commerce Commission (ICC) (today, the U.S. Department of Transportation).

In those days, the NFPA published its standards only in the Proceedings of the Annual Meeting (in May of each year) at which the standards were adopted. To make them more available to users, the National Board of Fire Underwriters (NBFU) (now American Insurance Association) obtained NFPA permission to publish many of them in pamphlet form. These were identified as standards of the NBFU "as recommended by the NFPA." Also, in those days, NFPA did not identify many of its standards by a number as well as a title. It was, however, the practice of the NBFU to do so and the standard was thus designated NBFU 52. Amended editions of the standard were adopted in 1928, 1933, and 1937.

Because the LP-Gas "bottles" had to be refilled — in those days only at a plant built for that purpsoe — which required the use of a much larger container for plant storage, new specifications had to be developed. Also, the major consumers found a container larger than an ICC cylinder to be advantageous. These containers required different design and siting criteria.

In 1931, NFPA tentatively adopted "Regulations for the Design, Installation and Construction of Containers and Pertinent Equipment for the Storage and Handling of Liquefied Petroleum Gases" and, at the next NFPA meeting in 1932, this standard was officially adopted. This standard covered the larger containers (the ASME containers of today). This standard was also published by the NBFU under the designation NBFU 58. It had 14 pages; ten more than NBFU 52.

In 1932 (and for many years before and after), the Committee on Gases was comprised of between 11 and 20 members and chaired by Harry E. Newell of the NBFU. In 1932 there were 15 members, representing the Associated Factory Mutual Fire Insurance Companies, American Gas Association, Board of Fire Underwriters of Alleghany County, Western Factory Insurance Association, Conference of Special Risk Underwriters, Manufacturing Chemists Association, Boston Board of Fire Underwriters, International Acetylene Association, Underwriters Laboratories Inc., Compressed Gas Manufacturers Association, American Petroleum Institute, Railway Fire Protection Association, and the U.S. Bureau of Standards, in addition to Chairman Newell who represented NBFU. With six of the 15 members (including the chairman) representing the insurance industry, the standard could be presumed to be rather conservative and this lack of balance of interests could not exist under later NFPA procedures.

Among these 15 members was H. Emerson Thomas, employed by the Phillips Petroleum Company and representative of the American Petroleum Institute. Mr. Thomas has served as a member of the Committee on Gases and its successors continuously ever since — a tenure unchallenged in NFPA history and for which he was honored by NFPA in 1982. He is recognized as a contributor to the first edition of this *Handbook*.

At the 1931 NFPA meeting, Chairman Newell noted in his report that the proposed standard had been prepared by a joint subcommittee of the Committee on Gases and the Committee on Flammable Liquids. He attributed this to the fact that "it was difficult to decide whether liquefied petroleum gases were flammable liquids or gases." Even today, confusion on this score exists. Many municipal ordinances (undoubtedly of considerable vintage) equate them and many communities have not adopted LP-Gas regulations under the erroneous impression that their flammable liquid regulations apply to LP-Gas.

The extreme versatility of "gas in a bottle" was now leading to more and more complex uses and the Committee on Gases was hard-pressed to keep up. Amended editions of this standard were adopted in 1934, 1937, 1938, and 1939.

In the early 1930s, the use of LP-Gas cargo vehicles become widespread. In 1935, a standard on these vehicles — called "tank trucks" at the time — was adopted and published by NBFU as NBFU 59.

By the middle 1930s, the use of LP-Gas as a fuel to power vehicles was prevalent enough to warrant a standard, which was adopted in 1937. However, in 1938, this became a part of the ASME Code container standard (NBFU 58).

By now it was apparent that the various standards contained considerable duplicate material and could be combined into a single standard. This was done in 1940. The 47-page standard (in the NBFU 58 pamphlet version) combined the 1937 edition of 52 and the 1939 editions of 58 and 59. The edition adopted in 1943 was the first edition to be designated NFPA 58.

The first pamphlet edition of NFPA 58 was published in 1950.

The dates (by year) of each edition of NFPA 58 are given in the "Origin and Development" section of each edition. To date, there have been 25 editions since the 1940 edition — an average of about one every two years. This is a very high rate and placed an impossible burden upon its use as a public safety regulatory instrument. From 1950 to 1961, in fact, a new edition was adopted each year. Since 1961, new editions have been adopted about every three years and this seems to be working well.

As early as the mid-1930s, regulatory and insurance interests viewed the standards as unnecessarily detailed for their purposes. This resulted in a *Liquefied Petroleum Gas Code,* adopted in 1937, which consisted of container siting and basic fabrication criteria. However, this proved to be inadequate from the industry's point of view and the Code was withdrawn when the 1940 combined standard was adopted.

The Committee on Gases itself developed NFPA 58 until 1956. By this time, the number and variety of NFPA standards covering various gases had grown to the point where the size of the committee was becoming difficult to manage. The solution was to establish a number of smaller committees (known as Sectional Committees) comprised of experts on the different gas applications. The developmental and interpretive responsibilities for NFPA 58 was thus assigned to the Sectional Committee On Liquefied Petroleum Gases. However, the Sectional Committee could only recommend amendments, Formal Interpretations, or Tentative Interim Amendments for adoption by the Committee on Gases.

Because of this lack of authority of the Sectional Committee, the membership of the Committee on Gases became essentially duplicative of the Sectional Committee and the purpose of the Sectional Committee was defeated. It was also becoming evident that the Committee on Gases seldom overturned a Sectional Committee recommendation.

In 1966, the Sectional Committees (four at the time) were organized into three full-fledged Technical Committees, and the Committee on Gases ceased to exist. NFPA 58 thus became the complete responsibility of only the Technical Committee on Liquefied Petroleum Gases.

From 1940 to and including the 1969 edition, NFPA 58 consisted of a chapter entitled "Basic Rules" followed by a number of chapters (entitled "Divisions") and appendices. For example, the 1969 edition was as follows:

Introduction
Basic Rules
Division I   Cylinder Systems
Division II   Systems Utilizing Containers Other than ICC
Division III   Truck Transportation of Liquefied Petroleum Gas
Division IV   Liquefied Petroleum Gas as a Motor Fuel
Division V   Storage of Containers Awaiting Use or Resale
Division VI   LP-Gas Exchangeable (Readily Portable Container) System Installations on Travel Trailers, Camper Trailers, Self-Propelled Campers, or Mobile Homes
Division VII   LP-Gas System Installations on Commercial Vehicles and Certain Self-Propelled or Trailer Type Mobile Living Units
Division VIII   Liquefied Petroleum Gas Service Stations
Appendices A-G   Covering Design of Relief Valves, Container Filling Volumes, a Pictorial Presentation of Container Siting, and Tubing Wall Thicknesses

The concept was that the introduction and "Basic Rules" would include all material common to all applications, e.g., scope statements, retroactivity, definitions, odorization, approval requirements, container fabrication, siting, design criteria for piping, valves, hoses and fittings, filling levels, and ignition source controls. The eight divisions amplified and sometimes modified the basic provisions for specific applications—some on the basis of kinds of containers (Divisions I, II, and III) and the others on the basis of the environment the system was used in.

Essentially, as a new application was developed it was treated as a package and added as a new divison. Not only was it easier to do it this way, but as a new application was always of great current interest to all users of the standard, there was corresponding pressure for timely inclusions in the standard. The problem with this approach was that very similar hazards were being treated differently and the basic rules were becoming less and less basic.

In the late 1960s, the Technology and Standards Committee of the National LP-Gas Association (now National Propane Gas Association) undertook an assignment to develop a format that could provide consistency in hazard evaluation, reduce the extent to which particular interests would have to study the entire standard, and would make the format more like that used in other NFPA standards.

The work of Mr. William D. Cook was the prime force in this reorganization, along with significant input from a review committee comprised of the three contributors to the first edition of this *Handbook*, Wilbur L. Walls, Walter H. Johnson, and H. Emerson Thomas.

The concepts in the current format are as follows:

1. Only material truly relevant to all applications — e.g., "basic" — is included in Chapter 1, "General Provisions".

2. Experience had shown that a fundamental breakdown of interest existed between manufacturers of equipment and those who assembled this equipment into systems and actually installed them. Fire experience had clearly revealed that the causes of incidents were nearly always due to failure to comply with one or more provisions and that many of these were due to confusion between manufacturers and installers as to who should do what. To help reduce this confusion, Chapter 2 is aimed at the manufacturer and Chapter 3 at the installer.

3. The same experience had also shown that the great majority of non-appliance related accidents occurred during operations where LP-Gas liquid was being transferred from one container to another. An essential aspect of such operation is the presence and performance of a person or persons conducting the transfer operation: an accident involves human behavior as well as equipment factors.

To define better the operational factors and their relation to equipment factors, these have been put together into Chapter 4. As the LP-Gas service station operation is essentially a liquid transfer function, former Divison VIII was largely absorbed into Chapter 4.

4. Chapter 5 ("Storage of Portable Containers Awaiting Use or Resale") and Chapter 6 ("Vehicular Transportation of LP-Gas") retain the old format (Divisions V and III).

5. Chapter 7 was essentially new in the 1972 edition. Earlier editions had not considered the role of room or building structural behavior in an explosion.

6. Chapter 8 was added in the 1989 edition to provide requirements for the new coverage of refrigerated storage containers. It was drawn largely from NFPA 59 and 59A.

7. The new appendix material contained a great deal of explanatory text for the first time — all designed to make it easier for the user of the standard to understand some of its major provisions. This is especially true of Appendices A, B, and C, which were the work of Mr. Cook.

While the 1972 edition went a long way towards bringing NFPA 58 into a format in line with other NFPA standards, it still did not properly segregate mandatory and non-mandatory provisions, and used a different paragraph numbering system. These differences were corrected in the 1983 edition.

Experience with the 1972 format — and subsequent editions — indicates that it has improved comprehension and application. However, it is more difficult to review existing provisions to see if a new application is in fact already covered than it is to start from scratch. Furthermore, the pressure is always great to include all provisions applicable to a new and "hot" application in one place. This temptation must be resisted lest the standard evolve into the conflicting and cumbersome document it once was.

The administration of the committees responsible for NFPA 58 has been remarkedly stable over its history. Mr. Newell remained chairman until 1956 — a span of over 32 years — until his retirement from NBFU. As an indication of the respect the Committee and NFPA held for him, the position of honorary chairman was created for him — the only time this position has existed. He served in this capacity until 1958.

Mr. Newell was succeeded as chairman of the Committee on Gases in 1956 by Franklin R. Fetherston. Initially representing the Liquefied Petroleum Gas Association (now the National Propane Gas Association), he represented the Compressed Gas Association until he retired in 1966. During Chairman Fetherston's tenure, the previously mentioned Sectional Committee on Liquefied Petroleum Gases was chaired by Harold L. DeCamp of the Fire Insurance Rating Organization of New Jersey (1956–1964) and Myron Snell of the Hartford Accident and Indemnity Company (1964–1966).

Hugh V. Keepers of the Fire Prevention and Engineering Bureau of Texas became the first chairman of the present Committee in 1966 and served for ten years until his retirement.

Connor L. Adams, a building official with the city of Miami, Florida, took over the reins in 1976. During his chairmanship, the Miami building and fire departments were combined, giving him a unique base of operations.

Chairman Newell's dedication was even more remarkable in that he also performed the chores of committee secretary—albeit in an anonymous manner. It wasn't until 1954 that NFPA was able to retain a staff person and assign him as committee secretary. This person was Clark F. Jones until his untimely death in 1962. Mr. Jones was succeeded by Wilbur L. Walls in September, 1962. Mr. Walls was the committee secretary until his retirement from NFPA in 1984. In May 1985, Theodore C. Lemoff became the NFPA staff secretary.

ABOUT THE EDITOR

After graduation from the City College of New York with a Bachelor of Engineering (Chemical) degree, Theodore C. Lemoff was employed in the chemical and petrochemical industry by the Procter and Gamble Company, Sun Chemical Corporation, and Badger Engineers, Inc. Since 1985, when he joined NFPA, he has served as the secretary to the Technical Committee on Liquefied Petroleum Gases, and as staff liaison to the other NFPA technical committees that are responsible for standards on flammable gases.

He represents NFPA on the ANSI Code for Pressure Piping (B31), Safety in Cutting and Welding (Z49), and Gas Utilization Equipment (Z21 and Z83) projects, and on the National Propane Gas Association Technology and Standards and Safety Committees.

Mr. Lemoff is a registered Professional Engineer in the Commonwealth of Massachusetts and is a member of the American Institute of Chemical Engineers.

ACKNOWLEDGMENTS

A book that compiles the history, intent, and application of an NFPA standard such as the *Liquefied Petroleum Gases Handbook* does, is the sum of the work of many contributors. It is a collection of the knowledge and opinions of many individuals, some of whom are members of the Technical Committee on Liquefied Petroleum Gases and all of whom work with LP-Gas on a daily basis. The work of the three contributors to the first edition, Wilbur L. Walls, editor, Walter H. Johnson, and H. Emerson Thomas, laid the firm groundwork on which this second edition is built.

In preparing the second edition it was recognized that the commentary on vaporizers, mixers, and other "industrial" LP-Gas equipment could be strengthened and Verle H. Brown agreed to supply commentary on this topic. Mr. Brown graduated from the University of Utah with a degree in Mechanical Engineering. His professional career has included work for a manufacturer of LP-Gas containers and equipment and consulting work. Since 1964 he has been the president of International Gas Systems, a manufacturer of gas processing equipment and electronic controls. He has been a member of the National Propane Gas Association's Technology and Standards Committee for many years and contributes to proposals that committee has presented to NFPA 58 in the industrial gas equipment area.

I also acknowledge the assistance of the many others who provided their expertise in the development of the second edition, especially Wilbur L. Walls who reviewed the manuscript and made many contributions.

A handbook is made usable with the addition of photographs and drawings and I wish to acknowledge the assistance of those who contributed the artwork listed under "Figure Credits," and to especially thank Mr. Mark Hood of Fisher Controls for his extensive contribution.

In addition to the above I wish to acknowledge my wife, Sharon, for her patience and understanding during the time I worked on this *Handbook*.

Theodore C. Lemoff, P.E.
Editor

DEDICATION

This *Handbook* is respectfully dedicated to Messrs. Wilbur L. Walls, Walter H. Johnson, and H. Emerson Thomas whose efforts led to the first edition of this handbook.

NOTE: The text and illustrations that make up the commentary on the various sections of the *Standard for the Storage and Handling of Liquefied Petroleum Gases* are printed in color and in a column narrower than the standard text. In the commentary, all photographs are printed in black for clarity. They can be identified as commentary by their figure captions, which are printed in color. The text of the standard itself is printed in black.

The Formal Interpretations included in this *Handbook* were issued as a result of questions raised on specific editions of the standard. They apply to all previous and subsequent editions in which the text remains substantially unchanged. Formal Interpretations are not part of the standard and therefore are printed in color to the full column width.

Information on referenced publications can be found in Chapter 9 and Appendix J.

NFPA 58

Standard for the Storage and Handling of

Liquefied Petroleum Gases

1989 Edition

Metric equivalents in this standard are approximate and shall not be used to lessen any provision.

NOTICE: An asterisk (*) following the number or letter designating a paragraph indicates explanatory material on that paragraph in Appendix A.

Appendix A was added in the 1989 Edition in conformance with the NFPA *Manual of Style*. Material formerly in footnotes was relocated to Appendix A.

Information on referenced publications can be found in Chapter 9 and Appendix J.

1

GENERAL PROVISIONS

1-1 Introduction.

1-1.1 General Properties of LP-Gas.

1-1.1.1 LP-Gases, as defined in this standard (*see 1-2.1*), are gases at normal room temperatures and atmospheric pressure. They liquefy under moderate pressure, readily vaporizing upon release of this pressure. It is this property which permits transporting and storing them in concentrated liquid form, while normally using them in vapor form. The potential fire hazard of LP-Gas vapor is comparable to that of natural or manufactured gas, except that LP-Gas vapors are heavier than air. The ranges of flammability are considerably narrower and lower than those of natural or manufactured gas. For example, the lower flammable limits of the more commonly used LP-Gases are: propane, 2.15 percent; butane, 1.55 percent. These figures represent volumetric percentages of gas in gas-air mixtures.

The sources of LP-Gases are natural gas and crude oil.

In a natural gas processing plant or a refinery, LP-Gases are removed from the base natural gas or crude oil and liquefied. They are then separated one from the other for particular uses. Impurities such as sulphur and water are removed.

LP-Gases, by nature, can be easily liquefied at moderate pressure for ease in storage and transportation. For instance, 1 cu ft (.028 m³) of liquid will expand to 270 cu ft (7.6 m³) of vapor at atmospheric pressure. The major uses of propane are as a domestic and commercial fuel for cooking, water heating, and building heating. Propane also has numerous uses on farms; e.g., power for irrigation pumping, crop drying, poultry incubating and brooding, fuel for tractors, and many other uses. It is also used in industry for heat treating, cutting and welding, building heating, as fuel for industrial lift trucks, etc. In addition, propane is widely used as a standby fuel in the case of natural gas shortages and interruption of natural gas supply.

In utility gas plants, propane is used for enrichment, for standby or, as a propane-air mixture, for the entire gas supply. Propane and butane are also used as feed stocks for chemical processing. Butane's uses are similar to those of propane but it is used to a much lesser degree. The use of butane is growing in consumer appliances such as cigarette lighters, hair curling

3

irons, and irons, as well as a replacement for freon in aerosol spray cans and as a foaming agent in foaming plastic products such as meat trays. It is also used in refinery gasoline to alter gasoline's volatility. In some operations a mixture of propane and butane is used.

1-1.1.2   The boiling point of pure normal butane is 31°F (-0.6°C); of pure propane, -44°F (-42°C). Both products are liquids at atmospheric pressure at temperatures lower than their boiling points. Vaporization is rapid at temperatures above the boiling point, thus liquid propane normally does not present a flammable liquid hazard. For additional information on these and other properties of the principal LP-Gases, see Appendix B.

As noted in Appendix B, commercial butane and propane contain other gaseous hydrocarbons (and, therefore, are not "pure") and their boiling points are usually somewhat lower than those of pure butane and propane. The commercial forms also tend to vaporize even more rapidly than do the pure products. These commercial grades of butane and propane are not unique mixtures, but will vary with their source, seasonally, and with the changing market value of their components.

This tendency to vaporize rapidly when they escape from containment is a primary difference in hazards and, consequently, in firesafety standards between LP-Gases and flammable liquids such as gasoline, alcohol, paint thinners, etc. As noted earlier, lack of appreciation of this vital difference has caused public safety enforcement problems.

1-1.2 Federal Regulations.

1-1.2.1   Regulations of the U.S. Department of Transportation (DOT) are referenced throughout this standard. Prior to April 1, 1967, these regulations were promulgated by the Interstate Commerce Commission (ICC).

A number of provisions of NFPA 58 rely upon certain regulations of DOT since their origin and continued development is traced to this agency. At the turn of the century ICC was charged with the duty of promulgating regulations for the safe transportation of hazardous commodities by rail — including shipping container specifications — and utilized the services of the Bureau of Explosives of the American Association of Railroads in this activity. These regulations were subsequently extended to other transportation modes. DOT was established in 1967 and took over this function of ICC. The Hazardous Materials Regulations issued by DOT and published in the *Code of Federal Regulations*, Title 49, Parts 100-199, are the most relevant to NFPA 58.

LP-Gas containers covered by these regulations are cylinders, portable tanks, cargo tanks, tank trucks, and tank cars. The use of LP-Gas cylinders as storage containers or fuel supply containers is recognized in the development of requirements for their use as shipping containers. DOT requirements for containers include specifications for construction, testing, marking, equipping with stipulated pressure relief devices, periodic requalifications, repair, method of filling, and maximum filling limits. In addition to regulations covering cylinders, portable tanks, tank and cylinder trucks, and tank cars, there are also regulations covering pipeline safety (Office of Pipeline Safety Operations) and motor carrier safety (Bureau of Motor Carrier Safety).

1-2 Scope.

1-2.1 Liquefied Petroleum Gas.

1-2.1.1 As used in this standard, the terms "liquefied petroleum gas(es)," "LP-Gas" and "LPG" are synonymous and shall mean and include any material having a vapor pressure not exceeding that allowed for commercial propane composed predominantly of the following hydrocarbons, either by themselves or as mixtures: propane, propylene, butane (normal butane or isobutane), and butylene (including isomers).

The maximum allowable vapor pressure of commercial propane is 208 psig at 100°F (1 462 kPa gauge at 38°C) as specified in ASTM D 1835, *Standard Specification for Liquefied Petroleum (LP) Gases*, and the Gas Processors Association Publication 2140, *Liquefied Petroleum Gas Specifications for Test Methods*. Propylene, one of the hydrocarbon components of LP-Gas listed in 1-2.1.1, has a vapor pressure of 225 psig at 100°F (1 582 kPa gauge at 38°C). Since design of equipment used in LP-Gas systems is based upon the maximum allowable vapor pressure of commercial propane, this criteria limits the amount of propylene that may be in LP-Gas as LP-Gas is defined in this standard. It also clarifies that 100 percent propylene fuel used for metal cutting operations, etc., is not covered by this standard.

1-2.1.2 LP-Gas stored or used in systems within the scope of this standard shall not contain ammonia. When such a possibility exists (such as may result from the dual use of transportation or storage equipment), the LP-Gas shall be tested as follows:

(a) Allow a moderate vapor stream of the product to be tested to escape from the container. A rotary, slip tube, or fixed level gauge is a convenient vapor source.

(b) Wet a piece of red litmus paper by pouring distilled water over it while holding it with clean tweezers.

(c) Hold the wetted litmus paper in the vapor stream from the container for 30 seconds.

(d) The appearance of any blue color on the litmus paper indicates that ammonia is present in the product.

NOTE 1:   Since the red litmus paper will turn blue when exposed to any basic (alkaline) solution, care in making the test and interpreting the results is required. Tap water, saliva, perspiration or hands that have been in contact with water having a pH greater than 7, or with any alkaline solution, will give erroneous results.

NOTE 2:   For additional information on the nature of this problem and conducting the test, see *Recommendations for Prevention of Ammonia Contamination of LP-Gas*, published by the National Propane Gas Association.

Since anhydrous ammonia and LP-Gas have about the same vapor pressure characteristics, transportation and storage equipment is sometimes used for ammonia when it is not needed for LP-Gas, and vice versa, due to economic reasons. It is very important when switching a tank from ammonia to LP-Gas service that the tank is cleaned properly. If not, very serious consequences can result. It is common to use brass fittings on tanks in LP-Gas service, particularly on the tanks of consumers and dealers. Under certain conditions ammonia can cause brass fittings to fail while in service through a process known as stress corrosion cracking. If ammonia gets into an LP-Gas distribution system, it will spread through the system and may cause damage to any brass fitting it contacts. Maximum concentrations of ammonia that can be tolerated in LP-Gas have been determined through a research project, "A Study of Cleaning Methods for LP-Gas Transports," conducted by Battelle Memorial Institute, Columbus, Ohio. This research indicates that the Red Litmus Paper Test for Ammonia in LP-Gas, if performed properly during a purging operation, is much more sensitive a test than the absence of ammonia odor and will ensure that tolerable limits are not exceeded. This test procedure is described in 1-2.1.2(a) through (d). Recommended procedures for purging storage and cargo tanks by steam cleaning, water flooding, and combination of water flooding and steaming methods when switching from anhydrous ammonia to LP-Gas service are contained in the National Propane Gas Association publication cited in Note 2.

1-2.2 Application of Standard.

1-2.2.1 This standard applies to the highway transportation of LP-Gas and to the design, construction, installation, and operation of all LP-Gas systems, including marine and pipeline terminals, except those designated by 1-2.3.

While the intent has been to make NFPA 58 applicable to all LP-Gas uses, it has not been feasible to do so for a number of reasons. Because of the great versatility of LP-Gas, the number and variety of its applications are very large and have steadily increased since its original commercial use as a residential fuel. This is especially true in industry, where use of LP-Gas is not only of great variety but also quite complex.

This paragraph was changed in the 1989 Edition by the addition of marine and pipeline terminals to the list. This change recognizes that these facilities are storage and transfer facilities which are similar to storage and transfer facilities already covered by the standard, and that, in many cases, authorities having jurisdiction and owners/operators were applying NFPA 58 and NFPA 59 to these facilities even though they were specifically excluded in paragraph 1-2.3.1 prior to the 1989 Edition.

It is vital that the Technical Committee and support staff responsible for NFPA 58 be technically qualified to develop and interpret its provisions. At the same time, the necessary balance of interests on the Committee must be maintained. The size of the Committee also must be kept manageable if the standard is to keep pace with the ever-growing variety of applications.

For these reasons, the applications covered by NFPA 58 are those which impact the largest number of common interest groups — e.g., the general public and employees and management of general commerce and industry — and those who must regulate, insure, and handle the emergencies associated with these applications. Obviously, these represent the bulk of all LP-Gas applications.

The applications not covered by NFPA 58 often are addressed by other standards-writing bodies or by technically qualified groups. Where the Technical Committee responsible for NFPA 58 considers these standards adequate, it has been the practice to call attention to them in NFPA 58. In most instances, this takes the form of stating that the referenced standard shall be used. Therefore, if NFPA 58 is adopted as, for example, a regulatory instrument, this gives the referenced document equal status (unless the adopting agency states otherwise, of course).

Whenever a standard or other document covers an LP-Gas application, it is important that its scope be carefully correlated with that of NFPA 58 to avoid conflicts. This is not always easy to do because of the large number of standards-writing bodies. The effort has been rather successful, however, as the spirit of cooperation among these bodies has always been high.

1-2.3 Nonapplication of Standard.

1-2.3.1   This standard does not apply to:

(a) Frozen ground containers used for the storage of LP-Gas.

This exclusion was changed in the 1989 Edition from "LP-Gas refrigerated storage systems." This change in the scope of the standard means that LP-Gas refrigerated storage systems are now covered under the 1989 Edition of this standard. (Note that LP-Gas refrigerated storage systems at utility gas plants continue to be covered under NFPA 59.) LP-Gas refrigerated storage systems are often located at marine and pipeline terminals, and may also be found at natural gas processing plants, refineries, and petrochemical plants. A new Chapter 8 has been added with requirements for refrigerated storage containers.

Frozen ground containers are a type of refrigerated LP-Gas container made by excavating an open pit and covering it with a gastight cap and freezing the ground to provide containment. Because of the differences between these and aboveground refrigerated containers, the fact that they are rarely used for LP-Gases, and the lack of standards that may be referenced, the Committee decided to continue their exclusion from this standard.

Refrigerated LP-Gas storage systems are those in which the liquid product is stored as a boiling liquid at or near atmospheric pressure. Refrigeration necessary to maintain the LP-Gas as a liquid is often provided by compressing and condensing the vapors which boil off the liquid. The storage containers are characterized by their large size [over a million gallons (3 785 m^3) is not unusual]. They are thermally insulated as a matter of operational necessity and to minimize the refrigeration load, and they can withstand only very low internal pressures. Such containers are not only very different from non-refrigerated LP-Gas containers, they are limited in number and found usually in specialized applications in LP-Gas production facilities, marine and pipeline terminals, and in natural gas peak-shaving plants.

(b) Natural gas processing plants, refineries, and petrochemical plants except at the discretion of the owner/operator or the authority having jurisdiction, this standard may be utilized with respect to the storage and transfer portions of such installations.

This paragraph was changed in the 1989 Edition to drop the exclusion of marine and pipeline terminals from the standard (see also commentary after 1-2.2.1) and to permit application to the storage and transfer portion of natural gas processing plants, refineries, and petrochemical plants subject to the owner/operator or the authority having jurisdiction. These facilities had been excluded from NFPA 58 prior to the 1989 Edition of the standard, and a footnote referred the user to API Standard No. 2510-1978, *The Design and Construction of Liquefied Petroleum Gas Installations at Marine and Pipeline Terminals, Natural Gas Processing Plants, Refineries, Petrochemical Plants and Tank Farms.*

(c) LP-Gas (including refrigerated storage) at utility gas plants. NFPA 59, *Standard for the Storage and Handling of Liquefied Petroleum Gases at Utility Gas Plants*, shall apply.

The LP-Gas application covered by this exclusion is essentially the storage (which can be both refrigerated and non-refrigerated) of LP-Gas, its vaporization, and often its mixing with air to form a mixture having a heating value similar to natural gas in a facility operated by a gas utility. A utility is a private or governmental body specifically designated as a public utility by an action of a governmental entity — e.g., state or local government.

Again, such facilities have limited firesafety impact upon general consumers and industry. However, because they are covered by an NFPA standard, NFPA 59 is adopted by reference in NFPA 58.

In recent years, the size and nature of an increasing number of industrial non-refrigerated installations have approached that of many utility gas plants. In 1966, a reorganization of the NFPA Technical Committee structure in the field of gases resulted in assignment of NFPA 59 to the same committee responsible for NFPA 58 (and the committee membership was augmented accordingly). As a result, both standards have been amended to bring their provisions closer together, and this process will continue. This development further supports the credibility of mandatory adoption of NFPA 59 in NFPA 58.

(d) Chemical plants where specific approval of construction and installation plans, based on substantially similar requirements, is obtained from the authority having jurisdiction.

This exclusion acknowledges the use of LP-Gas as a chemical reactant (feedstock), rather than as a fuel, in a chemical plant, and the unique and complex fire hazard problems that often exist. There are, however, many uses of LP-Gas as a fuel in such facilities and the storage hazards are often not unique. Moreover, there is no standard definition of a "chemical plant"; facilities in which few or no chemical reactions are carried out are often called chemical plants — e.g., aerosol product manufacturing facilities using LP-Gas only as a propellant. In practice, the chemical industry uses NFPA 58 extensively.

(e) LP-Gas used with oxygen. NFPA 51, *Standard for the Design and Installation of Oxygen-Fuel Gas Systems for Welding, Cutting, and Allied Processes*, and ANSI Z49.1, *Safety in Welding and Cutting*, shall apply.

The burning of LP-Gas with oxygen rather than air is most commonly associated with the flame cutting of metal in industry and construction. The use of oxygen introduces another dimension to the fire hazards. These operations are very widespread and the two standards referenced cover them. Because both NFPA 51 and ANSI Z49.1 are national consensus standards, their adoption in NFPA 58 is mandatory.

Unfortunately, the exclusion dealing with use with oxygen, and the specific scopes of NFPA 51 and ANSI Z49.1, technically have left some areas of firesafety uncovered. These include small jewelry and glass-forming operations (often in mercantile occupancies having considerable public exposure) and the use of oxy-propane torches by plumbers and "do-it-yourselfers." As manufacturers of these torches customarily obtain product listings, this has not proven to be a significant problem. However, regulatory authorities have had difficulty with small mercantile operations. Used with judgment, many of the provisions in NFPA 51, NFPA 58, and ANSI Z49.1 are appropriate, especially those concerned with LP-Gas and oxygen storage.

(f) Those portions of LP-Gas systems covered by NFPA 54 (ANSI Z223.1), *National Fuel Gas Code*.

NOTE: Several types of LP-Gas systems are not covered by the *National Fuel Gas Code* as noted in 1.1.1(b) therein. These include, but are not restricted to, most portable applications; many farm installations; vaporization, mixing, and gas manufacturing; temporary systems, e.g., in construction; and systems on vehicles. For those systems within its scope, the *National Fuel Gas Code* is applicable to those portions of a system downstream of the outlet of the first stage of pressure regulation.

The *National Fuel Gas Code* (NFPA 54/ANSI Z223.1) and its

predecessors (NFPA 54 and 54A and ANSI Z21.30 and Z83.1) were developed originally to cover the installation of utility gas piping and appliances in buildings. In the early days, utility gas was manufactured gas and the buildings were residential and commercial in character. After World War II, utility gas became natural gas predominantly, and by the early 1960s, industrial systems were included.

In the 1950s, NFPA 52, *Liquefied Petroleum Gas Piping and Appliance Installations in Buildings*, was developed to provide similar coverage for residential and commercial building undiluted LP-Gas piping and appliances. However, it was soon recognized that with the exception of the piping materials and testing of piping for leaks, NFPA 52 and 54 were similar. Differences were resolved in the 1959 edition of NFPA 54, NFPA 52 was withdrawn and NFPA 54 covered LP-Gas.

However, with very few minor exceptions, NFPA 54 is restricted by its scope to the installation of fixed-in-place appliances and other gas consuming equipment connected to a building gas piping system. Furthermore, many specialized types of farm equipment are not covered, even though they may be fixed in place in a building and connected to a piping system in that building. These systems are customarily LP-Gas systems rather than natural gas systems.

While the great majority of LP-Gas piping and appliance installations in completed buildings require the application of the *National Fuel Gas Code* to part of the system, it is necessary to refer to 1.1.1, "Applicability," in that code to determine this. In this respect, the Note to 1-2.3.1(f) should be considered only a rough guide.

To all LP-Gas systems covered by the *National Fuel Gas Code* that are supplied from on-site storage, it is necessary also to apply NFPA 58. The point of demarcation between these two documents is the outlet of the first stage pressure regulator. Most residential LP-Gas systems have only one regulator installed very close to the storage container(s) or a second regulator a few feet away. However, some installations involving multiple buildings — e.g., apartment complexes — may have second regulators downstream of the first that may be hundreds of feet away from the storage. The piping between them is within the scope of the *National Fuel Gas Code*.

(g) Transportation by air (including use in hot air balloons), rail, or water under the jurisdiction of the U.S. Department of Transportation.

This exclusion acknowledges the primary jurisdiction of the U.S. Department of Transportation in transportation of LP-Gas

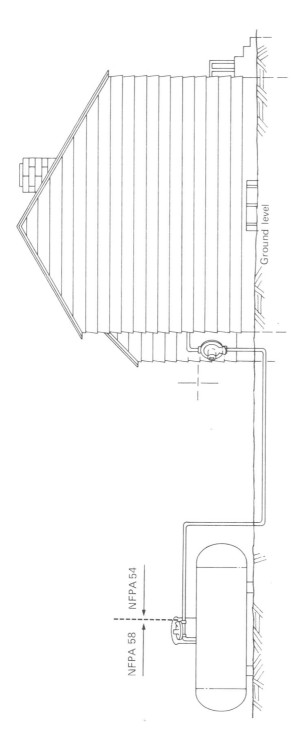

Figure 1.1 Split Between NFPA 54 and NFPA 58.

where that jurisdiction is valid. This is discussed in more detail in Chapter 6.

In recent years, the popularity of hot-air ballooning (in which LP-Gas is the source of the hot air) has resulted in fires and explosions involving the equipment used. Most of these have not occurred in flight but, rather, in storage, in ground transportation, or during refueling. While DOT (specifically the Federal Aviation Administration, FAA) regulations take precedence over NFPA 58 with respect to the system used to operate the balloon in flight, NFPA 58 is applicable at all other times.

(h) Marine fire protection. NFPA 302, *Fire Protection Standard for Pleasure and Commercial Motor Craft*, shall apply.

LP-Gas has become a popular fuel for galley stoves, cabin heaters, and other appliances on motor vessels within the scope of NFPA 302 [less than 300 gross tons (849 m³) and used for pleasure and commercial purposes]. This use is extensive enough to represent substantial public safety exposure and, therefore, its adoption in NFPA 58 is mandatory.

1-2.4 Retroactivity.

1 2.4.1 Unless otherwise stated, the provisions of this standard shall not be applied retroactively.

The matter of retroactivity in a standard that is as widely adopted by public safety regulatory agencies and used in litigation as NFPA 58, is of considerable importance. Over the more than 50 years of its existence, the Technical Committees responsible for NFPA 58 have consistently taken the approach that when an amendment should be applied retroactively the amendment will so state. Typically, this has taken the form of stipulating a certain date by which something must be accomplished (*see 3-2.7.9, for example*). The date is retained in subsequent editions of the standard for many years — even though it may predate the year of a specific edition — because of its potential importance in, for example, litigation.

Over the years, such retroactive provisions have been few. While many amendments do provide an additional degree of safety, the functional and economic burdens of complying with them are commensurate with a more gradual phasing-in, or no phasing-in, when compared with the additional degree of safety to be achieved. This principle is stated in more specific terms in 1-2.4.1(a) and (b).

The expression "distinct hazard to life or adjoining property" is rather obtuse and subject to a good deal of judgment. In practice, prior to an incident, what constitutes this hazard can be determined only by argument between parties or perhaps by regulatory directive. After an incident, especially a serious one, the opportunity for differing views is somewhat lessened. While many NFPA Committees have tried to express this thought in more specific terms, the above phrase remains the best they can come up with.

(a) Existing plants, appliances, equipment, buildings, structures, and installations for the storage, handling, or use of LP-Gas in compliance with the provisions of this standard in effect at the time of manufacture or installation may be continued in use provided that such continued use does not constitute a distinct hazard to life or adjoining property.

(b) The stocks of equipment and appliances on hand in such locations as manufacturer's storage, distribution warehouses, and dealer's storage and showrooms in compliance with the provisions of this standard in effect at the time of manufacture may be placed in use (provided such use does not constitute a distinct hazard to life or adjoining property), but all new equipment and appliances manufactured after the effective date of this standard shall comply with its provisions.

1-3 Acceptance of Equipment and Systems.

1-3.1 Method of Acceptance.

1-3.1.1 Systems, or components assembled to make up systems, shall be approved (*see Section 1-7, Approved*) as specified in Table 1-3.

1-3.1.2 Acceptance applies to the complete system, or to the individual components of which it is comprised, as specified in Table 1-3.

Approval of systems using DOT Cylinders includes:

(a) the cylinder itself, which must meet DOT specs and be labeled with a DOT number which indicates which spec it meets and date of manufacture or retest (*see Appendix C for information on requalification of DOT cylinders*)

(b) cylinder valves, which must be in accordance with DOT regulations and American National, Canadian, and CGA Standard Compressed Gas Cylinder Valve Outlet and Inlet Connections (*ANSI B57.1/CSA B 96/ CGA V-1*)

(c) manifold valve assemblies, regulators, and pressure relief devices must also be approved.

Table 1-3

Containers Used	Capacity in Water Gal (m³)	Approval Applies to:
DOT Cylinders	Up to 120 (0.454) (1,000 lb, 454 kg)	1. Container Valves and Connectors 2. Manifold Valve Assemblies 3. Regulators and Pressure Relief Devices
ASME Tanks	2,000 (7.6 m³) or less	1. Container System*, including Regulator, or 2. Container Assembly* and Regulator separately
ASME Tanks	Over 2,000 (7.6 m³)	1. Container Valves 2. Container Excess Flow Valves, Back Flow Check Valves, or alternate means of providing this protection such as remotely controlled Manual or Automatic Internal Valves 3. Container Gauging Devices 4. Regulators and Container Pressure Relief Devices

*Where necessary to alter or repair such systems or assemblies in the field in order to provide for different operating pressures, change from vapor to liquid withdrawal, or the like, such changes may be made by the use of approved components.

Approval of systems using ASME tanks (cylinders) of 2000 gal (7.6 m³) or less includes:

(a) the container (tank), which must be designed, fabricated, and stamped in accordance with the ASME Code, and its appurtenances may be approved as a unit, or

(b) the container may be approved separately in which case the pressure regulator must also be approved separately.

Approval of systems using ASME tanks (containers) over 2000 gal (7.6 m³) includes:

(a) container valves

(b) all other container appurtenances as listed in the table above. Note: On these larger systems utilizing containers over 2000 gal (7.6 m³), the entire system does not have to be approved as a system, rather the individual parts must be approved. The reason for this is that there are many application variations, and approval of the entire unit as a system would be very difficult, possibly requiring many different approvals to cover all possible applications.

1-4 LP-Gas Odorization.

1-4.1 LP-Gas to be Odorized.

1-4.1.1* All LP-Gases shall be odorized prior to delivery to a distributing plant by the addition of a warning agent of such character that they are detectable, by a distinct odor, down to a concentration in air of not over one-fifth the lower limit of flammability.

Exception: Odorization, however, is not required if harmful in the use of further processing of the LP-Gas, or if such odorization will serve no useful purpose as a warning agent in such further use or processing.

This provision was modified in the 1989 Edition to stipulate when odorization must take place — that is, at some point in its distribution chain prior to a distributing plant (see definition). This was done in recognition of research that has recently been carried out in the U.S. and Canada that has identified that odorants can be oxidized by oxygen or oxides (including iron oxide — rust) in LP-Gas containers, and a number of incidents where failure to smell leaking LP-Gas was a factor.

A-1-4.1.1 It is recognized that no odorant will be completely effective as a warning agent in every circumstance.

It is recommended that odorants be qualified as to compliance with 1-4.1.1 by tests or experience. Where qualifying is by tests, such tests should be certified by an approved laboratory not associated with the odorant manufacturer. Experience has shown that ethyl mercaptan in the ratio of 1.0 lb (0.45 kg) per 10,000 gal (37.9 m³) of liquid LP-Gas has been recognized as an effective odorant. Other odorants and quantities meeting the provisions of 1-4.1.1 may be used. Research* on odorants has shown that thiophane (tetrahydrothiophene) in a ratio of at least 6.4 lb (2.9 kg) per 10,000 gal (37.9 m³) of liquid LP-Gas may satisfy the requirements of 1-4.1.1.
   *"A New Look at Odorization Levels for Propane Gas," BERC/RI-77/1, United States Energy Research & Development Administration, Technical Information Center, September, 1977.

1-4.1.2 If odorization is required, the presence of such odorant shall be determined by sniff testing or other means and the results documented:

(a) Whenever LP-Gas is delivered to a distributing plant, and

(b) When shipments of LP-Gas bypass the distributing plant.

This provision, added in the 1989 Edition, resulted from a recommendation by the U.S. Consumer Product Safety Commission in their Status Report on LP-Gas Residential Heating Equipment, July 24, 1986 which stated:
   "NFPA Pamphlet 58 should be revised to assign responsibility for testing and certification of odorant testing at some point in

the distribution chain." This was based on technical work done for the CPSC. Testing for the presence of odorant is required to verify that the odorization has taken place, and that the odorant has not "faded" or been otherwise rendered ineffective.

The LP-Gas Committee, in considering this recommendation, evaluated the possible locations at which this testing might be required, including arrival at the distributing plant, departure from the distributing plant and upon delivery from a bobtail (delivery truck).

Additional information on odorization of LP-Gases may be found in the Fuel Gas Odorization supplement to the *National Fuel Gas Code Handbook.*

The odorization of gas, started as early as 1880 in Europe and shortly after World War I in the U.S., is used as a measure to prevent injuries or fatalities resulting from escaping gas. The first reported odorant used was ethyl mercaptan, which remains to this day a predominant compound used to odorize LP-Gas. This odorant has a characteristic odor frequently described as that of garlic, rotten eggs, or a skunk — essentially a sulfurous odor.

Previous wording in NFPA 58 specified that odorization was to indicate positively, by a distinctive odor, the presence of gas. The implication was that all must be warned by the odorant. This was an impossible requirement, for it is known that some persons have suffered complete loss of smell and would not be warned by any level of any odorant system. It is reported that about one person in 500 fits this category. On the other extreme, in certain medical cases such as Addison's disease, the nose is hundreds of times more sensitive than normal. Falling between these two extremes, about 96 percent of the population have sensitivities that vary widely but place them in a normal perception range. Even the normal olfactory response is subject to reduced perception caused by such common activities as smoking and eating, or by head colds and/or allergies. Age is also an important factor in olfactory responses, which are reportedly reduced by half for every 22 years of age between 20 and 70. Some persons also have specific nasal "blindness" to certain smells while retaining normal responses to most odors. It is estimated that about one person in a thousand cannot smell an odor such as the common skunk emits. Thus, regardless of dosage levels, it is not possible to warn 100 percent of the population of gas leaks by use of odorant systems. This is pointed out in A-1-4.1.1.

The most recent research on odorization of LP-Gas is that of the Bartlesville Energy Research Center (BERC) referenced in A-1-4.1.1. This research showed that in addition to defects of the nasal anatomy, psychological factors also probably affect olfactory responses, and that unfamiliarity with a given environment as well as anxieties or mental distractions can produce reduced awareness to odorants intended to warn individuals of the presence of LP-Gas.

1-5 Notification of Installations.

1-5.1 Fixed Installations.

1-5.1.1 Plans for fixed (stationary) installations utilizing storage containers of over 2,000 gal (7.6 m³) individual water capacity, or with aggregate water capacity exceeding 4,000 gal (15.1 m³), shall be submitted to the authority having jurisdiction before the installation is started. [*See also 3-4.9.1(e).*]

In terms of the number of installations actually made, installations of the sizes cited are much less numerous. Installations of lesser size comprise nearly all domestic installations and many commercial ones. Requiring plans for these installations to be submitted would place a burden upon the authority having jurisdiction (usually a fire department) that could not be justified for these small, relatively simple and standardized installations.

However, it is not uncommon for those authorities to require approval of small installations in heavily populated or congested areas. In such instances, they will either amend 1-5.1.1 or adopt specific wording in the enabling document.

1-5.2 Temporary Installations.

1-5.2.1 The authority having jurisdiction shall be notified of temporary (not to exceed six months) installations of the sizes covered in 1-5.1.1 before the installation is started.

This provision was added to the standard well after 1-5.1.1 when the use of temporary installations (primarily in building and road construction) became common. In retrospect, it could have just as easily been included by amending 1-5.1.1.

1-6 Personnel.

1-6.1 Qualification of Personnel.

1-6.1.1 In the interests of safety, all persons employed in handling LP-Gases shall be trained in proper handling and operating procedures.

The literal and original intent of this provision is directed at employees of LP-Gas distributing companies. However, in more recent times, LP-Gas is handled extensively by users of LP-Gas. Therefore, considerable educational information has been made available for users handling LP-Gas, as they should also receive proper training.

1-7 Definitions, Glossary of Terms and Abbreviations.

AGA. American Gas Association.

The American Gas Association is a trade association comprised largely of the gas utilities and gas transmission pipeline companies in the U.S. These utilities (the local "gas company") distribute primarily natural gas but often use LP-Gas in their supply and standby operations [*see commentary on 1-2.3.1(c)*]. The AGA is not specifically cited elsewhere in NFPA 58. The primary role of the AGA in the context of NFPA 58 is its sponsorship of the ANSI Z21 and Z83 series of standards covering residential, commercial, and industrial gas appliances and equipment. These standards cover both natural gas and LP-Gas appliances. (*See commentary on 2-6.2.1.*)

ANSI. American National Standards Institute.

The American National Standards Institute is the overall focus of all private sector standards making in the U.S. The institute does not develop its own standards, but does adopt standards prepared by others provided they are developed by ANSI procedures. Because these procedures are compatible with NFPA procedures, the three ANSI standards cited in Chapter 8 have been adopted as mandatory in NFPA 58. In fact, each edition of NFPA 58 is also submitted to ANSI for adoption both to support ANSI and for the additional status obtained from ANSI approval.

API. American Petroleum Institute.

The American Petroleum Institute is the trade association comprised largely of the oil companies in the U.S. Many oil companies also produce and distribute LP-Gas because it is a product of the refining of crude oil as well as the processing of natural gas. A number of oil companies have LP-Gas marketing subsidiaries that market LP-Gas and install and service LP-Gas installations.

The API produces many standards. While very useful, their

adoption is not considered mandatory in NFPA 58 unless they are adopted by ANSI.

API-ASME Container (or Tank).   A container constructed in accordance with the pressure vessel code jointly developed by the American Petroleum Institute and the American Society of Mechanical Engineers (*see Appendix D*).

Approved.   Acceptable to the "authority having jurisdiction."

NOTE: The National Fire Protection Association does not approve, inspect or certify any installations, procedures, equipment, or materials nor does it approve or evaluate testing laboratories. In determining the acceptability of installations or procedures, equipment or materials, the authority having jurisdiction may base acceptance on compliance with NFPA or other appropriate standards. In the absence of such standards, said authority may require evidence of proper installation, procedure or use. The authority having jurisdiction may also refer to the listings or labeling practices of an organization concerned with product evaluations which is in a position to determine compliance with appropriate standards for the current production of listed items.

ASME.   American Society of Mechanical Engineers.

ASME Code.   The *Boiler and Pressure Vessel Code* (Section VIII, "Rules for the Construction of Unfired Pressure Vessels") of the American Society of Mechanical Engineers. Only Division I of Section VIII of the ASME Code is applicable in this standard except UG-125 through UG-136 shall not apply.

The purpose of using the ASME Code was to adopt a nationally recognized code for LP-Gas containers other than those covered by DOT (ICC) regulations. As most states adopt the ASME Code, this results in national standardization which has a safety value as well as being practical from an economic standpoint.

There are two major exceptions to the ASME Code in the application of NFPA 58.

The Code contains a Division II which permits a smaller safety factor to be used provided an extensive and rigorous structural analysis is made. While a container fabricated in accordance with Division II could be suitable, the Committee is concerned about long-term safety—especially of the thousands of smaller 250-1000 gal (0.9 to 3.8 m³) water capacity ASME Code containers in domestic and commercial service which may not receive frequent inspections. Furthermore, for economic reasons, the manufacturers of these containers for LP-Gas service have not indicated any desire to use Division II.

As a broad document, the ASME Code applies to pressure vessels in many kinds of service with, as the membership of the ASME Committee indicates, emphasis on vessels in air, steam, and other non-flammable products service. Paragraphs UG-125 through UG-136 in Division I of the ASME Code cover pressure relief devices. The possible consequences of operation of the pressure relief devices discharging these products are obviously different from those discharging LP-Gas. Therefore, the provisions of NFPA 58 — e.g., start-to-discharge pressure settings — are different from those in the ASME Code.

Other differences between NFPA 58 and the ASME Code involve: the use of pilot-operated relief valves (UG-126) (*see commentary on 2-3.2.3*); the use of frangible disk devices (UG-127) (such devices discharge the entire contents of a container and reduce the pressure to atmospheric rather than control the pressure as a spring-loaded device does — a circumstance felt to be undesirable in installations covered by NFPA 58); marking (UG-129) (if functions are different, markings will be also); and certification (UG-131 and UG-132). NFPA 58 covers certification, in the context of "approved" and listing by Underwriters Laboratories Inc., Factory Mutual, or another agency acceptable to the authority having jurisdiction. These agencies, as noted in 2-3.2.3, have developed criteria especially for LP-Gas service which are believed to be more appropriate in NFPA 58 than are the ASME Code provisions.

ASME Container (or Tank)   A container constructed in accordance with the ASME Code. (*See Appendix D.*)

ASTM.   American Society for Testing and Materials.

The American Society for Testing and Materials promulgates the largest body of private sector standards in the U.S. and most are adopted as ANSI standards. As noted in Chapter 9, many ASTM/ANSI standards concerned with piping and castings are adopted as mandatory in NFPA 58.

Authority Having Jurisdiction.   The "authority having jurisdiction" is the organization, office or individual responsible for "approving" equipment, an installation or a procedure.

NOTE: The phrase "authority having jurisdiction" is used in NFPA documents in a broad manner since jurisdictions and "approval" agencies vary as do their responsibilities. Where public safety is primary, the "authority having jurisdiction" may be a federal, state, local or other regional department or individual such as a fire chief, fire marshal, chief of a fire prevention bureau, labor department, health department, building official, electrical inspector, or others having statutory authority. For insurance purposes, an insurance inspection department, rating bureau, or other insurance company representative may be the "authority having jurisdiction." In many circumstances the property owner or his designated agent assumes the role of the "authority having jurisdiction"; at government installations, the commanding officer or departmental official may be the "authority having jurisdiction."

NFPA 58 has been adopted as the basis of public safety regulations by most states in the U.S., by a number of federal agencies and in many other countries. In most states, the adopting and enforcing agency is the Office of the State Fire Marshal. The chief officer of each fire department in the state is considered an agent of the state fire marshal. In some states, an agency concerned only with LP-Gas has been established to adopt and enforce the regulations. In others, only the provisions concerned with certain applications have been adopted — for example, those concerned with vehicle applications or employee safety — and these involve other agencies, such as labor and industry departments and insurance commissions.

Since its formation, the U.S. Occupational Safety and Health Administration has adopted those provisions of NFPA 58 within its jurisdiction in its regulations [29 *Code of Federal Regulations* (CFR) 1910.110].

These authorities vary in their ability to update their adoptive documents with new editions of NFPA 58. It is rare that the edition of record is the current NFPA edition. This can present problems, although most authorities are amenable to use of the current edition, especially if the reason for an amendment is available to them.

As observed in the Note to the definition, non-governmental bodies can be the authority having jurisdiction by virtue of the conditions of a private contract — e.g., an insurance policy.

Bureau of Explosives (B of E).   An agency of the Association of American Railroads.

The Bureau of Explosives, an agency of the Association of American Railroads (AAR), functioned for many years in the 19th century as a source of safety requirements for the railroads. Its first activity related to railroad transportation of dynamite for blasting operations. With the turn of the century, the Interstate Commerce Commission was charged with the duty of promulgating regulations, including container specifications, for the safe transportation of hazardous commodities by rail. The Interstate Commerce Commission was authorized by Congress to use the services of the bureau. The bureau has also worked with the Canadian Transport Commission (CTC) in a similar capacity.

The bureau acted as a central point of coordination and communication for the railroads on the one hand, and DOT and CTC on the other. It also served all three groups as a central office for registration, licensing, and inspection. Its director and chief inspector held the title "Agent," and served in that capacity as the official agent and attorney for all AAR member rail carriers with DOT and CTC. Capacity of cylinder relief valves was certified through tests by the bureau. For many years the bureau published the ICC and DOT regulations under their "Tariff," which is still available through the bureau.

Today, DOT has assumed many of the functions previously delegated to the bureau and recognizes the bureau as one of several means to carry out and implement DOT regulations.

Cargo Tank.   (Primarily a DOT designation.) A container used to transport LP-Gas over the highway as liquid cargo, either mounted on a conventional truck chassis or as an integral part of a transporting vehicle in which the container constitutes in whole, or in part, the stress member used as a frame. Essentially a permanent part of the transporting vehicle.

CGA.   Compressed Gas Association, Inc.

In 1913 a nonprofit service organization was incorporated in New York as the Compressed Gas Manufacturers' Association to promote, develop, represent, and coordinate technical and standardization activities in the compressed gas industries in the interest of safety and efficiency. Among its members are companies and individuals producing compressed, liquefied, and cryogenic gases and their containers (including cargo containers), container appurtenances, and other system components.

From its inception, a major activity of CGA has been the development of standards through the efforts of over 40 technical committees. The 1932 Edition of what is now NFPA 58 (*see "History of NFPA 58"*) was based upon a draft prepared by CGA's Test and Specification Committee in 1930 and 1931. Although much of this work has now passed on to the National Propane Gas Association (NPGA), CGA has maintained membership on the NFPA Technical Committee and provides valuable technical assistance.

Charging.   See Filling.

Compressed Gas.   Any material or mixture having in the container an absolute pressure exceeding 40 psia (276 kPa absolute) at 70°F (21.1°C), or regardless of the pressure at 70°F (21.1°C), having an absolute pressure exceeding 104 psia (717 kPa absolute) at 130°F (54.4°C).

This is a very specific definition developed primarily to meet the needs of the U.S. Department of Transportation in regulating the safety of hazardous materials in transportation. All of the LP-Gases covered by NFPA 58 are compressed gases in accordance with this definition. In addition, they are liquefied compressed gases. Many compressed gases are stored and transported solely in the gaseous state.

Container.   Any vessel, including cylinders, tanks, portable tanks and cargo tanks, used for the transporting or storing of LP-Gases.

As noted in the definition, there are several kinds of pressure containers used to store and transport LP-Gas and each must comply with very specific fabrication criteria. The word "container" is a generic description and is used by itself in the standard whenever it is unnecessary to cite a specific fabrication type. Otherwise, it is qualified — e.g., DOT container, ASME container.

Container Appurtenances.   Items connected to container openings needed to make a container a gastight entity. These include, but are not limited to, pressure relief devices; shutoff, backflow check, excess flow check and internal valves; liquid level gauges; pressure gauges; and plugs.

Container Assembly.   An assembly consisting essentially of the container and fittings for all container openings. These include shutoff valves, excess flow valves, liquid level gauging devices, pressure relief devices and protective housings.

Cylinder.   A portable container constructed to DOT (formerly ICC) cylinder specifications or, in some cases, constructed in accordance with the ASME Code of a similar size and for similar service. The maximum size permitted under DOT specifications is 1,000 lb (454 kg) water capacity.

Direct Gas-Fired Tank Heater.   A gas-fired device which applies hot gas from the heater combustion chamber directly to a portion of the container surface in contact with LP-Gas liquid.

Dispensing Device (or Dispenser).   A device normally used to transfer and measure LP-Gas for engine fuel into a fuel container, serving the same purpose for an LP-Gas service station as that served by a gasoline dispenser in a gasoline service station.

Distributing Plant.   A facility, the primary purpose of which is the distribution of gas, and which receives LP-Gas in tank car, truck transport or truck lots, distributing this gas to the end user by portable container (package) delivery, by tank truck or through gas piping. Such plants have bulk storage [2,000 gal (7.6 m³) water capacity or more] and usually have container filling and truck loading facilities on the premises. So-called "bulk plants" are considered as being in this category. Normally no persons other than the plant management or plant employees have access to these facilities.

The terms "distributing plant" and "distributing point" were first included in the 1972 edition to indicate a basic distinction between two types of facilities with respect to public exposure. This distinction enables the organization of provisions in the standard that differ due to whether there is rather close involvement of the public with the facility. Particular attention is given for the protection of the public in such matters as ignition source control, protection against vehicular damage, container valve protection, separation distances, etc. This was found to be necessary due to a large increase in recent years of consumer-owned containers needing refilling, such as those used for gas-fired outdoor grills, recreational vehicles, vehicular propulsion fuel, etc. Generally, the public is expected to be in and around a distributing point facility, but not in and around liquid transfer areas in a distributing plant. The public may also go to bulk plants (a type of distributing plant) to have their cylinders filled, but public movement is controlled so that only employees are permitted in and around the filling operations. A distributing plant may also have a distributing point in connection with the plant facility.

Distributing plants have bulk storage facilities of 2,000 gal (7.6 m³) water capacity or more. A typical LP-Gas bulk plant distributing LP-Gas to its customers by delivering cylinders to them or by filling their storage tank is a distributing plant. A distributing plant may also be a facility serving a number of customers, such as in a residential area, by pipeline. However, if the latter is classified as a public utility it would be subject to NFPA 59, *Standard for the Storage and Handling of Liquefied Petroleum Gases at Utility Gas Plants*. A distributing plant serving customers from a pipeline would also be subject to the federal pipeline safety regulations, Part 192 CFR 49, if the system served 10 or more customers. This may not be the case, however, if the storage facility serves this type of customer from a master meter. In this case only that part of the system downstream from the master meter would be subject to federal regulations.

Distributing Point.   A facility, other than a distributing plant or industrial plant, which normally receives gas by tank truck, and which fills small containers or the engine fuel tanks of motor vehicles on the premises. Any such facility having LP-Gas storage of 100 gal (0.4 m³) or more water capacity, and to which persons other than the owner of the facility or his employees have access, is considered to be a distributing point. An LP-Gas service station is one type of distributing point.

The fundamental feature that differentiates this type of facility from a distributing plant or an industrial plant is that persons other than management or employees also have access to the facility (*see commentary on the definition of distributing plant*). As noted, an LP-Gas service station is one type of distributing point, but this type of facility is limited to dispensing LP-Gas into engine fuel tanks of highway vehicles. In the 1969 Edition of NFPA 58 there was a chapter devoted solely to service stations. The distinct provisions for dispensing fuel in this type of facility are now included in 3-2.10.6.

Other types of distributing points include cylinder dealers whose distribution is on a smaller scale and open to the public (for a facility at a distributing plant), cylinder refueling facilities at recreational vehicle parks, a facility at hardware, equipment rental, and sporting goods stores for filling gas grill cylinders, a facility at a marina, etc. Refueling of industrial trucks could be at a distributing plant where exchange cylinders are refilled or at an industrial plant where either exchange cylinders or containers mounted on the industrial truck are refilled. An installation where industrial trucks are refilled at an industrial plant [and the storage facilities are less than 2,000 gal (7.6 m³) water capacity] is not considered to be a distributing point since only employees have access to this operation.

DOT.   U.S. Department of Transportation.

DOT Cylinder.   See Cylinder.

Emergency Shutoff Valve.   A shutoff valve incorporating thermal and manual means of closing and providing for remote means of closing.

Excess-Flow Valve (also called Excess-Flow Check Valve).   A device designed to close when the liquid or vapor passing through it exceeds a prescribed flow rate as determined by pressure drop.

Fill, Filling.   Transferring liquid LP-Gas into a container.

Filling by Volume.   See Volumetric Filling.

Filling by Weight.   See Weight Filling.

Fixed Liquid Level Gauge.   A type of liquid level gauge using a relatively small positive shutoff valve and designed to indicate when the liquid level in a container being filled reaches the point at which this gauge or its connecting tube communicates with the interior of the container.

Fixed Maximum Liquid Level Gauge.   A fixed liquid level gauge which indicates the liquid level at which the container is filled to its maximum permitted filling density.

Flexible Connector. A short [not exceeding 36 in. (1 m) overall length] component of a piping system fabricated of flexible material (such as hose) and equipped with suitable connections on both ends. LP-Gas resistant rubber and fabric (or metal), or a combination of them, or all metal may be used. Flexible connectors are used where there is the need for, or the possibility of, greater relative movement between the points connected than is acceptable for rigid pipe.

Float Gauge. A gauge constructed with a float inside the container resting on the liquid surface which transmits its position through suitable leverage to a pointer and dial outside the container indicating the liquid level. Normally the motion is transmitted magnetically through a nonmagnetic plate so that no LP-Gas is released to the atmosphere.

Gallon. U.S. Standard. 1 U.S. gal = 0.833 Imperial gal = 231 cu in. = 3.785 liters.

Gas. Liquefied Petroleum Gas in either the liquid or vapor state. The more specific terms "liquid LP-Gas" or "vapor LP-Gas" are normally used for clarity.

Gas-Air Mixer. A device, or system of piping and controls, which mixes LP-Gas vapor with air to produce a mixed gas of a lower heating value than the LP-Gas. The mixture thus created is normally used in industrial or commercial facilities as a substitute for some other fuel gas. The mixture may replace another fuel gas completely, or may be mixed to produce similar characteristics and mixed with the basic fuel gas. Any gas-air mixer which is designed to produce a mixture containing more than 85 percent air is not subject to the provisions of this standard.

ICC. U.S. Interstate Commerce Commission.

ICC Cylinder. See Cylinder.

Ignition Source. See Sources of Ignition.

Industrial Plant. An industrial facility which utilizes gas incident to plant operations, with LP-Gas storage of 2,000 gal (7.6 m³) water capacity or more, and which receives gas in tank car, truck transport, or truck lots. Normally LP-Gas is used through piping systems in the plant, but may also be used to fill small containers, such as for engine fuel on industrial (i.e., forklift) trucks. Since only plant employees have access to these filling facilities, they are not considered to be distributing points.

As with distributing plants, the public does not have access to liquid transfer operations in an industrial plant facility since only employees are involved and the LP-Gas storage capacity is 2,000 gal (7.6 m³) water capacity or more. Actually, storage capacities can be rather large (in some instances these may be industrial tank farms) but not those facilities covered by API standard 2510. [*See commentary on 1-2.3.1(b).*]

Internal Valve. A primary shutoff valve for containers which has adequate means of actuation and which is constructed in such a manner that its seat is inside the container and that damage to parts exterior to the container or mating flange will not prevent effective seating of the valve.

Labeled. Equipment or materials to which has been attached a label, symbol or other identifying mark of an organization acceptable to the "authority having jurisdiction" and concerned with product evaluation, that maintains periodic inspection of production of labeled equipment or materials and by whose labeling the manufacturer indicates compliance with appropriate standards or performance in a specified manner.

See definition of Listed.

Liquefied Petroleum Gas (LP-Gas or LPG). Any material having a vapor pressure not exceeding that allowed for commercial propane composed predominantly of the following hydrocarbons, either by themselves or as mixtures: propane, propylene, butane (normal butane or isobutane) and butylenes.

Listed. Equipment or materials included in a list published by an organization acceptable to the "authority having jurisdiction" and concerned with product evaluation, that maintains periodic inspection of production of listed equipment or materials and whose listing states either that the equipment or material meets appropriate standards or has been tested and found suitable for use in a specified manner.

NOTE: The means for identifying listed equipment may vary for each organization concerned with product evaluation, some of which do not recognize equipment as listed unless it is also labeled. The "authority having jurisdiction" should utilize the system employed by the listing organization to identify a listed product.

The authorities having jurisdiction will customarily accept or even require (whether or not NFPA 58 requires) the use of listed equipment whenever it is available. In the U.S. and Canada, the major LP-Gas equipment listing agencies customarily recognized as such by the authorities having jurisdiction are Underwriters Laboratories Inc., Underwriters' Laboratories of Canada, Inc., the Factory Mutual Engineering Corporation, the American Gas Association Laboratories, and the Canadian Gas Association Laboratories. However, many other organizations perform such services, and equipment listed by them is acceptable provided the organization is acceptable to the authority having jurisdiction.

A point of some confusion is that not all the above organizations use the term "listed." For example, AGA and CGA Laboratories use the term "certified" and Factory Mutual uses the term "approved." Both are synonymous with "listed" in this definition.

Load, Loading.   See Filling.

LPG.   See Liquefied Petroleum Gas.

LP-Gas.   See Liquefied Petroleum Gas.

LP-Gas Service Station.   See Distributing Point. A facility open to the public which consists of LP-Gas storage containers, piping and pertinent equipment, including pumps and dispensing devices, and any buildings, and in which LP-Gas is stored and dispensed into engine fuel containers of highway vehicles.

LP-Gas System.   An assembly consisting of one or more containers with a means for conveying LP-Gas from the container(s) to dispensing or consuming devices (either continuously or intermittently) and which incorporates components intended to achieve control of quantity, flow, pressure, or state (either liquid or vapor).

Magnetic Gauge.   See Float Gauge.

Movable Fuel Storage Tenders or Farm Carts.   Containers not in excess of 1,200 gal (4.5 m³) water capacity, equipped with wheels to be towed from one location to another. They are basically non-highway vehicles, but may occasionally be moved over public roads or highways for short distances to be used as a fuel supply for farm tractors, construction machinery, and similar equipment.

Multipurpose Passenger Vehicle.   A motor vehicle with motive power, except a trailer, designed to carry 10 persons or fewer which is constructed on a truck chassis or with special features for occasional off-road operations.

This definition and amendment of 3-6.2.2(a)(6) comprised TIA 58-83-1 (issued October 8, 1982) originally. Such vehicles are identified by a label affixed to the vehicle in accordance with federal regulations.

NFPA.   National Fire Protection Association.

NPGA.   National Propane Gas Association.

The National Propane Gas Association (NPGA) (originally the Bottled Gas Association of America), is the national association for the liquefied petroleum gas industry. Among its membership are producers of LP-Gas, wholesale and retail marketers, gas appliance and equipment manufacturers and distributors, tank and cylinder fabricators, transport firms and others. NPGA has nearly 4,000 members throughout the U.S. and 29 other countries.

One of the primary reasons for the establishment of the association in 1932 was to afford all interested divisions of the industry full opportunity to aid in making the rules which safeguard their industry. The association was first affiliated with the Compressed Gas Manufacturers' Association (now the Compressed Gas Association). This was a natural beginning as the CGMA was the only technical association in existence which could cope with the problems involving gases in cylinders, the primary method of handling LP-Gases at that time. Today, the NPGA is a separate and distinct organization and maintains a liaison with CGA, but carries on its own technical activities through its Technology and Standards Committee.

The NPGA Technology and Standards Committee has engineering and technical representation from all phases of the industry. In addition, there are Advisory Members representing other related industry associations, testing laboratories, regulatory officials, insurance organizations, etc., who provide much assistance with the Technical Committee's standards development work. Many recommendations for new provisions and amendments of existing ones in NFPA 58 are initiated by this Technical Committee as industry sponsored recommendations. These in turn are considered by the NFPA Committee on LP-Gases which reflects a broader interest. Because the Technology and Standards Committee has at its disposal a large amount of technical talent, the NFPA Committee on LP-Gases will at times refer matters to this group for initial study and recommendations.

For many years prior to 1988, the association was known as the National LP-Gas Association (NLPGA).

Permanent Installation.   See Stationary Installation.

Piping, Piping Systems.   Pipe, tubing, hose, and flexible rubber or metallic hose connectors made up with valves and fittings into complete systems for conveying LP-Gas in either the liquid or vapor state at various pressures from one point to another.

Point of Transfer.   The location where connections and disconnections are made or where LP-Gas is vented to the atmosphere in the course of transfer operations.

The release of a small quantity of LP-Gas in either vapor or liquid form cannot be avoided during liquid transfer operations when filling a container. In all instances liquid is trapped between the valve on the end of the transfer hose and the filler valve on the container or piping. When the connection is disconnected, a small amount of liquid escapes to the atmosphere. If the filling level is being gauged with a fixed liquid level, fixed maximum liquid level, rotary, or slip tube gauge (*see definitions*), vapor and liquid are released during the operation. Therefore, the formation of a flammable mixture is unavoidable in the immediate vicinity of the points where the LP-Gas is released and it is necessary that there be space for the gas to dissipate without ignition.

In most residential and commercial fixed installations, the filling connection is fastened to the container (it is, in fact, a "container appurtenance") and the container siting distances are adequate to control the hazard of ignition. Where large containers are installed, it would be unsafe to require the delivery person to climb to the top to connect and disconnect the hose, so the connection is made to a filler pipe located near the ground some distance from the container. Even though container siting criteria would not necessarily address this hazard, established good practice in these large installations generally has provided an adequate safeguard.

With the growing general public ownership of portable containers associated with recreational applications, however, the location of the points where this release was occurring was no longer controlled by container siting criteria or by the expertise brought to bear in large installations. Moreover, the actual liquid transfer was being done by individuals not employed by LP-Gas industry firms and to whom this operation was incidental to their principal occupations. Experience soon showed that the standard needed to incorporate a means by which the location of these release points could be better controlled.

The point-of-transfer concept was incorporated into NFPA 58 for the first time in 1972.

Portable Container.   A container designed to be readily moved, as distinguished from containers designed for stationary installations. Portable containers designed for transportation filled to their maximum filling density include "cylinders," "cargo tanks" and "portable tanks," all three of which are separately defined. Containers designed to be readily moved from one usage location to another, but substantially empty of product, are "portable storage containers" and are separately defined.

Portable Storage Container.   A container similar to, but distinct from, those designed and constructed for stationary installation, designed so that it can be readily moved over the highways, substantially empty of liquid, from one usage location to another. Such containers either have legs or other supports attached, or are mounted on running gear (such as trailer or semitrailer chassis) with suitable supports, which may be of the fold-down type, permitting them to be placed or parked in a stable position on a reasonably firm and level surface. For large volume, limited duration product usage (such as at construction sites and normally for 12 months or less) portable storage containers function in lieu of permanently installed stationary containers.

Portable Tank (also called Skid Tank).   A container of more than 1,000 lb (454 kg) water capacity used to transport LP-Gas handled as a "package," that is, filled to its maximum permitted filling density. Such containers are mounted on skids or runners and have all container appurtenances protected in such a manner that they can be safely handled as a "package."

Pressure Relief Device.   A device designed to open to prevent a rise of internal fluid pressure in excess of a specified value due to emergency or abnormal conditions. (*See ANSI B95.1, Standard Terminology for Pressure Relief Devices.*)

Pressure Relief Valve.   A type of Pressure Relief Device designed to both open and close to maintain internal fluid pressure.

> **This definition was added in the 1989 Edition to clarify the different types of relief valves in use. Note that the standard contains only a list of the different pressure relief valves. The descriptions and drawings are included in Appendix A, which means that they are advisory material which is included for information purposes only. The use of Appendix A to provide informative material was begun in the 1989 Edition to conform with the NFPA Manual of Style.**

Pressure Relief Valves are further characterized as follows:

External Pressure Relief Valve*.   A relief valve in which the entire relief valve is outside the container connection except the threaded portion which is screwed into the container connection, and all of the parts are exposed to the atmosphere.

Flush Type Full Internal Pressure Relief Valve*.   A full internal relief valve in which the wrenching section is also within the container connection, except for pipe thread tolerances on make up.

Full Internal Pressure Relief Valve*.   A relief valve in which all working parts are recessed within the container connections, and the spring and guiding mechanism are not exposed to the atmosphere.

Internal Spring Type Pressure Relief Valve*.   A relief valve in which only the spring and stem are within the container connection and the spring and stem are not exposed to the atmosphere. The exposed parts of the relief valve have a low profile.

Sump Type Full Internal Pressure Relief Valve*.   A relief valve in which all working parts are recessed within the container connection, but the spring and guiding mechanism are exposed to the atmosphere.

A-1-7 External Pressure Relief Valve.   Describes the type relief valves used on older domestic tanks, relief valve manifolds and piping protection.

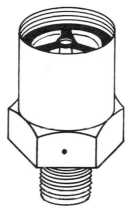

External Relief Valve.

Flush Type Full Internal Pressure Relief Valve.   Describes the type relief valve being required on cargo vehicles in most states.

Flush Type Full Internal Relief Valve.

Full Internal Pressure Relief Valve.   Describes the type relief valve being changed to for engine fuel use.

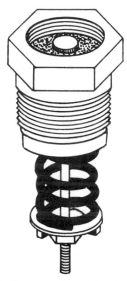

Full Internal Relief Valve.

Internal Spring Type Pressure Relief Valve.   Describes the type relief valves used on modern domestic tanks, looks similar to Full Internal Relief Valve, but has seat and poppet above the tank connection.

Internal Spring Type Relief Valve.

Sump Type Full Internal Pressure Relief Valve.   Describes the type relief valve used on older engine fuel tanks.

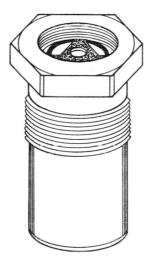

Sump Type Full Internal Relief Valve.

PSI, PSIG, and PSIA.   Pounds per square inch, pounds per square inch gauge, and pounds per square inch absolute, respectively.

Quick Connectors.   Devices used for quick connections of the acme thread or lever-cam types. This does not include devices used for cylinder-filling connections.

Quick connectors, in the context of this definition and as used in 2-4.6.2, are generally end fittings as hose couplings used on transfer hose to make connections for loading and unloading cargo tank vehicles, railroad tank cars, etc. They may also be used as connectors on swivel-type transfer piping. There are other quick connectors used in LP-Gas systems, such as a quick-closing coupling for engine fuel refueling and cylinder filling, quick-disconnect devices for gas-fired barbeque grills, connecting gas systems between two sections of mobile homes, etc. This definition, however, is limited only to the two basic types used for transfer of LP-Gas as cargo.

Rotary Gauge. A variable liquid level gauge consisting of a small positive shutoff valve located at the outer end of a tube, the bent inner end of which communicates with the container interior. The tube is installed in a fitting designed so that the tube can be rotated with a pointer on the outside to indicate the relative position of the bent inlet end. The length of the tube and the configuration to which it is bent is suitable for the range of liquid levels to be gauged. By a suitable outside scale, the level in the container at which the inner end begins to receive liquid can be determined by the pointer position on the scale at which a liquid-vapor mixture is observed to be discharged from the valve.

A rotary gauge is built with an open-ended bent tube within the container so that the open end of the tube can be rotated to determine the point of liquid level in the container. The gauge can be rotated 360 degrees and is generally made for containers up to 130 in. (3 300 mm) internal diameter of the container. It can be end or side-mounted, and can be used either on stationary or mobile storage containers. On straight-bodied delivery trucks, generally the rotary gauge is mounted in the head of the cargo tank because the length of the average unit is such that being at the end does not throw off the gauging accuracy to a great extent when the truck is not on a level surface. On the other hand, on longer semitrailer transports, rotary gauges must be side-mounted so as to read accurately the actual liquid level in such containers. As a result, the proper side location for the rotary gauge is at the center longitudinal point so as to gauge the liquid level at the average level when the transport unit is not on a level surface. The length of the tube within the container must be proper for the particular container in which it is installed.

To determine the liquid level in the container, the rotary gauge should be set with the tube in the vapor space and then slowly rotated toward the liquid level with the vent valve open. When in the vapor space there will be a vapor discharge from the vent but, when the gauge reaches the liquid level, liquid (a white fog) will start spewing out of the vent opening. It is then good practice to move the rotary gauge back to the vapor space slowly and again to the liquid state so as not to get an erroneous reading. The percentage full point is indicated on the dial on the outside of the gauge to determine the content level.

On large transports, a heavy metal hanger should be welded to the inside top of the container and attached to the rotary gauge stem to give it support so that in the movement over the road the bent tube will not be bent out of proper position. This also prevents or lessons vibration which could result in the eventual cracking or breaking of the tube.

The rotary gauge is not a gauge to be used for accurate determination of the contents of a container, but only for general use in determining the approximate quantity in the container. The rotary gauge should never be used to determine the maximum filling level of a container. This should be done by a fixed liquid level gauge set in accordance with 2-3.4.2.

Skid Tank. See Portable Tank.

Slip Tube Gauge. A variable liquid level gauge in which a relatively small positive shutoff valve is located at the outside end of a straight tube, normally installed vertically, and communicates with the container interior. The installation fitting for the tube is designed so that the tube can be slipped in and out of the container and the liquid level at the inner end determined by observing when the shutoff valve vents a liquid-vapor mixture.

Sources of Ignition. Devices or equipment which, because of their modes of use or operation, are capable of providing sufficient thermal energy to ignite flammable LP-Gas vapor-air mixtures when introduced into such a mixture or when such a mixture comes into contact with them, and which will permit propagation of flame away from them.

Although the standard has contained provisions to control ignition sources since its inception, this definition was adopted first in the 1979 Edition. It became necessary because of the existence of devices or equipment which could ignite a flammable LP-Gas/air mixture under certain conditions although not in the manner in which they were normally used. In addition, the definition clarifies the fact that other devices might produce ignition but would not permit a flame to propagate away from them; in which case, they are not a hazard. Certain Class I, Division 1 electrical devices using gap flame quenching principles and devices equipped with flame arrestors are examples of such devices.

Some sources of ignition are:

1. Devices or equipment, such as gas or oil-burning appliances, torches, matches and lighters, producing, or capable of producing, flame.

2. Devices or equipment, such as motors, generators, switches, lights, wiring and welding machines, producing, or capable of producing, electrical arcs or sparks of sufficient energy. Such devices or equipment listed for Class I, Group D locations are not sources of ignition when properly used and maintained in the classified areas prescribed in this standard.

3. Devices or equipment having, or capable of having, surfaces at temperatures exceeding 700°F (371°C).

4. Devices or equipment, such as grinders, metal saws, chipping hammers and flint lighters, producing, or capable of producing, mechanical (struck) sparks of sufficient energy. Unpowered hand tools used only by one individual at a time and for their intended purposes are not considered sources of ignition.

5. Cigarettes and other smoking materials.

6. Because they constitute a readily communicating pathway, gravity or mechanical air intakes and exhaust terminals of ventilation systems for structures housing the ignition sources described. Air intakes and exhaust terminals for gas or oil-burning appliances, including those for direct-vent appliances. Air intakes and exhaust terminals for air conditioners.

Operating spark-ignition internal combustion engines do not appear, based upon experience, to be an ignition source. However, they have been an ignition source during the act of starting them. Operating diesel engines have been identified as the ignition source in fire reports — usually as the result of burning carbon being blown out of their exhausts. Such engines tend to overspeed when LP-Gas is drawn into their intake manifolds.

Special Protection.   A means of limiting the temperature of an LP-Gas container for purposes of minimizing the possibility of failure of the container as the result of fire exposure.

When required in this standard, special protection consists of any of the following: applied insulating coatings, mounding, burial, water spray fixed systems or fixed monitor nozzles, meeting the criteria specified in this standard (*see 3-10.3*), or by any means listed (*see definition of Listed*) for this purpose.

In the early 1970s, the U.S. was subjected to an outbreak of Boiling Liquid-Expanding Vapor-Explosions (BLEVEs) of LP-Gas railroad tank cars as a result of derailments. Most of these were due to container failure resulting from overheating of the container metal (steel) through contact with flames from flammable or combustible materials released from other cars. While this hazard was beyond the scope of NFPA 58 and, in fact, the circumstances involved control problems not found in containers covered by the standard, these accidents resulted in widespread fear and apprehension of all large LP-Gas containers. Notable in this respect was an increasing reluctance by fire fighters to approach an LP-Gas container exposed to fire in order to limit the container temperature by the application of water from hose streams. As this procedure was, and is, a fundamental and universal BLEVE prevention safeguard, the NFPA Technical Committee on LP-Gases and the industry were concerned that a reduction in its effectiveness would increase the frequency and severity of BLEVEs.

A joint task force of the NFPA Technical Committee and the NPGA's Technology and Standards Committee (augmented by representatives of railroad and U.S. DOT groups studying the tank car problem) concluded that additional provisions were needed to reduce the chances for BLEVEs from fire exposure. Two new provisions were first included in the 1976 Edition of NFPA 58.

The first provision was for the emergency shutoff valves described in 2-4.5.4 and 3-2.7.9. These valves were considered important enough to require retroactive installation in all of the larger facilities as prescribed in 3-2.7.9.

The second provision was for special protection as stipulated in 3-10.2.5. It is noted, however, that the need for this protection is predicated upon the results of a required firesafety analysis.

The five modes of special protection cited in the above definition represent those currently recognized and for which standards exist. However, the definition is broad enough to recognize other means, provided they are listed for the purpose. There is ongoing research in this area.

Stationary Installation (also called "Fixed" or "Permanent" Installation).
An installation of LP-Gas containers, piping and equipment for use indefinitely at a particular location; an installation not normally expected to change in status, condition or place.

UL.   Underwriters Laboratories Inc.

Founded in 1894, Underwriters Laboratories Inc. is chartered as a not-for-profit, independent organization performing testing for public safety. It maintains laboratories for the examination and testing of devices, systems and materials to determine their relation to life, fire and casualty hazards and crime prevention.

UL-listed materials within the scope of NFPA 58 are included in the following UL directories: *Hazardous Location Equipment; Marine Products; Automotive, Burglary Protection, and Mechanical Equipment;* and *Gas and Oil Equipment.*

Universal Cylinder.   A DOT cylinder specification container, constructed and fitted with appurtenances in such a manner that it may be connected for service with its longitudinal axis in either the vertical or the horizontal position, and so that its fixed maximum liquid level gauge, pressure relief device(s) and withdrawal appurtenance will function properly in either position.

Vaporizer.   A device other than a container which receives LP-Gas in liquid form and adds sufficient heat to convert the liquid to a gaseous state.

This was editorially revised in the 1986 Edition to facilitate the use of standardized classifications for the many types of liquefied gas vaporizers in several NFPA standards.

Vaporizer, Direct-Fired.   A vaporizer in which heat furnished by a flame is directly applied to some form of heat exchange surface in contact with the liquid LP-Gas to be vaporized. This classification includes submerged-combustion vaporizers.

Some have considered submerged-combustion vaporizers to be indirect-fired because they contain a noncombustible heat transfer medium. However, for purposes of this standard, they are treated as direct-fired. The second sentence was added in the 1986 Edition to make this clear.

Vaporizer, Electric.   A unit using electricity as a source of heat.

Because electric vaporizers can be designed so that they are not a source of ignition (unlike a vaporizer using fuel-fired burners), they can be treated differently and need to be defined separately. Electric vaporizers are further separated into two types, each of which may be designed to not be a source of ignition.

1. *Direct immersion electric vaporizer.* A vaporizer wherein an electric element is immersed directly in the LP-Gas liquid and vapor.

2. *Indirect electric vaporizer.* An immersion type wherein the electric element heats an interface solution in which the LP-Gas heat exchanger is immersed or heats an intermediate heat sink.

Vaporizer, Indirect (also called Indirect-Fired).   A vaporizer in which heat furnished by steam, hot water, the ground, surrounding air or other heating medium is applied to a vaporizing chamber or to tubing, pipe coils, or other heat exchange surface containing the liquid LP-Gas to be vaporized; the heating of the medium used being at a point remote from the vaporizer.

This definition was changed in the 1989 Edition to clarify that ambient air or earth are sources of heat for some indirect vaporizers.

Vaporizer, Waterbath (also called Immersion Type).   A vaporizer in which a vaporizing chamber, tubing, pipe coils, or other heat exchange surface containing liquid LP-Gas to be vaporized is immersed in a temperature controlled bath of water, water-glycol combination, or other noncombustible heat transfer medium, which is heated by an immersion heater not in contact with the LP-Gas heat exchange surface.

Although the inference was that the heat transfer medium was not capable of burning, the word "noncombustible" was added in the 1986 Edition to make this clear.

Vaporizing-Burner (also called Vaporizer-Burner and Self-Vaporizing Liquid Burner).   A burner containing an integral vaporizer which receives LP-Gas in liquid form and which uses part of the heat generated by the burner to vaporize the liquid in the burner so that it is burned as a vapor.

Variable Liquid Level Gauge.   A device to indicate the liquid level in a container throughout a range of levels. See Float, Rotary, and Slip Tube Gauge.

Volumetric Filling.   Filling a container by determination of the volume of LP-Gas in the container. Unless a container is filled by a fixed maximum liquid level gauge, correction of the volume for liquid temperature is necessary.

Gauging the quantity of liquid in a container by means of its volume rather than its weight is a practical necessity on ASME containers, cargo vehicles, tank cars, and larger DOT containers, and is a convenience on smaller DOT containers.

Slip tube, rotary, and fixed tube gauges are used. Because these gauges must be manipulated during gauging and, with the exception of the fixed maximum liquid level gauge, the temperature of the liquid during filling must be taken into consideration, the volumetric method may be somewhat less precise than the weight method in practice. The degree of accuracy has seldom been a problem. However, the smaller the container, the greater the opportunity for overfilling, whether filling by volume or weight.

Appendix F discusses at considerable length the factors involved in volumetric filling.

Volumetric Loading.   See Volumetric Filling.

Water Capacity. The amount of water, in either lb or gal, at 60°F (15.6°C) required to fill a container liquid full of water.

Many years ago the U.S. Department of Transportation (ICC) drew up regulations that limit the amount of gases that may be charged into containers. NFPA 58 provisions are based on these regulations. With respect to LP-Gas, which is a liquefied compressed gas, the filling limitation is primarily for the purpose of preventing excess hydrostatic pressures if liquid expansion results in a liquid-full container. A certain outage is left in the container to account for this expansion. DOT regulations use a filling-density concept to designate the maximum amount of liquid LP-Gas that may be placed into a container and, thus, what the outage should be. Filling density, as defined in 4-5.2.2, is essentially a ratio of the quantity of LP-Gas in a container to the water capacity of the container. Water capacity of a container is a measure of its total volume and is the only way to express its gross capacity.

Water capacity may be expressed in gallons or pounds of water. Where expressed in pounds, it reflects both the feasibility and traditional practice of filling by weight using a scale. This is obviously limited to containers small enough to be moved by a person — usually up to about 240 lb (109 kg) water capacity. Where capacity is expressed in gallons, it usually applies to containers larger than 240 lb (109 kg) water capacity. In recent years, however, it is not unusual to find containers as small as 50 lb (23 kg) water capacity filled by volume.

Weight Filling. Filling containers by weighing the LP-Gas in the container. No temperature determination or correction is required as a unit of weight is a constant quantity regardless of temperature.

REFERENCES CITED IN COMMENTARY

The following publication is available from the National Fire Protection Association, Batterymarch Park, Quincy, MA 02269.

NFPA 54, *National Fuel Gas Code.*

The following publication is available from U.S. Government Printing Office, Washington, DC. Parts 171-190 are also available from the Association of American Railroads, American Railroads Bldg, 1920 L St., NW, Washington, DC 20036 and the American Trucking Associations, Inc., 2201 Mill Rd., Alexandria, VA 22314.

Code of Federal Regulations, Title 49, Parts 100-199.

The following publication is available from the Compressed Gas Association, Inc., 1235 Jefferson Davis Highway, Arlington, VA 22202.

ANSI B57.1/CGA V-1/CSA B 96, *Standard Compressed Gas Cylinder Valve Outlet and Inlet Connections.*

The following publication is available from the National LP-Gas Association, 1301 West 22nd St., Oak Brook, IL 60521.

A Study of Cleaning Methods for LP-Gas Transport.

The following publication is available from the American Society for Testing and Materials, 1916 Race St., Philadelphia, PA 19103.

ASTM D 1835, *Standard Specification for Liquefied Petroleum (LP) Gases.*

The following publication is available from the Gas Processors Association, 1812 First National Building, Tulsa, OK 74103.

Liquefied Petroleum Gas Specifications for Test Methods Standard 2140.

The following publications are available from Underwriters Laboratories Inc., 333 Pfingsten Rd., Northbrook, IL 60062.

Automotive, Burglary Protection, and Mechanical Equipment.

Gas and Oil Equipment.

Hazardous Location Equipment.

Marine Products.

For information on the ANSI Z21 and Z83 Series, see the reference list following Chapter 2.

2

LP-GAS EQUIPMENT AND APPLIANCES

2-1 Scope.

2-1.1 Application.

2-1.1.1 This chapter includes the basic provisions for individual components, or for such components shop-fabricated into subassemblies, container assemblies, or complete container systems.

The intent of Chapter 2 is to include all information needed by the designer and fabricator of LP-Gas system components, with the exception of cargo tank vehicles which are covered by Chapter 6.

2-1.1.2 The field assembly of components, subassemblies, container assemblies or complete container systems into complete LP-Gas systems is covered by Chapter 3. (*See Definition of LP-Gas System.*)

2-2 Containers.

2-2.1 General.

2-2.1.1 This section includes design, fabrication, and marking provisions for containers, and features normally associated with container fabrication, such as container openings, appurtenances required for these openings to make the containers gastight entities, physical damage protecting devices, and container supports attached to, or furnished with the container by the manufacturer.

2-2.1.2 Nonrefrigerated containers shall comply with 2-2.1.3 or shall be designed, fabricated, tested, and marked using criteria which incorporate an investigation to determine that it is safe and suitable for the proposed service, is recommended for that service by the manufacturer, and is acceptable to the authority having jurisdiction. Refrigerated containers shall comply with Chapter 8.

To the editors' knowledge, all pressurized LP-Gas containers in the U.S., Canada, and other countries served by marketers based in the U.S. and Canada comply with 2-2.1.3. Other countries that use NFPA 58 have their own container require-

ments. Furthermore, a safe container can be built using other criteria and this flexibility is desirable in the standard.

2-2.1.3* Containers shall be designed, fabricated, tested, and marked (or stamped) in accordance with the Regulations of the U.S. Department of Transportation (DOT), the "Rules for the Construction of Unfired Pressure Vessels," Section VIII, Division 1, ASME *Boiler and Pressure Vessel Code*, or the API-ASME *Code for Unfired Pressure Vessels for Petroleum Liquids and Gases* applicable at the date of manufacture; and as follows:

A-2-2.1.3 Prior to April 1, 1967, these regulations were promulgated by the Interstate Commerce Commission. In Canada, the regulations of the Canadian Transport Commission apply.

Construction of containers to the API-ASME Code has not been authorized after July 1, 1961.

Containers originally used for LP-Gas were Interstate Commerce Commission (ICC) cylinders. DOT/ICC restricts cylinders to 1,000 lb (454 kg) of water capacity and less. In the early days there were some ICC specifications, such as ICC-26 and others, but these have not been used for many years. The current basic container for reusable service is a DOT 4BA-240 cylinder (steel construction) and a DOT 4E specification (aluminum construction). There are also "throw-away" cylinders, such as the hand torch cylinders, which have different DOT specifications (or old ICC specifications) — currently DOT 39.

Even though the responsibilities of the U.S. Department of Transportation and Canadian Transport Commission are restricted to the transportation environment and much of the application of NFPA 58 is concerned with container usage and storage, there are compelling reasons why the standard must recognize these containers. From a safety point of view, it would be extremely hazardous to require that the LP-Gas be transferred from a DOT shipping container to another use or storage container on the premises of the user or storer. Such a procedure would also be adverse from an economic standpoint. However, the NFPA Technical Committee continuously monitors the DOT and CTC standards to assure that they are adequate for the applications covered by NFPA 58. For example, the Committee did not permit the 4E (aluminum) cylinder for several years until its use and storage safety had been demonstrated by tests and experience.

While all DOT/ICC/CTC containers (cylinders) are portable in the sense that they are designed to be transportable full of product, when it is considered that such containers commonly hold from 1 lb (0.5 kg) to 420 lb (190 kg) of propane, the degree

Table 2.1 DOT/ICC/CTC Cylinder Applications.

Type of Service	Typical Use	Propane Capacity (lb) (kg)	Propane Capacity (gal)	Water Capacity (lb) (kg)	Common DOT Mfg. Code
Stationary	Homes, Business	420 (191)	99	1,000 (454)	4B, 4BA, 4BW
Stationary	Homes, Business	300 (136)	71	715 (324)	4B, 4BA, 4BW
Stationary	Homes, Business	200 (91)	47	477 (216)	4B, 4BA, 4BW
Stationary	Homes, Business	150 (68)	35	357 (162)	4B
Exchange	Homes, Business	100 (45)	24	239 (108)	4B, 4BA, 4BW
Exchange	Homes, Business	60 (27)	14	144 (65)	4B, 4BA, 4BW
Motor Fuel	Tractor	100 (45)	24	239 (108)	4B, 4BA, 4BW
Motor Fuel	Tractor	60 (27)	14	144 (65)	4B, 4BA, 4BW
Motor Fuel	Forklift	43.5 (19.7)	10	104 (47)	4B, 4BA, 4BW, 4E
Motor Fuel	Forklift	33.5 (15.2)	8	80 (36)	4B, 4BA, 4BW, 4E
Motor Fuel	Forklift	20 (9)	4.7	48 (22)	4B, 4BA, 4BW, 4E
Motor Fuel	Forklift	14 (6.4)	3.3	34 (15.4)	4B, 4BA, 4BW, 4E
Portable	Rec. Vehicles	40 (18)	9.5	95 (43)	4B, 4BA, 4BW, 4E
Portable	Rec. Vehicles	30 (13.6)	7.1	72 (32.7)	4B, 4BA, 4BW, 4E
Portable	Rec. Vehicles	25 (11.3)	5.9	59.5 (27)	4B, 4BA, 4BW
Portable	Rec. Vehicles, Grills	20 (9)	4.7	48 (22)	4B, 4BA, 4BW, 4E
Portable	Rec. Vehicles & Sm. Ind.	10 (4.5)	2.4	23.8 (10.8)	4B, 4BA, 4BW, 4E
Portable	Indoors, Trailers	5 (2.3)	1.2	12 (5.4)	4B, 4BA, 4BW, 4E
Portable	Torches, RV	.93 lb (1 lb) (.42)	0.2	2.2 (1)	39 (disposable) 4B240 (refillable)

of portability varies widely. In practice, cylinders in portable service are those that can be moved with reasonable ease by a strong individual, or those that contain up to about 100 lb (45 kg) of LP-Gas [about 200 lb (90 kg) total weight of cylinder and product]. These containers are usually filled at a location other than where they are installed or used [the 100 lb (45 kg) size is often filled at the installed location as well]. Containers larger than these usually are filled at the installed location and are considered to be in stationary service.

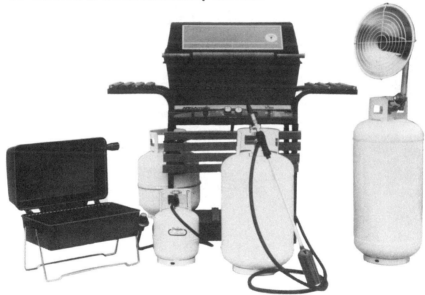

Figure 2.1 DOT/ICC/CTC Portable Cylinders.

A DOT/ICC cylinder is required to be marked with certain information — usually on the cylinder itself or its neck ring or collar. This information is especially vital in determining the suitability of the cylinder for continued service indicated by the retest date.

The second type of container is the "bulk container," as exemplified by the ASME unfired pressure vessel. The original LP-Gas bulk containers were riveted but there was great difficulty in keeping them leak free around the rivets and at the seams. The industry then began to forge weld containers, and in the 1930s began to use the fusion welded pressure vessels. At that time the ASME Code specifications were U-68 and U-69, which had a factor of safety of 5 to 1 (design operating pressure to theoretical burst pressure). The normal working pressure of these containers was 200 psig (1.4 MPa gauge). At that time,

Figure 2.2 DOT/ICC/CTC Industrial Truck Motor (Engine) Fuel Cylinders.

NFPA 58 allowed the pressure relief valve setting to be 125 percent of the working pressure, or 250 psig (1.7 MPa gauge). This was done to ensure that there would be no pressure relief of the flammable product into the atmosphere as a result of ambient air temperatures or solar radiation, which would create a hazard. Prior to 1949, there were also ASME specifications U-200 and U-201, which were developed in conjunction with the American Petroleum Institute. These were known as the API-ASME Code specifications. Such containers, however, were restricted in use to installations such as refineries, gas processing plants, and tank farms.

In 1949 ASME adopted the current ASME Code precept with a factor of safety of 4 to 1. At that time the requirement for the pressure relief valve setting was changed to the working pressure of a container which generally was 250 psig (1.7 MPa

Figure 2.3 DOT/ICC/CTC Stationary and Portable Cylinders.

gauge). In effect, the previous 4 to 1 factor of safety and the 5 to 1 factor of safety were comparable from an ultimate burst standpoint because 200 × 5 and 250 × 4 both equal 1,000.

With the exception of those used in vehicular systems, ASME Code containers are usually found in stationary installations.

At times there are ASME Code Case Interpretations and Addenda to the base code; they are also applicable to ASME containers. See Appendices C and D for further description of DOT/CTC and ASME Code containers.

(a) Adherence to applicable ASME Code Case Interpretations and Addenda shall be considered as compliance with the ASME Code.

(b) Containers fabricated to earlier editions of regulations, rules or codes listed in 2-2.1.3 and the ICC *Rules for Construction of Unfired Pressure Vessels*, prior to April 1, 1967, may continue in use in accordance with 1-2.4.1. (*See Appendices C and D.*)

2-2.1.4 Containers complying with 2-2.1.3 may be reused, reinstalled, or continued in use as follows:

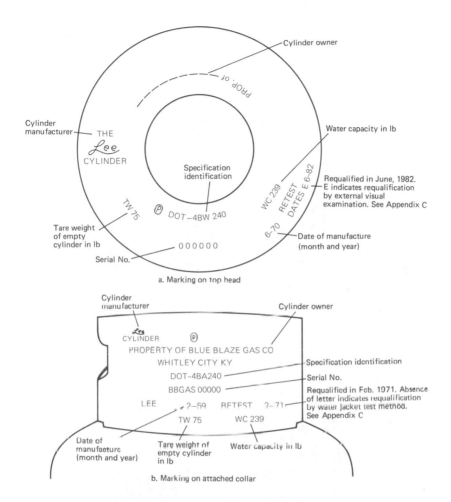

Figure 2.4 Typical DOT Cylinder Marking.

Maintenance of the container is extremely important. Containers may be subjected to conditions at the point of installation which can create corrosion. In addition, the handling of cylinders can result in cuts, gouges, and dents which can weaken them. It is the responsibility of the owner of the container to see that it is properly maintained and requalified, that the container is repaired if such can be done in a safe manner and, if not, that the container is withdrawn from service and scrapped.

In regard to DOT and CTC cylinders, the U.S. Department of Transportation and Canadian Transport Commission have regulations for retest or requalification of such cylinders (*see Appendix C*).

Corrosion is a major factor in the failure of DOT cylinders,

Table 2.2   Typical Stationary ASME Container Applications.

Type of Service	Water Capacity (in Gallons)	LP-Gas Capacity (in Gallons*)	LP-Gas Capacity (in Pounds)
Domestic	100 (379 L)	80 (301 L)	338
Domestic	125 (473 L)	100 (379 L)	423
Domestic	150 (568 L)	120 (454 L)	508
Domestic	250 (946 L)	200 (757 L)	848
Domestic	325 (1 230 L)	260 (984 L)	
Domestic	500 (1 893 L)	400 (1 514 L)	
Domestic	1,000 (3.8 m³)	800 (3 m³)	
Industrial/Agricultural/Commercial	1,000–5,000 (3.8–19 m³)	800–4,500 (3–17 m³)	
Service Stations	1,000–6,500 (3.8–24.6 m³)	800–5,850 (3–22 m³)	
Bulk Plant or Standby Storage	12,000–18,000 (45.4–68 m³)	10,800–16,200 (41–61 m³)	
Bulk Plant or Standby Storage	20,000–30,000 (76–114 m³)	18,000–27,000 (45.4–102 m³)	
Bulk Plant or Standby Storage	30,000–60,000 (114–227 m³)	27,000–54,000 (102–204 m³)	
Bulk Plant or Standby Storage	60,000–120,000 (227–454 m³)	48,000–96,000 (182–364 m³)	

* Based on propane specific gravity of 0.508 at 60°F (15.6°C). Actual quantity depends upon actual specific gravity.

Dome

Barrel or Shell

Seam

Head

Foot

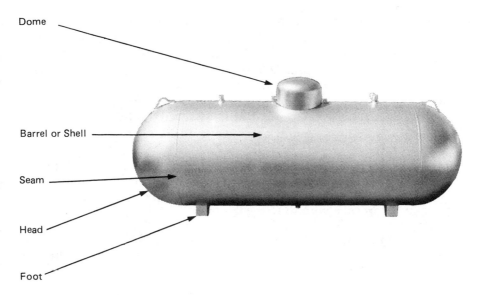

Figure 2.5 Small ASME Container. Used commonly in residential and commercial applications.

especially where the wearing ring is attached to the bottom of the cylinder. This area is especially affected by the elements and there will be cycles of drying and wetting which accelerate corrosion. The cylinder has to be upended in order for this area to be checked, and it has to be done very thoroughly or small areas of heavy corrosion can be missed. Many times there will be deep corrosion under scale on the surface and the deep corrosive spot will not be observed. Then when the cylinder is put back into use, it can fail at such a point. All scale, rust, and corrosion must be thoroughly removed at the time of retest or requalification. This is especially the case in regard to the area where the wearing ring is attached to the bottom of the container. After the rusted or corroded area is properly cleaned, the cylinder surface should be well coated with a protective coating to prevent future rust and corrosion.

When a cylinder is removed from service and discarded, it should be cut up so that it cannot be repaired by a junkyard or others and put back into pressure service. Care should be used when it is cut up to ensure that it has been properly purged of all product, including vapors. Compressed Gas Association Pamphlet C-2, *Recommendations for the Disposition of Unserviceable Compressed Gas Cylinders*, is useful in these circumstances.

The retesting and requalification of DOT and CTC cylinders are best done by a qualified test organization or, if not, then by a thoroughly qualified and trained individual.

If a cylinder has been subjected to fire, the steel or the aluminum can be weakened and it can be requalified only by a properly qualified repair facility authorized by DOT or CTC to do such requalification. When an ASME Code container has been subjected to fire, it must be retested using the hydrostatic test procedure applicable at the time it was originally fabricated, because the fire heat can alter the properties of the steel and result in weakness from a pressure holding standpoint.

(a) The owner of a container shall be responsible for its suitability for continued service. DOT cylinders shall not be refilled, continued in service or transported unless they are properly qualified or requalified for LP-Gas service in accordance with DOT regulations.

(b) Containers which have been involved in a fire and showing no distortion shall be requalified for continued service before being used or reinstalled as follows:

(1)   DOT containers shall be requalified by a manufacturer of the type of cylinder to be requalified or by a repair facility approved by DOT.

Exception: DOT 4E specification (aluminum) cylinders shall be permanently removed from service.

(2)   ASME or API-ASME containers shall be retested, using the hydrostatic test procedure applicable at the time of original fabrication.

2-2.1.5   Containers showing serious denting, bulging, gouging, or excessive corrosion shall be removed from service.

This includes ASME and API/ASME containers as subject to inspection. Prior to the 1986 Edition, only DOT/CTC containers were addressed in this context. ASME containers are being used more — especially in vehicular applications where they are more subject to physical damage and corrosion.

2-2.1.6   Repair or alteration of containers shall comply with the Regulations, Rules or Code under which the container was fabricated. Other welding is permitted only on saddle plates, lugs, or brackets attached to the container by the container manufacturer.

The heat of welding directly to the container can affect the strength of the metal and change its characteristics so that it does not meet the Code. Such heating can set up localized stressing which can affect materially the strength of the metal.

2-2.1.7   Containers for general use shall not have individual water capacities greater than 120,000 gal (454 m³). Containers in service stations shall not have individual water capacities greater than 30,000 gal (114 m³).

This limits the size of containers that may be installed for general use and in service stations. The 30,000 gallon (114 m³) limit on container size in service stations is a practical one as this common container size is the largest that can be easily transported by truck. The term "general use," which is not defined in the standard, is intended to mean installations where LP-Gas is not the primary or sole product or raw material used. Thus an LP-Gas distribution plant where most employees can be expected to be aware of the proper procedures for handling LP-Gas is *not* general use, while a container used as an alternate fuel supply at a manufacturing plant whose product is not related to LP-Gas *is* general use.

2-2.1.8 Heating or cooling coils shall not be installed inside storage containers.

If there is leakage from heating or cooling coils installed inside of the storage container, the gases can get out into the piping of the heating or cooling system because the LP-Gas pressure is normally higher. This can result in a pressure that can rupture or create leaks in the heating or cooling system. Even if this doesn't occur, a flammable product will be present in the heating or cooling system.

2-2.2 Container Design or Service Pressure.

2-2.2.1 The minimum design, or service, pressure of DOT specification containers shall be in accordance with the appropriate DOT regulations.

See Appendix C.

2-2.2.2 The minimum design pressure for ASME containers shall be in accordance with Table 2-2.2.2.

See Appendix D.

2-2.2.3 In addition to the applicable provisions for horizontal ASME storage containers, vertical ASME storage containers over 125 gal (0.5 m³) water capacity shall comply with 2-2.2.3(a) through (e).

When a container is installed in a vertical rather than a horizontal position, in addition to the pressure factors involved there is also an increased weight load in the vertical container.

Table 2-2.2.2

For Gases with Vapor Pressure in psig (MPa gauge) at 100°F (37.8°C) not to Exceed	Minimum Design Pressure in psig (MPa gauge) ASME Code, Section VIII, Division 1, 1986 Edition (Note 1)
80 (0.6)	100 (0.7) (Note 2)
100 (0.7)	125 (0.9)
125 (0.9)	156 (1.1)
150 (1.0)	187 (1.3)
175 (1.2)	219 (1.5)
215 (1.5)	250 (1.7)
215 (1.5)	312.5 (2.2) (Note 3)

Note 1:   See Appendix D for information on earlier ASME or API-ASME Code.

Note 2:   New containers for 100 psig (0.7 MPa gauge) design pressure (or equivalent under earlier codes) not authorized after December 31, 1947.

Note 3:   See 3-6.2.2 for certain service conditions which require a higher pressure relief valve start-to-leak setting.

As a result, some changes may have to be made in the design specifications for the vertical container as opposed to those of the horizontal container. In addition, the methods of support of the vertical container must be designed to take into account the greater concentration of stresses. These methods are addressed further in 2-2.5.3.

Lifting lugs are not designed to carry the load of lifting the container when it is full of LP-Gas or water (the latter for testing or purging).

This provision was changed in the 1989 Edition to reduce the minimum size of vertical containers covered from 2,000 gal (7.6 m³) to 125 gal (0.5m³) to correct an omission of containers of 125 to 2000 gal (0.5 m³ to 7.6 m³) made when the standard was reorganized in 1972, and to clarify that only shop-fabricated containers require lifting lugs.

(a) Containers shall be designed to be self-supporting without the use of guy wires and shall satisfy proper design criteria taking into account wind, seismic (earthquake) forces, and hydrostatic test loads.

(b) Design pressure (see Table 2-2.2.2) shall be interpreted as the pressure at the top head with allowance made for increased pressure on lower shell sections and bottom head due to the static pressure of the product.

(c) Wind loading on containers shall be based on wind pressures on the projected area at various height zones aboveground in accordance with *Design Loads for Buildings and Other Structures*, ANSI A58.1. Wind speeds shall be based on a Mean Occurrence Interval of 100 years.

(d) Seismic loading on containers shall be based on forces recommended in the ICBO *Uniform Building Code*. In those areas identified as zones 3 and 4 on the Seismic Risk Map of the United States, Figures 1, 2 and 3 of Chapter 23 of the UBC, a seismic analysis of the proposed installation shall be made which meets the approval of the authority having jurisdiction.

(e) Shop-fabricated containers shall be fabricated with lifting lugs or some other suitable means to facilitate erection in the field.

2-2.3 Container Openings.

2-2.3.1 Containers shall be equipped with openings suitable for the service in which the container is to be used. Such openings may be either in the container proper or in the manhole cover, or part in one and part in the other.

The number of openings in the container is not limited by the standard. Prior to the 1989 Edition of the standard no more than two plugged openings were permitted in a 30 to 1,999 gal (0.1 to 7.6 m³) container. The provision—2-2.3.3 in the 1986 Edition—was deleted because the Technical Committee could see no reason why a properly installed plug would not provide an effective seal and could find no other reason to continue the restriction.

2-2.3.2 Containers of more than 30 gal (0.1 m³) and less than 2,000 gal (7.6 m³) water capacity, designed to be filled volumetrically, and manufactured after December 1, 1963, shall be equipped for filling into the vapor space.

The reason for filling into the vapor space only and not through the liquid is to achieve equilibrium more rapidly between the vapor and liquid in the container. It has been demonstrated repeatedly that lower container pressure results when filling into the vapor space, compared to filling into the liquid space. This is a matter of real, practical interest as the time it takes to fill cylinders, especially in the summer, can be lengthy. A valve manufacturer has reported a 27 percent increase in the filling rate of a 100 lb (45 kg) cylinder when the cylinder valve was equipped with a spray filler that sprayed the liquid into the cylinder rather than admitting a solid stream of liquid.

2-2.3.3 Containers of 125 gal (0.5 m³) or more water capacity manufactured after July 1, 1961, shall be provided with a connection for liquid evacuation, not smaller than ¾ in. National Pipe Thread. A plugged opening will not comply with this provision.

The provision for a liquid evacuation connection is both safety and utility oriented. From the safety standpoint a tank which has been damaged can become a severe hazard if there is no method of removing the liquid. Liquid can, of course, be removed from a vapor connection by rolling or tipping the tank but this can be a hazardous procedure.

Although the provision does not specifically state, the intent is that the connection cited allows evacuation of the tank without moving the tank and with a device such as shown in Figure 2.6 (a liquid withdrawal valve) specifically designed for this purpose installed in the connection. This is the reason for stating that a plugged opening will not comply with the provision. There are other reasons for evacuating the tank other than those caused by an accident. Note that paragraph 6-5.2.1 requires containers of 125 gal (0.5 m³) or more water capacity, which are designed for stationary service, must contain no more then 5 percent of their capacity in liquid during transportation. The selection of 5 percent was for a number of reasons, not the least of which was that most float gauges on domestic size tanks, that are often transported with some liquid in them, would not read below 5 percent. However, it was believed that 5 percent left in the tank during transportation did not create a hazard. This enabled safe evacuation of the tank when the tank was to be moved from one location to another. Providing a method of removing most of the liquid in the tank prior to moving it reduces the hazard of possible accident during such operation.

Liquid withdrawal valves are combination check and excess-flow valves that are actuated by threading an adapter or pipe nipple into the exposed threaded portion of the valve. This opens a #80 drill size hole and allows liquid propane to escape. When a valve is threaded into the adapter or nipple (and the valve is closed) pressure will build up and the excess-flow valve will open. The valve will remain open until the flow through it exceeds the valve design flow rate, or the threaded valve or adapter is removed.

A frequently asked question regards whether liquid withdrawal valves may be used for liquid withdrawal from containers in liquid withdrawal service. The liquid withdrawal valve may be used for normal service if it meets all the requirements of the standard for container appurtenances (see 2-3 and 3-2.4). If in doubt, the manufacturer of the liquid withdrawal valve should be contacted to determine if the valve is designed for continuous service, or only for occasional use.

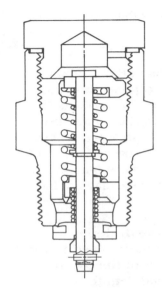

Figure 2.6 Special Fitting Used to Accomplish the Provision of 2-2.3.3 on Containers not Fitted for Liquid Withdrawal for Operational Purposes. Valve is a combination check valve and excess-flow check valve. Check valve is opened by insertion of an adapter (in place of caps) which opens the check valve against spring tension. The same functions can be achieved by using separate fittings.

2-2.3.4   Containers of more than 2,000 gal (7.6 m³) water capacity and all containers installed in LP-Gas service stations shall be provided with an opening for a pressure gauge (*see 2-3.5.2*).

   Knowledge of container pressure is useful on larger containers, especially during liquid transfer operations. The pressure in containers on vehicles can be quite varied; knowledge of storage container pressure is valuable regardless of container size in service stations.

2-2.3.5   Connections for pressure relief valves shall be located and installed in such a way as to have direct communication with the vapor space, whether the container is in storage or in use.

   (a) If located in a well inside the container with piping to the vapor space, the design of the well and piping shall permit sufficient pressure relief valve relieving capacity.

   (b) If located in a protecting enclosure, design shall be such as to permit this enclosure to be protected against corrosion and to permit inspection.

   (c) If located in any position other than uppermost point of the container, it shall be internally piped to the uppermost point practical in the vapor space of the container.

This provision has two purposes. Because of the liquid-to-vapor expansion ratio, a liquid discharge represents a much greater quantity of gas in the atmosphere. If vapor is discharged, the liquid is vaporizing inside the container and using the heat in the liquid to do so. This limits the pressure, and the relieving flow capacity of the relief valve can take advantage of this. If liquid is discharged, this effect takes place outside the container, is lost to the relief valve, and the valve may not have an adequate capacity. This is particularly critical if the valve is discharging as a result of fire exposure.

Note that the relief valve must be located either at the uppermost point of the container, or internally piped to the uppermost point practical in the vapor space of the container. This was added as a Tentative Interim Amendment to the 1986 Edition of the standard, and added in the 1989 Edition when the Committee became aware that some containers (in motor fuel and recreational vehicle applications) were being manufactured with the relief valve located just above the 80 percent liquid level. Incidents had occurred where these containers had released liquid LP-Gas upon relief valve operation resulting in a fire in at least one case. The provision that the relief valve be piped to the uppermost point practical in the container recognizes that in some containers, such as portable lift truck cylinders which are stored vertically and used horizontally, the relief valve cannot be piped to the uppermost point for both use positions.

2-2.3.6   Containers to be filled on a volumetric basis manufactured after December 31, 1965, shall be fabricated so that they can be equipped with a fixed liquid level gauge(s) capable of indicating the maximum permitted filling level(s) in accordance with 4-5.2.3.

The variable types of liquid level gauges are not 100 percent reliable. The float may develop a leak, the float arm may sag, the mechanism may bind, or the gauge may otherwise indicate an inaccurate reading. Therefore, a fixed liquid level gauge is required to be used in filling the container. [Note that paragraph 4-5.3.3(c)(1) requires that the variable gauge be checked for accuracy by comparison to the fixed maximum liquid level gauge.] In this way, overfilling can be prevented. If containers are filled only by weight, a fixed liquid level gauge is not required.

2-2.4 Portable Container Appurtenance Physical Damage Protection.

2-2.4.1 Portable containers of 1,000 lb (454 kg) [nominal 120 gal (0.5 m³)] water capacity or less shall incorporate protection against physical damage to container appurtenances and immediate connections to these while in transit, storage, while being moved into position for use, and when in use except in residential and commercial installations, by:

There have been a number of serious accidents because the valving on portable cylinders has not been properly protected both in transportation and in moving the container into place for use. The major cause of these accidents has been failure to have the valves properly protected with a cap or attachable collar. These should be in place at all times except when the container is properly installed and connected for use, or while it is being filled in a proper filling plant. The valve on the cylinder should not be used for lifting or moving the cylinder because this puts a strain on it which it is not designed to endure.

(a) Recessing connections into the container so that valves will not be struck if the container is dropped on a flat surface, or,

(b) A ventilated cap or collar designed to permit adequate pressure relief valve discharge and capable of withstanding a blow from any direction equivalent to that of a 30 lb (14 kg) weight dropped 4 ft (1.2 m). Construction shall be such that the force of the blow will not be transmitted to the valve. Collars shall be designed so that they do not interfere with the free operation of the cylinder valve.

2-2.4.2 Portable containers of more than 1,000 lb (454 kg) [nominal 120 gal (0.5 m³)] water capacity, including skid tanks or for use as cargo containers, shall incorporate protection against physical damage to container appurtenances by recessing, protective housings, or by location on the vehicle. Such protection shall comply with the provisions under which the tanks are fabricated, and shall be designed to withstand static loadings in any direction equal to twice the weight of the container and attachments when filled with LP-Gas, using a safety factor of not less than four, based on the ultimate strength of the material to be used. (*See Chapters 3 and 6 for additional provisions applying to the LP-Gas system used.*)

2-2.5 Containers with Attached Supports.

2-2.5.1 Horizontal containers of more than 2,000 gal (7.6 m³) water capacity designed for permanent installation in stationary service may be provided with steel saddles designed to permit mounting the containers on flat topped concrete foundations. The total height of the outside bottom of the container shell above the top of the concrete foundation shall not exceed 6 in. (152 mm).

2-2.5.2 Horizontal containers of 2,000 gal (7.6 m³) water capacity or less, designed for permanent installation in stationary service, may be equipped with nonfireproofed structural steel supports and designed to permit mounting on firm foundations in accordance with 2-2.5.2 (a) or (b).

(a) For installation on concrete foundations raised above the ground level by more than 12 in. (305 mm), the structural steel supports shall be designed so that the bottoms of the horizontal members are not less than 2 in. (51 mm), nor more than 12 in. (305 mm) below the outside bottom of the container shell.

(b) For installation on paved surfaces or concrete pads within 4 in. (102 mm) of ground level, the structural steel supports may be designed so that the bottoms of the structural members are not more than 24 in. (610 mm) below the outside bottom of the container shell. [*See 3-2.3.2(a)(3) for installation provisions for such containers which are customarily used as components of prefabricated container-pump assemblies.*]

2-2.5.3 Vertical ASME containers over 125 gal (0.5 m³) water capacity designed for permanent installation in stationary service shall be designed with steel supports designed to permit mounting the container on, and fastening it to, concrete foundations or supports. Such steel supports shall be designed to make the container self-supporting without guy wires and shall satisfy proper design criteria, taking into account wind, seismic (earthquake) forces, and hydrostatic test load criteria established in 2-2.2.3.

(a) The steel supports shall be protected against fire exposure with a material having a fire resistance rating of at least two hours. Continuous steel skirts having only one opening 18 in. (457 m) or less in diameter need such fire protection applied only to the outside of the skirt.

> **This provision was changed in the 1989 Edition to reduce the minimum size of vertical containers covered from 2,000 gal (7.6 m³) to 125 gal (0.5 m³) to correct an omission of containers of 125 to 2000 gal (0.5 to 7.6 m³) made when the standard was reorganized in 1972.**

2-2.5.4 Containers to be used as portable storage containers (*see definition*) for temporary stationary service (normally less than 12 months at any given location) and to be moved only when substantially empty of liquid shall comply with 2-2.5.4(a) and (b).

(a) If mounted on legs or supports, such supports shall be of steel, and shall either be welded to the container by the manufacturer at the time of fabrication or shall be attached to lugs which have been so welded to the container. The legs or supports or the lugs for the attachment of these legs or supports shall be secured to the container in accordance with the code or rule under which the container is designed and built, with a minimum factor of safety of four, to withstand loading in any direction equal to twice the weight of the empty container and attachments.

(b) If the container is mounted on a trailer or semitrailer running gear so that the unit can be moved by a conventional over-the-road tractor, attachment to the vehicle, or attachments to the container to make it a vehicle, shall comply with the appropriate DOT requirements for cargo tank service; except that stress calculations shall be based on twice the weight of the empty container. The unit shall also comply with applicable State and DOT motor carrier regulations and shall be approved by the authority having jurisdiction.

2-2.5.5 Portable tanks (*see definition*) shall comply with DOT portable tank container specifications as to container design and construction, securing of skids or lugs for the attachment of skids and protection of fittings. In addition, the bottom of the skids shall be not less than 2 in. (51 mm) or more than 12 in. (305 mm) below the outside bottom of the container shell.

2-2.6 Container Markings.

2-2.6.1 Containers shall be marked as provided in the Regulations, Rules or Code under which they are fabricated and in accordance with 2-2.6.2 through 2-2.6.5 as applicable.

2-2.6.2 When LP-Gas and one or more other compressed gases are to be stored or used in the same area, the containers shall be be marked "Flammable" and either "LP-Gas," "LPG," "Propane" or "Butane." Compliance with marking requirements of Title 49 of the *Code of Federal Regulations* shall meet this provision.

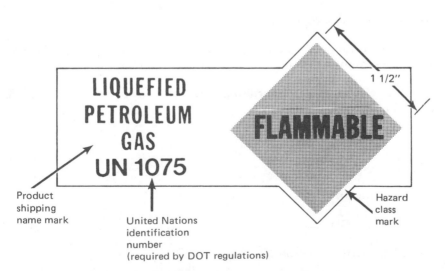

Figure 2.7 Label Marking for DOT Container.

2-2.6.3 When being transported, portable DOT containers shall be marked and labeled in accordance with Title 49 of the *Code of Federal Regulations*.

2-2.6.4 Portable DOT containers designed to be filled by weight, including those optionally filled volumetrically but which may require check weighing, shall be marked with:

(a) The water capacity of the container in lb.

(b) The tare weight of the container in lb, fitted for service. The tare weight is the container weight plus the weight of all permanently attached valves and other fittings, but does not include the weight of protecting devices removed in order to load the container.

See Figure 2.4.

2-2.6.5 ASME containers shall be marked in accordance with 2-2.6.5(a) through (1). The marking specified shall be on a stainless steel metal nameplate attached to the container, so located as to remain visible after the container is installed. The nameplate shall be attached in such a way to minimize corrosion of the nameplate or its fastening means and not contribute to corrosion of the container.

(a) Service for which the container is designed; i.e. underground, aboveground, or both.

(b) Name and address of container supplier or trade name of container.

(c) Water capacity of container in lb or U.S. Gallons.

(d) Design pressure in psig.

(e) The wording "This container shall not contain a product having a vapor pressure in excess of _____ psig at 100°F." (*See Table 2-2.2.2.*)

(f) Tare weight of container fitted for service for containers to be filled by weight.

(g) Outside surface area in sq ft.

(h) Year of manufacture.

(i) Shell thickness _____ head thickness.

(j) OL _____ OD _____ HD _____.

(k) Manufacturer's Serial Number.

(l) ASME Code Symbol.

Note that the container nameplate must be stainless steel, and must be attached to the container to eliminate the possibility of corrosion to the nameplate, fasteners, or the container. Since under the ASME Code the nameplate must be attached to a container (or it does not meet the ASME Code), the need for this provision is consistent with the long service life of propane containers.

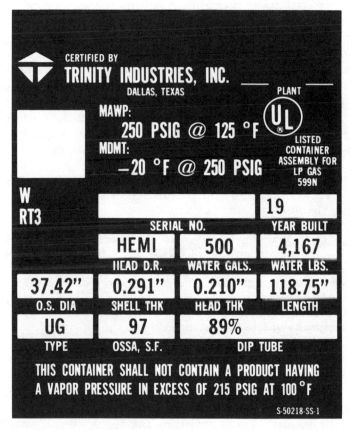

CERTIFIED BY
TRINITY INDUSTRIES, INC.
DALLAS, TEXAS   PLANT

MAWP:
250 PSIG @ 125 °F
MDMT:
−20 °F @ 250 PSIG

LISTED CONTAINER ASSEMBLY FOR LP GAS 599N

W RT3

			19
	SERIAL NO.		YEAR BUILT
	HEMI	500	4,167
	HEAD D.R.	WATER GALS.	WATER LBS.
37.42"	0.291"	0.210"	118.75"
O.S. DIA	SHELL THK	HEAD THK	LENGTH
UG	97	89%	
TYPE	OSSA, S.F.	DIP TUBE	

THIS CONTAINER SHALL NOT CONTAIN A PRODUCT HAVING A VAPOR PRESSURE IN EXCESS OF 215 PSIG AT 100 °F

S-50218-SS-1

Figure 2.8 Marking Required by 2-2.6.5 as Given on a Nameplate. Container listing [in this example, by Underwriters Laboratories Inc. (UL)] is optional.

2-3 Container Appurtenances.

2-3.1 General.

2-3.1.1 This section includes fabrication and performance provisions for container appurtenances, such as pressure relief devices, container shutoff valves, backflow check valves, internal valves, excess-flow check valves, plugs, liquid level gauges and pressure gauges connected directly into the container openings described in 2-2.3. Shop installation of such appurtenances in containers listed as container assemblies or container systems in accordance with 1-3.1.1 is a responsibility of the fabricator under the listing. Field installation of such appurtenances is covered in Chapter 3.

In the early days of distribution of LP-Gas in bulk quantities to an ASME container located on the consumer's premises, each container assembly of 1,200 gal (4.5 m³) [and subsequently 2,000 gal (7.6 m³)] water capacity or less had to be tested and listed by a nationally recognized testing laboratory or inspected and approved by the authority having jurisdiction. On larger containers each appurtenance and regulator had to be so tested and listed. Many LP-Gas distributing companies and tank fabricators had their own systems listed by Underwriters Laboratories Inc. Today, the individual appurtenances are tested and listed and either shop installed — e.g., by container fabricators — or field installed. This section sets out general, fundamental criteria for these appurtenances, which are used as a basis for testing laboratories, manufacturers, and installers.

2-3.1.2 Container appurtenances shall be fabricated of materials suitable for LP-Gas service and resistant to the action of LP-Gas under service conditions. The following shall also apply:

(a) Pressure containing metal parts of appurtenances, such as those listed in 2-3.1.1, except fusible elements, shall have a minimum melting point of 1500°F (816°C) such as steel, ductile (nodular) iron, malleable iron, or brass. Ductile iron shall meet the requirements of ASTM A 395 or equivalent and malleable iron the requirements of ASTM A 47 or equivalent. Approved or listed liquid level gauges used in containers of 3500 gal (13.2 m³) water capacity or less are exempted from this provision.

These criteria are intended to provide a degree of structural integrity in the event of fire exposure and application of water for fire control. It is recognized, however, that appurtenance configuration, mass, and location on the container affect the degree of hazard. The exception for liquid level gauges on smaller containers reflects this consideration.

Note that only metals are recognized for pressure containing parts.

(b) Cast iron shall not be used.

Cast iron is subject to cracking from severe thermal shock under fire conditions when used in container appurtenances and pipe fittings.

(c) Non-metallic materials shall not be used for parts such as bonnets or bodies.

Although there has been considerable research and development in plastic materials, at this time nonmetallic materials should not be used for bonnets and bodies of appurtenances which may or may not be subject to pressures of LP-Gas.

2-3.1.3 Container appurtenances shall have a rated working pressure of at least 250 psig (1.7 MPa gauge).

2-3.1.4 Gaskets used to retain LP-Gas in containers shall be resistant to the action of LP-Gas. They shall be made of metal or other suitable material confined in metal having a melting point over 1500°F (816°C) or shall be protected against fire exposure, except that aluminum O-rings and spiral wound metal gaskets are acceptable and gaskets for use with approved or listed liquid level gauges for installation on a container of 3500 gal (13.2 m³) water capacity or less are exempted from this provision. When a flange is opened, the gasket shall be replaced.

The exemption for gaskets used with approved or listed liquid level gauges on containers of 3,500 gal (13.2 m³) water capacity or less is attributed to the long and successful experience with this equipment which has been listed for many years by Underwriters Laboratories Inc. [*See also commentary on 2-3.1.2(a).*]

2-3.2 Pressure Relief Devices. (*See 2-4.7 for hydrostatic relief valves.*)

Editions of the standard prior to 1983 referred to "safety relief devices" to distinguish between those pressure relief devices normally in service with nonhazardous materials. The 1983 Edition was amended to refer to "pressure relief devices" to bring the standard in agreement with ANSI B95.1, *Standard Terminology for Pressure Relief Devices.*

2-3.2.1 Containers shall be equipped with one or more pressure relief devices which, except as otherwise provided for in 2-3.2.2, shall be designed to relieve vapor.

Vapor discharge permits smaller valves to be used to obtain the needed relieving capacity under fire exposure conditions because some of the heat is used to vaporize the liquid inside the container. Also, the discharged vapor represents a lesser hazard than does discharged liquid.

2-3.2.2 DOT containers shall be equipped with pressure relief valves or fusible plug devices as required by DOT Regulations. (*See Appendix E for additional information.*)

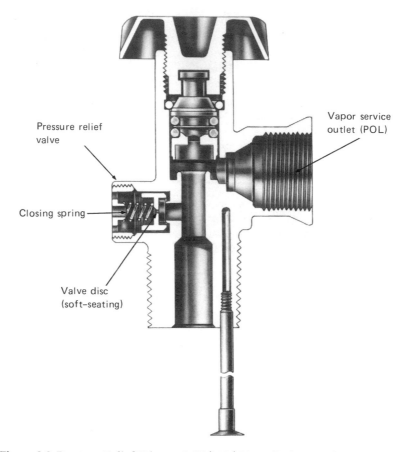

Pressure relief
valve

Closing spring

Valve disc
(soft–seating)

Vapor service
outlet (POL)

Figure 2.9 Pressure Relief Valve on DOT/ICC/CTC Cylinder. On these containers the relief valve is part of a fitting which includes the manual shutoff valve (also known as the "service valve"). Note that the relief valve is unaffected by the position of the manual valve and is, therefore, able to function whether the manual valve is open or closed.

DOT regulations [CFR 49, 173.34(d)] currently require that DOT containers be equipped with one or more pressure relief devices selected and tested in accordance with CGA Pamphlet S-1-1, *Pressure Relief Device Standards.* The pressure relief device system must be capable of preventing rupture of the normally charged cylinder when subjected to a fire test conducted in accordance with CGA Pamphlet C-14, *Procedures for Fire Testing of DOT Cylinder Safety Relief Device Systems.* For many years this testing was done by the Bureau of Explosives of the Association of American Railroads but manufacturers may now conduct their own testing.

2-3.2.3   ASME containers for LP-Gas shall be equipped with direct spring-loaded pressure relief valves conforming with applicable requirements of the *Standard on Safety Relief Valves for Anhydrous Ammonia and LP-Gas*, UL 132; or other equivalent pressure relief valve standards. The start-to-leak setting of such pressure relief valves, with relation to the design pressure of the container, shall be in accordance with Table 2-3.2.3.

Exception: On containers of 40,000 gal (151 m³) water capacity or more, a pilot operated pressure relief valve in which the relief device is combined with and is controlled by a self-actuated, direct, spring-loaded pilot valve may be used provided it complies with Table 2-3.2.3, is approved (see definition), is inspected and maintained by persons with appropriate training and experience, and is tested for proper operation at intervals not exceeding 5 years.

Table 2-3.2.3

Containers	Minimum	Maximum
All ASME Codes prior to the 1949 Edition, and the 1949 Edition, paragraphs U-68 and U-69	110%	125%*
ASME Code, 1949 Edition, Paragraphs U-200 and U-201, and all ASME Codes later than 1949	100%	100%*

*Manufacturers of pressure relief valves are allowed a plus tolerance not exceeding 10 percent of the set pressure marked on the valve.

This provision now specifically states that these valves must comply with UL or FM standards, or the equivalent. This provision relates to the definition of ASME Code in Section 1-7 stating that Division I of Section VIII of that Code is applicable to NFPA 58 except for the relief valve provisions in Sections UG-125 through UG-136 of the Code. The exception was made in light of differences in this area between the ASME Code and longstanding practice within the LP-Gas and petroleum industry. Recognition is given to UL standard 132 and FM standards as the basic requirements for pressure relief valves installed on ASME containers. One of the basic differences between the ASME Code and other standards was the tolerances for the start-to-leak setting of relief valves. The ASME Code permitted a maximum of ± 3 percent tolerance, whereas NFPA 58 set out a plus tolerance only, and one not exceeding 10 percent. The plus tolerance is based on the opinion that there should be no premature discharge of the valve and that the pressure relief valve should function only as a last resort. The ASME Code has subsequently been changed to agree with this concept.

There is apparently one remaining difference and that is the official capacity of the valves. The ASME Code requires 10

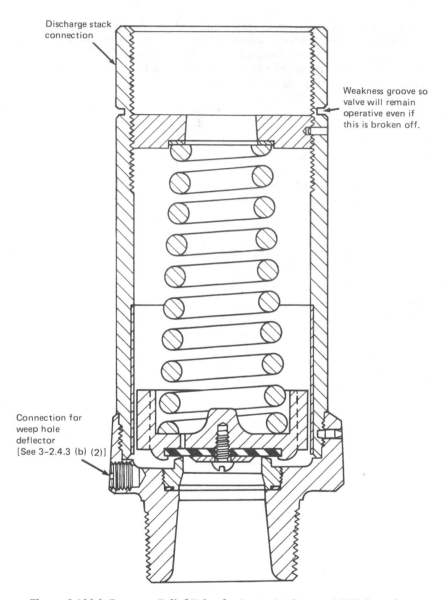

Discharge stack connection

Weakness groove so valve will remain operative even if this is broken off.

Connection for weep hole deflector [See 3-2.4.3 (b) (2)]

Figure 2.10(a) Pressure Relief Valve for Large Stationary ASME Container.

percent derating of the average test capacities of three samples as the official capacity of the valve. NFPA 58, UL and FM do not require such derating.

Reference is made to "direct spring-loaded pressure relief valves" to describe accurately the type of relief valve desired on ASME containers used in installations covered by NFPA 58. Other, more complex types are not functionally needed, are

Weakness groove

Adapter for stack
connection

Rain cap

Weep hole deflector

Figure 2.10(b) Pressure Relief Valve for Small ASME Container.

unlikely to receive adequate maintenance and could result in
undesirable frequent premature operation in facilities covered
by NFPA 58.

One exception to this provision is given for pilot operated
valves on 40,000 gal (151 m³) water capacity or larger containers under certain conditions. Large installations utilizing this
size storage container are more likely to be able to comply with
the conditions specified.

2-3.2.4 Pressure relief valves for ASME containers shall also comply with
2-3.2.4(a) through (e).

(a) Pressure relief valves shall be of sufficient individual or aggregate
capacity as to provide the relieving capacity in accordance with Appendix E
for the container on which they are installed, and to relieve at not less than
the rate indicated before the pressure is in excess of 120 percent of the
maximum (not including the 10 percent referred to in the footnote of Table
2-3.2.3) permitted start-to-leak pressure setting of the device. This provision is applicable to all containers (including containers installed partially
aboveground) except containers installed wholly underground in accordance with E-2.3.1.

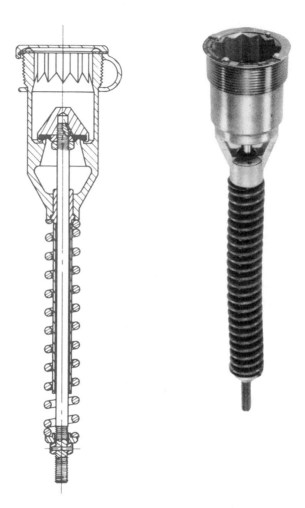

Figure 2.10(c) Internal Type Pressure Relief Valves Used on Cargo Containers. All working parts are within the containers or below container shell. Stacks are not used.

While a pressure relief valve can function for other reasons, the maximum relieving capacity needed is the result of fire exposure. The CGA documents cited in E-2.1.1 describe the factors involved.

As specified in E-2.3.1, underground or mounded container pressure relief valve capacities may be as small as 30 percent of those specified in Table E-2.2.2 if certain conditions are met.

(b) Each pressure relief valve shall be plainly and permanently marked with: (1) the pressure in psig at which the valve is set to start-to-leak; (2) rated relieving capacity in cu ft per minute of air at 60°F (16°C) and 14.7 psia (0.1 MPa absolute); and (3) the manufacturer's name and catalog number. Example: A pressure relief valve is marked 250-4050 AIR. This

Connection for deflector or pipeaway system.

Figure 2.10(d) Internal Pressure Relief Valve for Motor Fuel Containers (with Pipeway Adapter). See 3-6.2.3(a)(4).

indicates that the valve is set to start-to-leak at 250 psig (1.7 MPa gauge); and that its rated relieving capacity is 4050 cfm (1.9 m³/s) of air.

(c) Shutoff valves shall not be located between a pressure relief device and the container, unless the arrangement is such that the relief device relieving capacity flow specified in 2-3.2.4(a) will be achieved through additional pressure relief devices which remain operative.

(d) Pressure relief valves shall be so designed that the possibility of

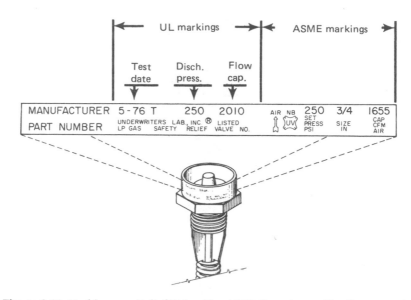

Figure 2.11 Markings on Relief Valves for ASME Containers. The flow capacity marked on the relief valve is the volume of air that the valve can discharge when it is fully open. Again, this marking is based on a factory test when the relief valve is built. This capacity is usually measured in cubic feet of air per minute (cfm-air). Two different flow capacities are marked on the relief valve in Figure 2.11. On the left is the UL flow capacity rating (2,010). On the right is the ASME flow capacity rating (1,655). The UL rating is higher than the ASME rating because the UL capacity is measured under a higher test pressure (UL: 300 psig; ASME: 275 psig) and the ASME rating is at 90 percent of actual flow.

tampering will be minimized. Externally set or adjusted valves shall be provided with an approved means of sealing the adjustment.

(e) Fusible plug devices, with a yield point of 208°F (98°C) minimum and 220°F (104°C) maximum, with a total discharge area not exceeding 0.25 sq in. (1.6 cm²), and which communicate directly with the vapor space of the container, may be used in addition to the spring-loaded pressure relief valves (as specified in Table 2-3.2.3) for aboveground containers of 1,200 gal (4.5 m³) water capacity or less.

2-3.2.5 All containers used in industrial truck (including forklift truck cylinders) service shall have the container pressure relief valve replaced by a new or unused valve within 12 years of the date of manufacture of the container and each 10 years thereafter.

Studies have shown that the most critical serviceability factor relative to relief valves on DOT or ASME containers involves those in industrial truck use. Many of these valves have been found to be inoperative due to plugging, etc., attributed to the

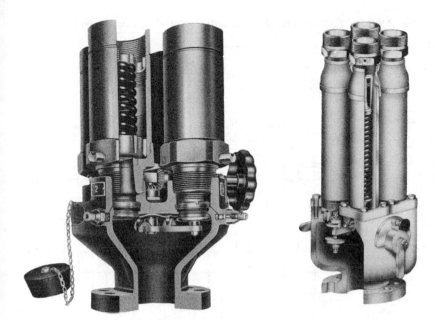

Figure 2.12 ASME Container Relief Valve Manifolds. The container requires three relief valves. The manifold contains four. By manipulating the handwheel or lever, an internal clapper-type valve can be rotated to isolate any one of the four relief valves for testing, maintenance, or replacement.

dust and corrosive environment in which some of the trucks operate.

2-3.3 Connections for Flow Control (Filling, Withdrawal, Equalizing).

2-3.3.1 Shutoff valves, excess-flow check valves, backflow check valves and quick closing internal valves, used individually or in suitable combinations, at container filling, withdrawal, and equalizing connections, shall comply with 2-3.1.2 and 2-3.1.3.

2-3.3.2 Filling, withdrawal and equalizing connections shall be equipped with the appurtenances for the appropriate type and capacity of container and the service in which they are to be used in accordance with Table 2-3.3.2. Cylinder valve outlet connections on all DOT cylinders except those used for engine fuel, from which vapor can be withdrawn, shall not be interchangeable with those used for liquid withdrawal.

(a) If the loading or transfer point is not on the container, it shall be equipped as specified for filling connections on the container.

Table 2-3.3.2 summarizes all the container appurtenances used for filling, withdrawal, and vapor equalization together

Table 2-3.3.2 Filling Withdrawal and Equalizing Connections

Type of Use	General Uses												Used as Fuel on Vehicles			Portable Tanks			Cargo Tanks		
Type of Container	DOT Cylinder Specifications									ASME			DOT or ASME			ASME			ASME		
Water Capacity of Containers — Pounds (kg)	1 (0.5) to 1,000 (454)			50 (23) to 1,000 (454)			2.5 (1) to 245 (110)			—			Any			Any			Any		
Gallons (m³)	—			—			—			Up to 120,000 (454 m³)			Any			Any			Any		
Conditions under which Container is Used	Replacement or Exchange Outdoors			Filled at Point of Use Outdoors			When Used Inside Buildings			Filled at Point of Use Outdoors			Replacement or Fixed			Transportation of LP-Gas			Transportation of LP-Gas		
Connection Use (Note 1): "F"—Filling, "W"—Withdrawal, "E"—Equalizing	F	W	E	F	W	E	F	W	E	F	W	E	F	W	E	F	W	E	F	W	E
Appurtenances to be provided (Note 2):																					
1. Positive (Manual) Shutoff Valve	√ 3*	√ 3*	√	5*	5*						√ 6*										
2. Positive (Manual) Shutoff & Internal Excess Flow Check Valve		√	√	√	√	√	√	√	√	√	√	√	√ 8*	√	√	√	√	√		√	
3. Positive (Manual) Shutoff & External Excess Flow Check Valve (Note 4)				√	√	√	√	√	√	√	√	√				√	√	√	√		
4. Single Back Flow Check Valve			√																		
5. Positive (Manual) Shutoff & Internal Back Flow Check Valve											√			√			√			√	
6. Excess Flow Check Valve & Back Flow Check Valve		√	√	√						√	√					√	√		√	√	
7. Double Back Flow Check Valve			√								√			√			√			√	
8. Quick Closing Internal Valve										√	√	√				√	√		√	√ 7*	√
Column Number	1			2			3			4			5			6			7		

*Note Number.

Notes to Table 2-3.3.2

Note 1:   Containers are not required to be equipped with all three connections, but if used, appurtenances shall be those shown. Suitably fitted multipurpose valves may be used.

Note 2:   If more than one appurtenance, or combination of appurtenances, is shown for any connection use, any one of the appurtenances or combinations shown will comply. (*See also DOT regulations for cargo and portable containers.*)

Note 3:   Single manual shutoff valve normally used for both filling and withdrawal.

Note 4:   External excess-flow check valves shall be installed in such a way that any undue strain beyond them will not cause breakage between the container and the excess-flow check valves.

Note 5:   Containers of less than 50 lb (23 kg) water capacity need only be equipped with a positive shutoff valve for filling at the point of use.

Note 6:   An excess-flow check valve is not required in the withdrawal connection provided the following are all complied with:

   (a) Container water capacity does not exceed 2,000 gal (7.6 m³)
   (b) Withdrawal outlet is equipped with a manually operable (having a handwheel or the equivalent) shutoff valve, which is:

      (1) threaded directly into the container outlet, or

      (2) an integral part of a substantial fitting which is threaded directly into or on the container outlet, or

      (3) threaded directly into a substantial fitting which is threaded directly into or on the container outlet.

   (c) The controlling orifice between the container contents and the shutoff valve outlet does not exceed 5⁄16 in. (8 mm) in diameter for vapor withdrawal or 1⁄8 in. (3 mm) for liquid withdrawal.

   (d) An approved pressure-reducing regulator is directly attached to the outlet of the shutoff valve and is rigidly supported, or is adequately supported and properly protected on or at the container, and is connected to the shutoff valve by means of a suitable flexible connection.

Note 7:   See 6-3.2.1(a) for special requirements for containers constructed to DOT cargo tank specifications.

Note 8:   Authorized for exchangeable (removable) containers only.

with certain conditions set out in the notes to this table. As indicated in Note 1, it is not required that containers be equipped with all three connections, but if used they must be as shown. Multipurpose valves are available to accomplish several functions in a single fitting.

DOT cylinder valve vapor outlet connections, except those for engine fuel, must not be interchangeable with liquid withdrawal. CGA Standard V-1/ANSI B57.1, *Compressed Gas Cylinder Valve Outlet and Inlet Connections*, sets out Connection No. 510 for vapor withdrawal and Connection No. 555 for liquid withdrawal. (There is also a 600 Connection for small torch type cylinders.)

2-3.3.3 The appurtenances specified in Table 2-3.3.2 shall comply with 2-3.3.3(a) through (d).

(a) Manual shutoff valves shall be designed to provide positive closure under service conditions.

(b) Excess-flow check valves shall be designed to close automatically at the rated flows of vapor or liquid specified by the manufacturer. Excess flow valves shall be designed with a bypass, not to exceed a No. 60 drill size opening, to allow equalization of pressure.

(c) Backflow check valves, which may be of spring-loaded or weight-loaded type with in-line or swing operation, shall close when flow is either

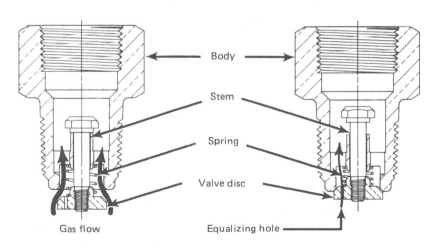

a. Open position (normal)   b. Closed position "slugged"

Figure 2.13(a) Operation of Excess-flow Check Valve. After flow has been stopped by closing the downstream valve (or other means), pressure on both sides will equalize through equalizing hole and spring will cause valve to reopen.

Figure 2.13(b) Various Excess-flow Check Valves.

stopped or reversed. Both valves of double backflow check valves shall comply with this provision.

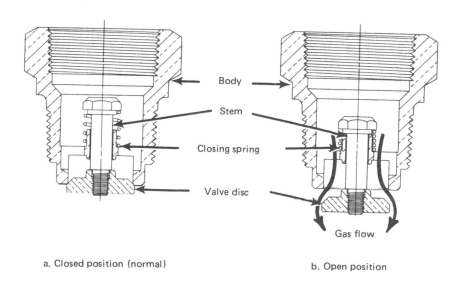

a. Closed position (normal)   b. Open position

Figure 2.14 Operation of Backflow Check Valve.

(d) Internal valves (*see definition*), either manually or automatically operated and designed to remain closed except during operating periods, shall be considered positive shutoff valves. [*See 6-3.2.1(a) for special requirements for such valves used on cargo units.*]

Figure 2.15(a) Lever Operated Internal Valve for Threaded Installation.

Figure 2.15(b) Lever Operated Internal Valve for Flanged Installation.

Figure 2.15(c) Pneumatically Operated Internal Valve.

An internal valve can be considered a substitute for an excess-flow check valve with the added feature that it remains in the closed position except when liquid is being transferred. An excess-flow check valve is in the open position except when a flow large enough to close the valve occurs.

Internal valves are categorized by the means by which they are opened. Some are opened manually by means of a lever. Others are opened by liquid or gas pressure downstream of the valve. In the event of leakage, they close automatically — either by operation of an incorporated excess-flow check valve or by lowering of downstream pressure. They also can be arranged for automatic closing through operation of fusible elements (fire exposure) and for remote manual closing.

2-3.3.4 The appurtenances specified in Table 2-3.3.2 may be installed as individual components or as combinations completely assembled by the appurtenance manufacturer.

2-3.4 Liquid Level Gauging Devices.

2-3.4.1 Liquid level gauging devices shall be provided on all containers filled by volume. Fixed level gauges or variable gauges of the slip tube, rotary tube or float types (or combinations of such gauges) may be used to comply with this provision.

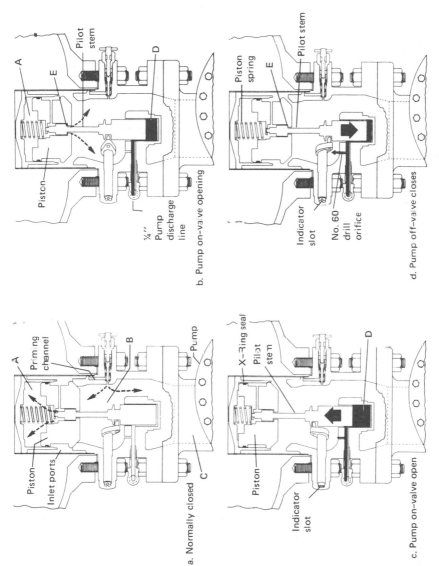

Figure 2.15(d) Operation of Pump Pressure Operated Internal Valve.

2-3.4.2 Every container constructed after December 31, 1965, designed to be filled on a volumetric basis, shall be equipped with a fixed liquid level gauge(s) to indicate the maximum filling level(s) for the service(s) in which the container is to be used (*see 4-5.3.3*). This may be accomplished either by using a dip tube of appropriate length, or by the position of the gauging device in the container. The following shall apply:

> Because of the possibility of inaccuracy of rotary or magnetic gauges, it is necessary to have a fixed liquid level gauge available on all tanks to determine maximum filling levels in emergencies and to check the accuracy of these variable gauges. This provision was incorporated in the 1965 Edition of the standard. Fixed liquid level gauges are dip tubes of certain lengths installed in the container. The gauge discharge is invisible if vapor is emitted but a fog of condensed water vapor is created by refrigeration when vaporizing liquid is discharged. The criteria for their length is based on the specific gravity of the liquid at 40°F (4°C). Subparagraphs (a) through (d) provide marking requirements for fixed liquid level gauges.

(a) ASME containers manufactured after December 31, 1969, shall have permanently attached to the container adjacent to the fixed liquid level gauge, or on the container nameplate, markings showing the percentage full that is indicated by that gauge.

(b) Containers constructed to DOT cylinder specifications shall have stamped on the container the letters "DT" followed by the vertical distance (to the nearest tenth inch) from the top of the boss or coupling into which the gauge, or the container valve of which it is a part, is made up, to the end of the dip tube. [*See 2-3.4.2(c)(2) for DOT containers designed for loading in either the vertical or horizontal position.*]

(c) Each container manufactured after December 31, 1972, equipped with a fixed liquid level gauge for which the tube is not welded in place shall be permanently marked adjacent to such gauge or on container nameplate as follows:

(1)   Containers designed to be filled in one position shall be marked with the letters "DT" followed by the vertical distance (to the nearest tenth inch) measured from the top center of the container boss or coupling into which the gauge is installed to the maximum permitted filling level.

(2)   Portable universal type containers that may be filled in either vertical or horizontal position shall be marked as follows:

a. *For Vertical Filling*: With the letters "VDT" followed by the vertical distance (to the nearest tenth inch), measured from the top center of the container boss or coupling into which the gauge is installed to the maximum permitted filling level.

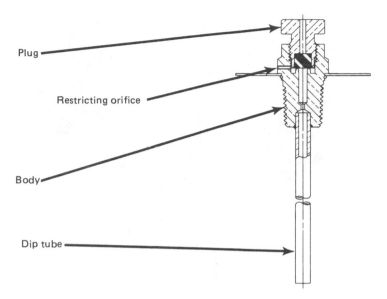

Plug

Restricting orifice

Body

Dip tube

Figure 2.16(a) Fixed Liquid Level Gauge.

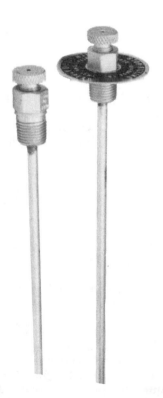

Figure 2.16(b) Fixed Liquid Level Gauge. Marking on disk reads, "STOP FILLING WHEN LIQUID APPEARS."

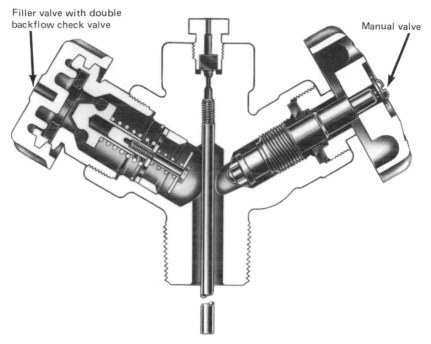

Filler valve with double
backflow check valve

Manual valve

Figure 2.16(c) Fixed Liquid Level Gauge Incorporated into Multiple Function Container Valve Assembly.

b. *For Horizontal Filling*: With the letters "HDT" followed by the vertical distance (to the nearest tenth inch), measured from the top centerline of the container boss or coupling opening into which the gauge is installed to the inside top of the container when in the horizontal position.

(d) Cargo tanks having several fixed level gauges positioned at different levels shall have stamped adjacent to each gauge the loading percentage (to the nearest ⅗₀ percent) of the container content which that particular gauge indicates.

2-3.4.3 The intent of 2-3.4.2 may be achieved by other methods acceptable to the authority having jurisdiction.

2-3.4.4 Variable liquid level gauges shall comply with 2-3.4.4(a) through (e).

(a) Variable liquid level gauges shall be so marked that the maximum liquid level, in in. or percent of capacity of the container in which they are to be installed, is readily determinable. These markings shall indicate the maximum liquid level for propane, for 50/50 butane-propane mixtures, and for butane at liquid temperatures from 20°F (−6.7°C) to 130°F (54.4°C) and in increments not greater than 20 Fahrenheit degrees.

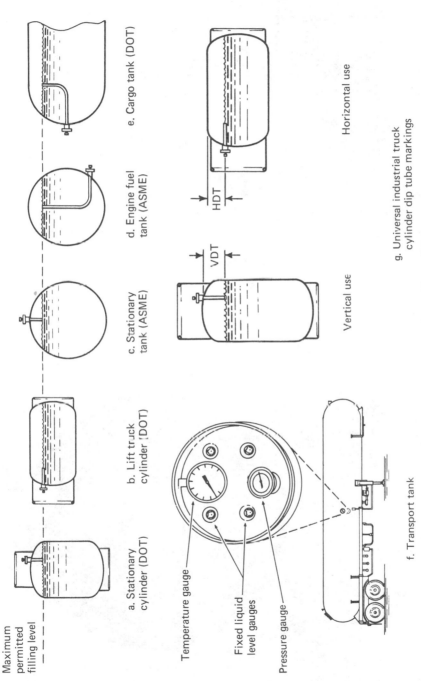

Maximum
permitted
filling level

a. Stationary
cylinder (DOT)

b. Lift truck
cylinder (DOT)

c. Stationary
tank (ASME)

d. Engine fuel
tank (ASME)

e. Cargo tank (DOT)

HDT

Horizontal use

VDT

Vertical use

g. Universal industrial truck
cylinder dip tube markings

Temperature gauge

Fixed liquid
level gauges

Pressure gauge

f. Transport tank

Figure 2.16(d) Location of Fixed Liquid Level Gauges.

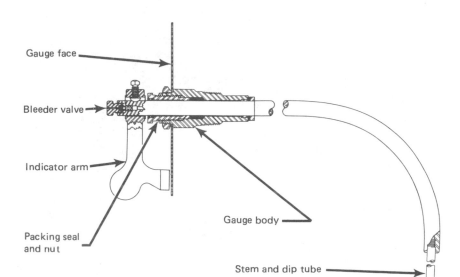

Gauge face

Bleeder valve

Indicator arm

Packing seal
and nut

Gauge body

Stem and dip tube

Figure 2.17(a) Rotary Type of Variable Liquid Level Gauge.

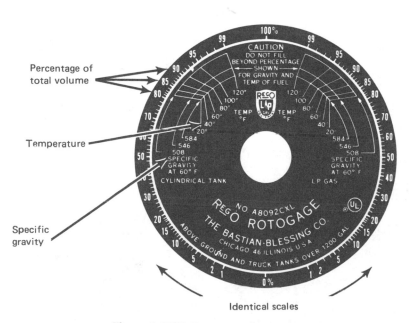

Percentage of
total volume

Temperature

Specific
gravity

Identical scales

Figure 2.17(b) Rotary Gauge Face.

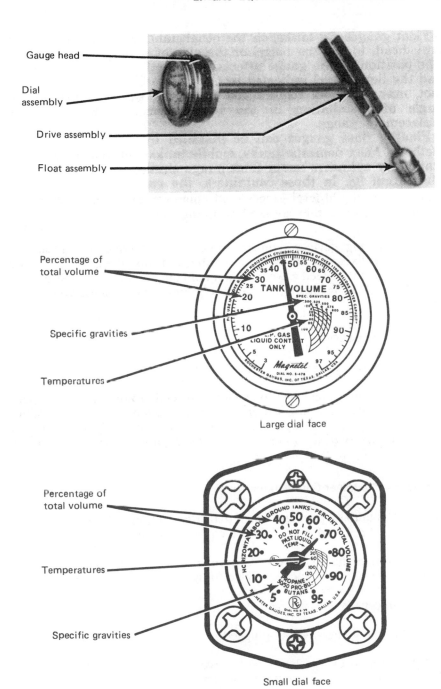

Figure 2.17(c) Float Type of Variable Liquid Level Gauge.

(b) The markings indicating the various liquid levels from empty to full shall either be directly on the system nameplate or on the gauging device or on both.

(c) Dials of magnetic float or rotary gauges shall show whether they are for cylindrical or spherical containers, and whether for aboveground or underground service.

(d) The dials of gauges for use only on aboveground containers of over 1,200 gal (4.5 m³) water capacity shall be so marked.

(e) Variable liquid level gauges shall comply with the accuracy provisions of 4-5.3.3(b) if they are used for filling containers.

2-3.4.5 Gauging devices requiring bleeding of product to the atmosphere, such as fixed liquid level, rotary tube, and slip tube gauges, shall be designed so that the bleed valve maximum opening to the atmosphere is not larger than a No. 54 drill size, unless equipped with excess-flow check valves.

2-3.5 Pressure Gauges.

2-3.5.1 Pressure gauges shall comply with 2-3.1.2 and 2-3.1.3.

2-3.5.2 Pressure gauges shall be attached directly to the container opening or to a valve or fitting which is directly attached to the container opening. If the effective opening into the container will permit a flow greater than that of a No. 54 drill size, an excess-flow check valve shall be provided.

2-3.6 Other Container Connections.

2-3.6.1 Container openings, other than those equipped as provided in 2-3.2, 2-3.3, 2-3.4, and 2-3.5, shall be equipped with one of the following:

(a) A positive shutoff valve in combination with either an excess-flow check valve or a backflow check valve, plugged.

(b) An internal valve, plugged.

(c) A backflow check valve, plugged.

(d) An internal excess-flow check valve, normally closed and plugged, with provision to allow for external actuation.

(e) A plug, blind flange, or plugged companion flange.

This provides a complete list of appurtenances which may by installed in container openings. The appurtenances referred to in 2-3.2 through 2-3.5 are:

2-3.2 Pressure relief devices.

2-3.3 Connections for flow control (shutoff, excess-flow, back-flow check and quick closing internal valves).

2-3.4 Liquid level gauging devices.

2-3.5 Pressure gauges.

2-4 Piping (Including Hose), Fittings, and Valves.

The term "piping," as used in NFPA 58, includes pipe, tubing, hose, and the valves and fittings used in the piping system. Provisions for the basic design of this equipment are given in this section for use by manufacturers in producing piping system components. Guidelines for both testing laboratories to list this equipment for possible acceptance by authorities having jurisdiction and for users to select proper materials in making a piping installation are also included.

2-4.1 General.

2-4.1.1   This section includes basic design provisions and material specifications for pipe, tubing, pipe and tubing fittings, valves (including hydrostatic relief valves), hose, hose connections and flexible connectors used to connect container appurtenances with the balance of the LP-Gas system in accordance with the installation provisions of Chapter 3.

Pressure ratings of fittings and valves are based on the concept that the maximum normal operating discharge pressure of a liquid pump or vapor compressor be designed for 350 psig (2413 kPa), while other piping above 125 psig (862 kPa) be designed for 250 psig (1724 kPa). Note that the metallic pipe used for service above 125 psig (862 kPa) will withstand 350 psig (2413 kPa) and a pressure limit is not specified in the standard.

2-4.1.2   Piping, pipe and tubing fittings and valves used to supply utilization equipment within the scope of NFPA 54, *National Fuel Gas Code*, shall comply with that Code.

Most LP-Gas piping systems from the outlet of the first stage regulator are subject to the provisions of NFPA 54, *National Fuel Gas Code*. Therefore, reference should be made to NFPA 54 first, to determine whether the particular installation (primarily one of a permanent piping system) is subject to that standard; and second, to Part 2, "Gas Piping System Design, Materials and Components."

In general, when an LP-Gas system serves a building, piping

between the container and the first stage regulator is in the scope of NFPA 58; and piping downstream of the first stage regulator is in the scope of NFPA 54.

2-4.1.3 Pipe and tubing shall comply with 2-4.2.1 and 2-4.3.1 or shall be of material which has been investigated and tested to determine that it is safe and suitable for the proposed service and is recommended for that service by the manufacturer, and be acceptable to the authority having jurisdiction.

Materials specified for pipe (2-4.2), tubing (2-4.3), and fittings (2-4.4) have been investigated and tested for recommended use. If materials are not referenced they should not be used unless given special approval by the authority having jurisdiction. For example, prior to the 1972 Edition of NFPA 58 provisions were included for the use of aluminum alloy pipe and tubing. However, the provisions were so restrictive due to adverse experience under a variety of conditions that their use was eliminated from the standard. Provisions for plastic pipe and tubing were first incorporated in the 1979 Edition. However, they too are quite restrictive. Pipe, tubing, and fittings are limited to polyethylene types meeting certain ASTM standards and have to be listed or approved. Joints are to be made by heat fusion, not mechanically (i.e., by threads). The service is limited to vapor (no liquid) and even then not exceeding 30 psig (208 kPa gauge). The system can only be installed outside and underground. With reference to copper tubing specifications, Type M is not included since experience has indicated that a minimum wall thickness of 0.032 in. (0.81 mm) should be used to avoid kinking.

2-4.2 Pipe.

2-4.2.1 Pipe shall be wrought iron or steel (black or galvanized), brass, copper, or polyethylene (*see 3-2.7.6*) and shall comply with 2-4.2.1(a) through (g).

(a) Wrought iron pipe; ANSI B36.10, *Welded and Seamless Wrought Steel Pipe.*

(b) Steel Pipe; *Specification for Pipe, Steel, Black and Hot-Dipped, Zinc-Coated Welded and Seamless* (ANSI/ASTM A 53).

(c) Steel pipe; *Specification for Seamless Carbon Steel Pipe for High Temperature Service* (ANSI/ASTM A 106).

(d) Steel pipe; *Specification for Pipe, Steel, Black and Hot-Dipped*

Zinc-Coated (Galvanized) Welded and Seamless, for Ordinary Uses (ANSI/ ASTM A 120).

(e) Brass Pipe; *Specification for Seamless Red Brass Pipe, Standard Sizes* (ANSI/ASTM B 43).

(f) Copper Pipe; *Specification for Seamless Copper Pipe, Standard Sizes* (ANSI/ASTM B 42).

(g) Polyethylene Pipe; *Specification for Thermoplastic Gas Pressure Pipe, Tubing and Fittings* (ANSI/ASTM D 2513) and be listed or approved.

The service pressure of polyethylene pipe (and tubing) is limited to 30 psig (207 kPa gauge). [*See 3-2.6.1(b)*.]

2-4.3 Tubing.

2-4.3.1 Tubing shall be steel, brass, copper, or polyethylene (*see 3-2.7.6*) and shall comply with 2-4.3.1(a) through (d):

(a) Steel tubing; *Specification for Electric-Resistance-Welded Coiled Steel Tubing for Gas Fuel Oil Lines* (ANSI/ASTM A 539).

(b) Brass tubing [*see 3-2.6.1(d)(3)*]; *Specification for Seamless Brass Tube* (ANSI/ASTM B 135).

(c) Copper tubing [*see 3-2.6.1(d)(3)*]:

(1) Type K or L, *Specification for Seamless Copper Water Tube* (ANSI/ASTM B 88).

(2) *Specification for Seamless Copper Tube for Air Conditioning and Refrigeration Field Service* (ANSI/ASTM B 280).

(d) Polyethylene tubing; *Specification for Thermoplastic Gas Pressure Pipe, Tubing and Fittings* (ASTM D 2513) and be listed or approved.

The service pressure of polyethylene tubing (and pipe) is limited to 30 psig (207 kPa gauge). [*See 3-2.6.1(b)*].

2-4.4 Pipe and Tubing Fittings.

2-4.4.1 Fittings shall be steel, brass, copper, malleable iron, ductile (nodular) iron or polyethylene, and shall comply with 2-4.4.1(a) through (c). Cast iron pipe fittings (ells, tees, crosses, couplings, unions, flanges, or plugs) shall not be used.

(a) Pipe joints in wrought iron, steel, brass or copper pipe may be screwed, welded or brazed.

(1)   Fittings used at pressures higher than container pressure, such as on the discharge of liquid transfer pumps, shall be suitable for a working pressure of at least 350 psig (2.4 MPa gauge).

(2)   Except as provided in 2-4.4.1(a)(1), fittings used with liquid LP-Gas, or with vapor LP-Gas at operating pressures over 125 psig (0.9 MPa gauge), shall be suitable for a working pressure of 250 psig (1.7 MPa gauge).

(3)   Fittings for use with vapor LP-Gas at pressures not exceeding 125 psig (0.9 MPa gauge) shall be suitable for a working pressure of 125 psig (0.9 MPa gauge).

(4)   Brazing filler material shall have a melting point exceeding 1,000°F (538°C).

Prior to the 1989 Edition of the standard soldering was permitted so long as the filler material had a melting point exceeding 1000°F (538°C). This is impossible as solder is a low melting point material, and the provision has been changed to allow brazing only. While not specified in this standard, it should be recognized that NFPA 54, *National Fuel Gas Code,* **specifies that brazing alloys may not contain phosphorous as this can lead to a deterioration of the joint.**

(b) Tubing joints in steel, brass, or copper tubing shall be flared, brazed, or made up with approved gas tubing fittings.

It should be recognized that this does not specify that flared fittings be approved. Flared fittings essentially are fabricated when installed in the field and do not need to be specially approved.

(1)   Fittings used at pressures higher than container pressure, such as on the discharge of liquid transfer pumps, shall be suitable for a working pressure of at least 350 psig (2.4 MPa gauge).

(2)   Except as provided in 2-4.4.1(b)(1), fittings used with liquid LP-Gas, or with vapor LP-Gas at operating pressures over 125 psig (0.9 MPa gauge), shall be suitable for a working pressure of 250 psig (1.7 MPa gauge).

(3)   Fittings for use with vapor LP-Gas at pressures not exceeding 125 psig (0.9 MPa gauge) shall be suitable for a working pressure of 125 psig (0.9 MPa gauge).

(4)   Brazing filler material shall have a melting point exceeding 1,000°F (538°C).

See commentary following 2-4.4.1(a)(4).

(c) Joints in polyethylene pipe and tubing shall be made by heat fusion in accordance with the manufacturer's instructions.

(1) Polyethylene fittings shall conform to ANSI/ASTM D 2683, *Specification for Socket-Type Polyethylene (PE) Fittings for Outside Diameter — Controlled Polyethylene Pipe*, or ASTM D 3261, *Specification for Butt Heat Fusion Polyethylene (PE) Plastic Fittings for Polyethylene (PE) Plastic Pipe and Tubing*, and be listed or approved.

2-4.5 Valves, Other than Container Valves.

2-4.5.1 Pressure containing metal parts of valves (except appliance valves), including manual positive shutoff valves, excess-flow check valves, backflow check valves, emergency shutoff valves (*see 2-4.5.4*), and remotely controlled valves (either manually or automatically operated), used in piping systems shall be of steel, ductile (nodular) iron, malleable iron, or brass. Ductile iron shall meet the requirements of ANSI/ASTM A 395 or equivalent and malleable iron shall meet the requirements of ANSI/ASTM A 47 or equivalent. All materials used, including valve seat discs, packing, seals, and diaphragms shall be resistant to the action of LP-Gas under service conditions.

2-4.5.2 Valves shall be suitable for the appropriate working pressure, as follows:

(a) Valves used at pressures higher than container pressure, such as on the discharge of liquid transfer pumps, shall be suitable for a working pressure of at least 350 psig (2.4 MPa gauge). [400 psig (2.8 MPa gauge) WOG valves comply with this provision.]

(b) Valves to be used with liquid LP-Gas, or with vapor LP-Gas at pressures in excess of 125 psig (0.9 MPa gauge), but not to exceed 250 psig (1.7 MPa gauge), shall be suitable for a working pressure of at least 250 psig (1.7 MPa gauge).

Exception: Valves used at higher pressure as specified in 2-4.5.2(a).

(c) Valves (except appliance valves) to be used with vapor LP-Gas at pressures not to exceed 125 psig (0.9 MPa gauge) shall be suitable for a working pressure of at least 125 psig (0.9 MPa gauge).

2-4.5.3 Manual shutoff valves, emergency shutoff valves (*see 2-4.5.4*), excess-flow check valves, and backflow check valves used in piping systems shall comply with the provisions for container valves. [*See 2-3.3.3(a), (b) and (c).*]

2-4.5.4 Emergency shutoff valves shall be approved and incorporate all of the following means of closing (*see 3-2.7.9 and 3-3.3.4*):

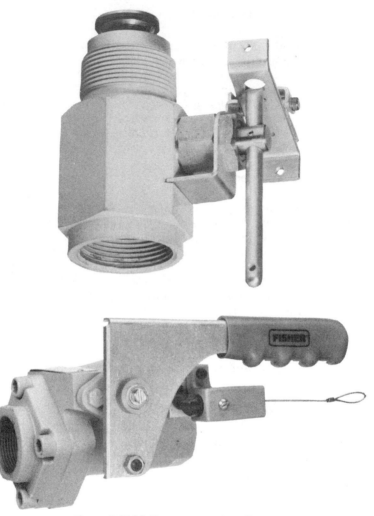

Figure 2.18(a) Emergency Shutoff Valves.

(a) Automatic shutoff through thermal (fire) actuation. When fusible elements are used they shall have a melting point not exceeding 250°F (121°C).

(b) Manual shutoff from a remote location.

(c) Manual shutoff at the installed location.

This provision sets forth the basic criteria for the emergency shutoff valve, a key valve in the protection of many liquid transfer operations. Actuating means for remote control may be electrical, mechanical, or pneumatic. Many systems use a pneu-

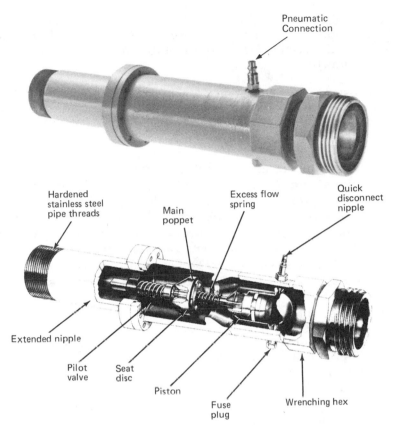

Figure 2.18(b) Emergency Shutoff Valve/Tank Car Unloading Adapter Combination.

matic plastic tubing system, where the tubing itself acts as a fusible element releasing the pressure holding the valve open. With respect to the feature of manual shutoff at the installed location, it is recommended that this valve be operated occasionally. Also, the system should be tested periodically to determine that it will function properly. See commentary on 3-2.7.9.

2-4.6 Hose, Quick Connectors, Hose Connections, and Flexible Connectors.

2-4.6.1   Hose, hose connections, and flexible connectors (*see definition*) shall be fabricated of materials resistant to the action of LP-Gas both as liquid and vapor. If wire braid is used for reinforcement it shall be of corrosion resistant material such as stainless steel.

Note that carbon steel may not be used in hose or hose connections. This prohibition is based on experience with hoses that weakened as the carbon steel braid corroded.

2-4.6.2   Hose and quick connectors shall be approved (*see Section 1-7, Approved*).

2-4.6.3 Hose, hose connections, and flexible connectors used for conveying LP-Gas liquid or vapor at pressures in excess of 5 psig (34.5 kPa gauge), and as provided in Section 3-4 regardless of the pressure, shall comply with 2-4.6.3(a) and (b):

(a) Hose shall be designed for a working pressure of 350 psi (240 MPa) with a safety factor of 5 to 1 and be continuously marked "LP-GAS," "PROPANE," "350 PSI WORKING PRESSURE," and the manufacturer's name or trademark.

> It is important to note that these provisions for hose, hose connections, and flexible connectors apply only to those involved with pressures in excess of 5 psig (34.5 kPa gauge). Requiring continuous marking, which is current industry practice, permits identification of short connectors. The term "propane" has been added to carry out the intent for industry identification internationally. The current provision is consistent with standards of the Rubber Manufacturers Association.

(b) Hose assemblies, after the application of connections, shall have a design capability of withstanding a pressure of not less than 700 psig (4.8 MPa gauge). If a test is made, such assemblies shall not be leak tested at pressures higher than the working pressure [350 psig (2.4 MPa gauge) minimum] of the hose.

> It is far more important to specify that the assembly must have a "design capability" of withstanding a pressure of 700 psig (4.8 MPa gauge) than to pressure test the assembly to this pressure to assure that the connector fittings have been properly assembled to the hose. Past experience with hydrostatic testing of hose at this pressure has resulted in injury to the hose. An assembly procedure can be verified to determine that the design capability has been met.

2-4.6.4 Hoses or flexible connectors used to supply LP-Gas to utilization equipment or appliances shall be installed in accordance with the provisions of 3-2.7.8 and 3-2.7.10.

2-4.7 Hydrostatic Relief Valves.

2-4.7.1 Hydrostatic relief valves designed to relieve the hydrostatic pressure which might develop in sections of liquid piping between closed shutoff valves shall have pressure settings not less than 400 psi (2.8 MPa gauge) or more than 500 psi (3.5 MPa gauge) unless installed in systems designed to operate above 350 psi (2.4 MPa gauge). Hydrostatic relief valves for use in systems designed to operate above 350 psi (2.4 MPa gauge) shall have settings not less than 110 percent or more than 125 percent of the system design pressure.

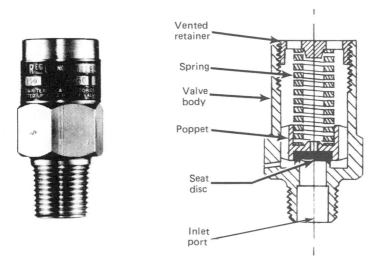

Vented retainer
Spring
Valve body
Poppet
Seat disc
Inlet port

Figure 2.19 Hydrostatic Relief Valve.

The number of types of hydrostatic relief valves has been reduced and the possibility of using the wrong valve eliminated through a rational consolidation of pressure settings. Small-diameter pipe and tubing usually used in smaller systems can easily handle 500 psig (3.5 MPa gauge) without danger. In large industrial or commercial systems where higher pressures might be used, the specified 110 to 125 percent of design pressure is reasonable. Liquid systems connected to DOT containers are rare and require no special consideration.

2-5 Equipment.

2-5.1 General.

2-5.1.1 This section includes fabrication and performance provisions for the pressure containing metal parts of LP-Gas equipment such as pumps, compressors, vaporizers, strainers, meters, sight flow glasses and regulators. Containers are not subject to the provisions of this section.

2-5.1.2 Equipment shall be suitable for the appropriate working pressure as follows:

(a) Equipment to be used at pressures higher than container pressure, such as on the discharge of a liquid pump, shall be suitable for a working pressure of at least 350 psig (2.4 MPa gauge). If pressures above 350 psig (2.4 MPa gauge) are necessary, the pump and all equipment under pressure from the pump shall be suitable for the pump discharge pressure.

(b) Equipment to be used with liquid LP-Gas, or vapor LP-Gas at pressures over 125 psig (0.9 MPa gauge) but not to exceed 250 psig (1.7 MPa gauge), shall be suitable for a working pressure of at least 250 psig (1.7 MPa gauge).

(c) Equipment to be used with vapor LP-Gas at pressures over 20 psig (138 kPa gauge), but not to exceed 125 psig (0.9 MPa gauge), shall be suitable for a working pressure of at least 125 psig (0.9 MPa gauge).

(d) Equipment to be used with vapor LP-Gas at pressures not to exceed 20 psig (138 kPa gauge) shall be suitable for a working pressure adequate for the service in which it is to be used.

2-5.1.3 Equipment shall be fabricated of materials suitable for LP-Gas service and resistant to the action of LP-Gas under service conditions. The following shall also apply:

It is important that the materials of components in equipment be suitable for LP-Gas; equipment should not be purchased until this has been ascertained from the manufacturer. Restrictions on the use of cast iron reflect its tendency to crack under low temperature conditions and when heated and suddenly cooled under fire control conditions. Behavior when subjected to fire is also reflected in the melting points of metals. The use of metals which melt at less than 1,000°F (538°C) under the service pressure stresses—e.g., aluminum and zinc—are restricted to situations in which their failure will not constitute an undue hazard. Subparagraph 2-5.1.3(e) was added in the 1986 Edition and reflects growing interest in the use of plastic materials.

(a) Pressure containing metal parts shall be of steel, ductile (nodular) iron (ASTM A 395-77 or A 536-77 Grade 60-40-18 or 65-45-12), malleable iron (ASTM A 47-77), higher strength gray iron (ASTM A 48-76, Class 40B), brass, or the equivalent.

(b) Cast iron shall not be used for strainers or flow indicators which shall comply with provisions for materials for construction of valves (*see 2-4.5.1*).

(c) Aluminum may be used for approved meters.

(d) Aluminum or zinc may be used for approved regulators. Zinc used for regulators shall comply with ASTM B 86-76.

(e) Non-metallic materials shall not be used for upper and lower casings of regulators.

2-5.2 Pumps.

2-5.2.1   Pumps shall be designed for LP-Gas service and may be of rotary, centrifugal, turbine or reciprocating type.

Internal relief valve

Figure 2.20(a) Sliding Vane Positive Displacement Pump.

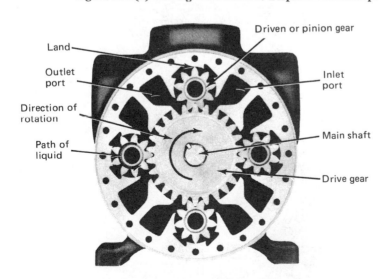

Figure 2.20(b) Gear Pump.

 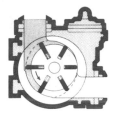 

a. Vanes move out, trapping liquid at the pump inlet.

b. Liquid is transferred toward the outlet between the vanes.

c. As the vanes move back into their slots, liquid is discharged through the outlet.

Figure 2.20(c) Operation of Sliding Vane Pump.

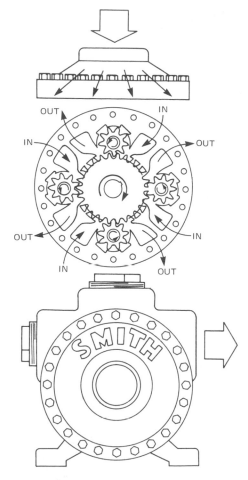

Figure 2.20(d) Operation of Gear Pump.

2-5.2.2   The maximum discharge pressure of a liquid pump under normal operating conditions shall be limited to 350 psig (2.4 MPa gauge).

It is important that pumps be designed and built for use with LP-Gas. Nonspecialized pumps may not work satisfactorily and some may be hazardous. A check should be made with the manufacturer to see whether or not the pump is capable of a discharge pressure higher than 350 psig (2.4 MPa gauge). If so, provisions must be made to limit the pressure to 350 psig (2.4 MPa gauge), or the system must be designed for the higher pressure.

Because they are pumping a liquefied gas which will vaporize if the pressure drops even slightly, LP-Gas pumps are eigher positive displacement pumps or special types of centrifugal pumps (regenerative turbine pumps). Ordinary centrifugal pumps will "vapor lock" rather easily and be rendered ineffective.

2-5.3 Compressors.

2-5.3.1   Compressors shall be designed for LP-Gas service and may be of the rotary or reciprocating type and shall be equipped with suitable glands or seals to minimize any release of LP-Gas.

Care should be taken to select compressors properly designed for LP-Gas service. Compressors designed for noncombustible gases may not provide adequate leak protection required for LP-Gases.

2-5.3.2   Means shall be provided to limit the suction pressure to the maximum for which the compressor is designed.

2-5.3.3   Means shall be provided to prevent the entrance of LP-Gas liquid into the compressor suction, either integral with the compressor, or installed externally in the suction piping [see 3-2.10.2(b)]. Portable compressors used with temporary connections are excluded from this requirement.

Because all liquids, including liquid LP-Gas, are not compressible, they must not be allowed to enter an operating compressor. A number of compressors manufactured for this service include relief devices built into the cylinder head to prevent destruction of the compressor in case small amounts of liquid are allowed to enter. Continuously taking advantage of this feature, however, will eventually destroy the compressor. Several float operated devices are designed to prevent liquid from entering the compressor. Some manufacturers use this type of device while others simply use a large receiver as a liquid trap on the suction side of the compressor. On these types the compressor is usually mounted on the top of the receiver. Portable compressors are exempted from this type of protection because the temporary connections used are empty at the time they are connected which usually prevents liquid from entering the compressor. Also, portable units are operated under continuous surveillance by an operator.

Compressors are used to move liquid by pumping vapor from the receiving storage into the supply transport car or truck, thereby increasing the pressure in the transport device and forcing liquid from the transport to the receiver tank. This type of loading or unloading operation is necessary when it is impractical or impossible to use liquid pumps, such as when unloading railroad tank cars. Tank cars have all fittings at the top of the car in the dome. The use of pumps is impractical because of the extended length of piping to the pump suction which results in pump cavitation and reduced capacity. When compressors are used to unload tank cars it is common practice to reverse the compressor, after all liquid is removed from the car, to recover the vapor. Normally, this continues until the pressure in the car is about 30 psi.

2-5.3.4   Engines used to drive portable compressors shall be equipped with exhaust system spark arrestors and shielded ignition systems.

This applies to portable pumps as well as portable compressors. The fact that these devices are portable indicates that they are going to be operated by an internal combustion engine of some type, in all probability. An obvious source of ignition of gas in the atmosphere is the possibility of sparks from an exhaust system of an internal combustion engine or sparks from the engine ignition system. Most manufacturers of these

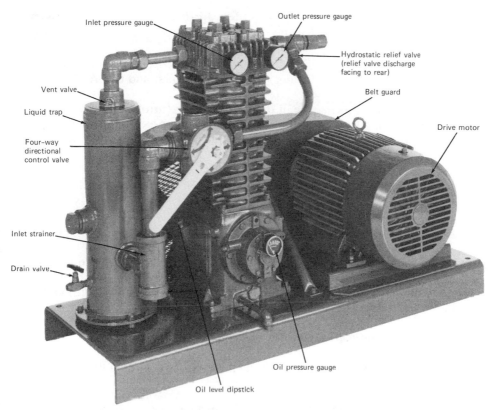

Figure 2.21(a) Compressor with Protection Devices.

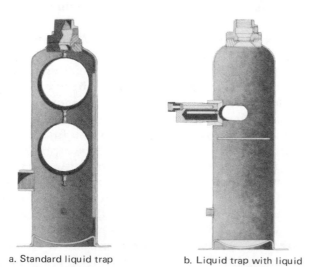

a. Standard liquid trap   b. Liquid trap with liquid

Figure 2.21(b) Compressor Liquid Traps. In part (a), the floats rise with the liquid level and close the valve at the top. In part (b), the float actuates a switch which shuts off the compressor motor.

type of engines supply shielded ignition systems and spark arrestors as an option for units which are going to be used in classified areas.

2-5.4 Vaporizers, Tank Heaters, Vaporizing-Burners, and Gas-Air Mixers.

2-5.4.1 Vaporizers may be of the indirect type (utilizing steam, hot water, or other heating medium), or direct fired. This subsection does not apply to engine fuel vaporizers or to integral vaporizer-burners such as those used with weed burners or tar kettles.

See definitions in Section 1-7 for description of various types of vaporizers.

2-5.4.2 Indirect vaporizers shall comply with 2-5.4.2(a) through (e):

(a) Indirect vaporizers with an inside diameter of more than 6 in. (152 mm) shall be constructed in accordance with the applicable provision of the ASME Code for a design pressure of 250 psig (1.7 MPa gauge) and shall be permanently and legibly marked with:

(1)   The marking required by the code.

(2)   The allowable working pressure and temperature for which designed.

(3)   The sum of the outside surface area and the inside heat exchange surface area in sq ft.

(4)   The name or symbol of the manufacturer.

With respect to the reference in 2-5.4.2(a) to indirect vaporizers with an inside diameter of more than 6 in. (152 mm) and the exemption in (b) for those units smaller than 6-in. (152-mm) inside diameter, the ASME Code offers two major exemptions: (1) for any size vessel which is designed for less than 15 psig (103 kPa gauge) operating pressure, and (2) for vessels at any operating pressure but which have an inside diameter of 6 in. (152 mm) or less. The latter is the exemption in 2-5.4.2(b). However, the intent is that the vessel be designed in accordance with the ASME Code but need not be inspected by an ASME approved inspector or marked with the Code marking. The second sentence in (b), which states that these vaporizers shall be constructed for a minimum of 250 psig (1.7 MPa gauge) design pressure, is intended to mean that this construction be in accordance with the ASME Code but exempt from inspection and marking as allowed by the ASME Code. Also, with regard to the ASME Code exception, some jurisdictions do not allow this exemption, e.g., the State of California requires any vaporizer with an internal volume that would hold more than 1 gal (4 liters) of liquid be built, inspected, and marked in accordance with the ASME Code.

(b) Indirect vaporizers having an inside diameter of 6 in. (152 mm) or less are exempt from the ASME Code and need not be marked. They shall be constructed for a minimum 250 psig (1.7 MPa gauge) design pressure.

(c) Indirect vaporizers shall be provided with a suitable automatic means to prevent liquid passing through the vaporizer to the vapor discharge piping. This means may be integral with the vaporizer, or otherwise provided in the external piping (*see 3-7.2.6*).

The provision for preventing liquid from leaving the vaporizer through vapor discharge piping is an important one. If the vaporizer is sized properly and is functioning properly, liquid will not be present in the vapor outlet. This provision is intended to protect against malfunction where the heat source of the vaporizer failed or the unit was overloaded intentionally or accidentally. The hazard of liquid in the vapor piping is that none of the vapor control gear, such as regulators, control valves, and burners, is designed to handle liquid. Liquid passing through vapor regulators will flash and can cause a supercooling situation where liquid actually is expelled through appliances creating a fire hazard.

(d) Indirect vaporizers, including atmospheric-type vaporizers using heat from the surrounding air or the ground, and of more than one quart (0.9 L) capacity, shall be equipped, at or near the discharge, with a spring-loaded pressure relief valve providing a relieving capacity in accordance with 2-5.4.5. Fusible plug devices shall not be used.

The prohibition of fusible plugs (instead of spring-loaded pressure relief valves) is because fusible plugs, as well as rupture discs, will cause the entire contents of the protected device to be discharged into the atmosphere Spring-loaded relief valves will reduce the quantity discharged because they close when the excessive pressure has been relieved.

(e) Indirect atmospheric-type vaporizers of less than one quart (0.9 L) capacity need not be equipped with pressure relief valves, but shall be installed in accordance with 3-7.2.9.

2-5.4.3 Direct-fired vaporizers shall comply with 2-5.4.3(a) through (f).

Note that all direct-fired vaporizers must be constructed in accordance with the ASME Code, which includes inspection and

marking. Unlike provisions for indirect vaporizers, there is no exemption from inspection and marking regardless of internal diameter. The prohibition against fusible plugs and/or rupture discs as a means to prevent liquid passing into the vapor discharge piping is common with the provisions for indirect vaporizers. [*See commentary on 2-5.4.2(d)*.] Additional provisions for direct-fired vaporizers include a provision for the relief valve to be located or protected in such a manner as to prevent excessive temperature because a direct-fired unit contains a burner that can create temperatures, near the burner itself or in the path of the products of combustion, that would destroy the seats and mechanism of a relief valve that might be subjected to those excessive temperatures. Marking required by the Code refers to the ASME Code. Paragraphs 2-5.4.3 (a)(2), (a)(3), (a)(4), and (a)(5) are markings that are not stipulated in the ASME Code. Note that (a)(2) and (a)(3) specify outside surface and heat exchange surface as separate items, which differs from the use of the sum of these in 2-5.4.2(a)(3) for indirect vaporizers. This is because heat exchange and outside surface are usually the same surface in a direct-fired unit, while they are not the same in an indirect-fired unit. The "outside surface" is not the surface of the outside of the cabinet or enclosure, but is that surface in contact with LP-Gas inside and atmosphere on the outside, which could add to the vaporization rate of the vaporizer if subjected to an external fire. Referring to Figure 2.22(b), the outside surface includes all of the heat exchange surface plus the top of the vessel. However, the marking for (a)(2) should be for the top only and (a)(3) should include the bottom and sides.

Paragraph 2-5.4.3 (e) reiterates the requirement for 100 percent shutoff to a gas burner and is not unique to a direct-fired vaporizer burner.

While 2-5.4.3 (f) infers that the limit control should be a pressure operated device, temperature sensing devices are commonly used for this purpose. Controlling the temperature of the output of the vaporizer is the only practical method to prevent the heater from raising the product pressure (the heater being the burner of the vaporizer). The vaporizer cannot produce a pressure higher than provided to the liquid inlet of the vaporizer since, if the pressure in the vaporizer were to exceed the inlet pressure, flow of liquid into the vaporizer would stop and, in fact, flow backward.

The reference in 2-5.4.3(f) to raising the pressure within the storage container is descriptive of a type of system seldom used in present day systems. This was a system where liquid ran by gravity from a storage tank into a vaporizer and vapor was returned from the vaporizer back into the top of the tank. Vapor was then removed from the tank through a pressure regulator into the distribution system. With this type of system, it is theoretically possible, in extremely hot weather, for a vaporizer connected in this manner to raise the pressure in the container. This, of course, would cause the container pressure relief valves to operate, discharging vapor which, because of the existence of the direct-fired vaporizer, would almost certainly become ignited. Again, temperature controls in the vaporizer would prevent this type of malfunction from taking place. Accordingly, a temperature control is the only practical type of control to satisfy this requirement.

Figure 2.22(a) Direct-fired Vaporizer.

(a) Design and construction shall be in accordance with the applicable requirements of the ASME Code for the working conditions to which the vaporizer will be subjected, and it shall be permanently and legibly marked with:

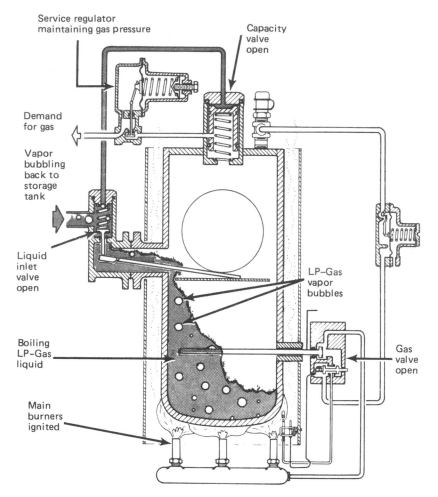

Service regulator
maintaining gas pressure

Capacity
valve
open

Demand
for gas

Vapor
bubbling
back to
storage
tank

Liquid
inlet
valve
open

LP–Gas
vapor
bubbles

Boiling
LP–Gas
liquid

Gas
valve
open

Main
burners
ignited

Figure 2.22(b) Direct-fired Vaporizer Operating on Demand.

(1)   The markings required by the code.

(2)   The outside surface area in sq ft.

(3)   The area of the heat exchange surface in sq ft.

(4)   The maximum vaporizing capacity in gal per hour.

(5)   The rated heat input in Btuh.

(6)   The name or symbol of the manufacturer.

(b) Direct-fired vaporizers shall be equipped, at or near the discharge, with a spring-loaded pressure relief valve providing a relieving capacity in accordance with 2-5.4.5. The relief valve shall be located so as not to be subject to temperatures in excess of 140°F (60°C). Fusible plug devices shall not be used.

(c) Direct-fired vaporizers shall be provided with suitable automatic means to prevent liquid passing from the vaporizer to its vapor discharge piping.

(d) A means for manually turning off the gas to the main burner and pilot shall be provided.

(e) Direct-fired vaporizers shall be equipped with an automatic safety device to shut off the flow of gas to the main burner if the pilot light is extinguished. If the pilot flow exceeds 2,000 Btuh (2 MJ/h), the safety device shall shut off the flow of gas to the pilot also.

(f) Direct-fired vaporizers shall be equipped with a limit control to prevent the heater from raising the product pressure above the design pressure of the vaporizer equipment, and to prevent raising the pressure within the storage container above the pressure shown in the first column of Table 2-2.2.2 corresponding with the design pressure of the container (or its ASME Code equivalent — *see Note 1 of Table 2-2.2.2*).

2-5.4.4 Waterbath vaporizers shall comply with 2-5.4.4(a) through (j).

(a) The vaporizing chamber, tubing, pipe coils, or other heat exchange surface containing the LP Gas to be vaporized, hereinafter referred to as "heat exchanger," shall be constructed in accordance with the applicable provisions of the ASME Code for a minimum design pressure of 250 psig (1.7 MPa gauge) and shall be permanently and legibly marked with:

(1) The marking required by the code.

(2) The allowable working pressure and temperature for which designed.

(3) The sum of the outside surface and the inside heat exchange surface area in sq ft.

(4) The name or symbol of the manufacturer.

(b) Heat exchangers for waterbath vaporizers having an inside diameter of 6 in. (152 mm) or less are exempt from the ASME Code and need not be marked. They shall be constructed for a 250 psig (1.7 MPa gauge) minimum design pressure.

(c) Heat exchangers for waterbath vaporizers shall be provided with a suitable automatic control to prevent liquid passing through the heat exchanger to the vapor discharge piping. This control shall be integral with the vaporizer.

(d) Heat exchangers for waterbath vaporizers shall be equipped at or near the discharge with a spring loaded pressure relief valve providing a relieving capacity in accordance with 2-5.4.5. Fusible plug devices shall not be used.

(e) Waterbath sections of waterbath vaporizers shall be designed to eliminate a pressure buildup above the design pressure.

(f) The immersion heater which provides heat to the waterbath shall be installed so as not to contact the heat exchanger and may be electric or gas-fired.

(g) A control to limit the temperature of the waterbath shall be provided.

(h) Gas-fired immersion heaters shall be equipped with an automatic safety device to shut off the flow of gas to the main burner and pilot in the event of flame failure.

(i) Gas-fired immersion heaters with an input of 400,000 Btu (422 mJ/h) per hour or more shall be equipped with an electronic flame safeguard and programming to provide for pre-purge prior to ignition, proof of pilot before main burner valve opens, and full shutdown of main gas and pilot upon flame failure.

(j) A means shall be provided to shut off the source of heat in case the level of the heat transfer medium falls below the top of the heat exchanger.

The waterbath vaporizer differs from a direct-fired vaporizer in that there is no flame impingement on the vaporizer chamber itself. It is to some extent a cross between a direct-fired vaporizer and an indirect-fired vaporizer. The portion of such a unit that contains and vaporizes the LP-Gas must be built in accordance with the provisions of the ASME Code for a minimum of 250 psig (1.7 MPa gauge) and be marked as stipulated. Again, those under 6 in. (152 mm) in diameter are exempt from the ASME Code, but must be constructed to a minimum of 250 psig (1.7 MPa gauge). [See also commentary following 2-5.4.2(a)(4).] The same automatic safety controls used with indirect and direct-fired vaporizers to prevent liquid passing into the vapor discharge piping, etc., must be provided. The immersion heater that provides heat to the waterbath must not contact the LP-Gas heat exchanger surface to avoid overheating and it must be properly equipped with an automatic shutoff device to prevent overheating of the liquid bath temperature should the level of the bath liquid fall below the top of the heat exchanger.

Paragraph 2-5.4.4 (f) states that the immersion heater which heats the waterbath may be electric or gas-fired. The intent is

not to prevent any other form of heat source, but at this writing, the only two that seem practical are electric or gas-fired immersion heaters. There is no provision preventing the use of steam coils or some form of waste heat coils to heat the waterbath. Such forms of heat would be better applied to indirect vaporizers as covered by paragraph 2-5.4.2, rather than adding the unnecessary additional dimension of a second interface between heat source and vaporizing coil. Although this type of an operation would be perfectly acceptable, the economics would probably preclude its use. There is no apparent safety advantage over the use of indirect units when the source of heat would be the same as that required by the indirect unit.

2-5.4.5 The minimum rate of discharge in cubic-feet-of-air-per-minute for pressure relief valves for LP-Gas vaporizers, either of the indirect type or direct fired, shall be determined as follows:

(a) The surface area of that part of the vaporizer shell directly in contact with LP-Gas shall be added to the heat exchange surface area directly in contact with LP-Gas to obtain the total surface area in sq ft.

(b) Refer to Table E-2.2.2 to obtain the rate of discharge in cu ft of air per minute (Flow Rate CFM Air) for the total surface area in sq ft for the vaporizer computed in accordance with 2-5.4.5(a).

In the event of fire exposure to a vaporizer, the pressure relief valve must be sized to handle vapor produced as a result of heating of the liquid by both the normal heat source and the exposure fire.

2-5.4.6 Direct gas-fired tank heaters shall be designed exclusively for outdoor aboveground use and so that there is no direct flame impingement upon the container. The provisions of 2-5.4.6(a) through (f) shall also apply.

Direct-fired tank heaters are no longer widely used. They are heaters which attach to the bottom of the tank to heat the tank shell at the point of application, which then heats the liquid product inside. However, there can be no flame impingement upon the container itself to avoid loss of container strength. Most of these units are constructed so that they can be easily removed for inspection for possible corrosion of the tank at the point of installation. These units have to be carefully engineered so that there will be automatic means to prevent overheating the product in the tank, to shut off the heater in case the tank goes empty of liquid, to shut off the gas flow to the heater in case the pilot light goes out, etc.

(a) Tank heaters shall be approved and be permanently and legibly marked with:

   (1)   The rated input to the burner in Btuh.

   (2)   The maximum vaporizing capacity in gal per hour.

   (3)   The name or symbol of the manufacturer.

   (b) The heater shall be designed so that it can be readily removed for inspection of the entire container.

   (c) The fuel gas supply connection to the tank heater shall originate in the vapor space of the container being heated and shall be provided with a manually operated shutoff valve at the heater.

   (d) The heater control system shall be equipped with an automatic safety shutoff valve of the manual-reset type arranged to shut off the flow of gas to both the main and pilot burners if the pilot flame is extinguished.

   (e) When installed on a container exceeding 1,000 gal (3.8 m^3) water capacity, the heater control system shall include a valve to automatically shut off the flow of gas to both the main and pilot burners if the container becomes empty of liquid.

   (f) Direct gas-fired tank heaters shall be equipped with a limit control to prevent the heater from raising the pressure in the storage container to more than 75 percent of the pressure shown in the first column of Table 2-2.2.2 corresponding with the design pressure of the container (or its ASME Code equivalent — *see Note 1 of Table 2-2.2.2*).

2-5.4.7   Vaporizing-burners shall be constructed with a minimum design pressure of 250 psig (1.7 MPa gauge) with a factor of safety of 5, and shall comply with 2-5.4.7(a) through (h):

   (a) The vaporizing-burner, or the appliance in which it is installed, shall be permanently and legibly marked with:

   (1)   The maximum burner input in Btuh.

   (2)   The name or symbol of the manufacturer.

   (b) Vaporizing coils or jackets shall be made of ferrous metals or high temperature alloys.

   (c) The vaporizing section shall be protected by a hydrostatic relief valve, located where it will not be subjected to temperatures in excess of 140°F (60°C), and with a pressure setting such as to protect the components involved but not lower than 250 psig (1.7 MPa gauge). The relief valve discharge shall be directed upward and away from the component parts of the vaporizing burner. Fusible plug devices shall not be used.

(d) A means shall be provided for manually turning off the gas to the main burner and the pilot.

(e) Vaporizing-burners shall be provided with an automatic safety device to shut off the flow of gas to the main burner and pilot in the event the pilot is extinguished.

(f) Dehydrators and dryers utilizing vaporizing-burners shall be equipped with automatic devices both upstream and downstream of the vaporizing section. These devices shall be installed and connected to shut off in the event of excessive temperature, flame failure, and if applicable, insufficient air flow. See NFPA 61B, *Standard for the Prevention of Fires and Explosions in Grain Elevators and Facilities Handling Bulk Raw Agricultural Commodities*, for ignition and combustion controls applicable to vaporizing-burners associated with grain dryers.

(g) Pressure regulating and control equipment shall be so located or so protected as not to be subject to temperatures above 140°F (60°C), unless it is designed and recommended for use by the manufacturer for a higher temperature.

(h) Pressure regulating and control equipment located downstream of the vaporizing section shall be designed to withstand the maximum discharge temperature of the hot vapor.

These burners actually are liquid fed and the vaporization takes place in the burner. Therefore, the minimum design pressure has to be 250 psig (1.7 MPa gauge) (the same as the container) and, because of the high temperature created, the coils and jackets must be made of ferrous metals or high-temperature alloys. Because liquid is fed to the unit, there must be a hydrostatic relief valve in the unit to protect it when the burner is shut off and the liquid is trapped in the burner and piping. If this were not done, an increase in the temperature of the liquid would create a pressure that could burst the heater or the attached piping. These units are often used for temporary heating during construction or during emergencies and are unattended. As a result, the careful use of automatic equipment is extremely important. The units are also used for agricultural product dehydrators and dryers and should be properly equipped with automatic devices both on the inlet and outlet of the vaporizer section of the burner. As these units are generally portable, extreme care must be used in the placement of them and in ensuring that the fuel connection to them is not subjected to mechanical damage.

2-5.4.8 Gas-air mixers shall comply with 2-5.4.8(a) through (e).

(a) Gas-air mixers shall be designed for the air, vapor and mixture pressures to which they are subjected. Piping materials shall comply with applicable portions of this standard.

(b) Gas-air mixers shall be designed so as to prevent the formation of a combustible mixture. Gas-air mixers which are capable of producing combustible mixtures shall be equipped with safety interlocks on both the LP-Gas and air supply lines to shut down the system if combustible limits are approached.

(c) In addition to the interlocks provided for in 2-5.4.8(b), a method shall be provided to prevent air from accidentally entering gas distribution lines without LP-Gas being present. Check valves shall be installed in the air and LP-Gas supply lines close to the mixer to minimize the possibility of backflow of gas into the air supply lines or of air into the LP-Gas system. Gas mixing control valves in the LP-Gas and air supply lines which are arranged to fail closed when actuated by safety interlock trip devices shall be considered as acceptable shutdown devices.

(d) Where it is possible for condensation to take place between the vaporizer and the gas-air mixer, an interlock shall be provided to prevent LP-Gas liquid from entering the gas-air mixer.

(e) Gas-air mixers which utilize the kinetic energy of the LP-Gas vapor to entrain air from the atmosphere, and are so designed that maximum air entrained is less than 85 percent of the mixture, need not include the interlocks specified in 2-5.4.8(b), (c) and (d), but shall be equipped with a check valve at the air intake to prevent the escape of gas to atmosphere when shut down. Gas-air mixers of this type receiving air from a blower, compressor or any source of air other than directly from the atmosphere, shall include a method of preventing air without LP-Gas, or mixtures of air and LP-Gas within the flammable range, from entering the gas distribution system accidentally.

The flammable limits for propane in air are approximately 2.15 percent to 9.6 percent (*see Appendix B*). In a venturi type blender or mixer where the kinetic energy of the LP-Gas vapor is used to entrain air from the atmosphere, the physical geometry of the venturi design is such that the quantity of air that can be drawn in cannot exceed 85 percent of the mixed gas formed. Thus, it is impossible to have a flammable LP-Gas/air mixture, and the interlock provisions of 2-5.4.8(b) can be eliminated. A check valve at each air inlet venturi will prevent any LP-Gas from escaping to the atmosphere.

To prevent malfunction, most venturi mixers have high and low vapor pressure interlocks, high and low mixed LP-Gas/air pressure interlocks and a low incoming vapor temperature interlock. The latter interlock prevents any LP-Gas liquid from entering the gas-air mixer [2-5.4.8(d)].

Figure 2.23 Venturi Type Gas-Air Mixer with Five (5) Venturis. This type of gas-mixer utilizes the kinetic energy of LP-Gas vapor to entrain air from the atmosphere to provide a desired LP-Gas/air mixture. Each venturi includes a solenoid valve on the gas inlet side of the venturi. A controller monitors the outlet manifold pressure of the mixer and energizes each individual solenoid valve to provide a constant flow to the process at the required pressure. The air inlet of the venturi includes a backflow check valve to prevent the escape of LP-Gas to the atmosphere. In addition, the controller is interlocked with a low vapor inlet temperature switch and high and low pressure switches on the gas inlet at gas-air outlet manifolds.

To prevent any possible vapor condensation problems, it is advisable that the mixer be located as close as possible to the vaporizer, and that the mixer vapor inlet connection be sloped back toward the vaporizer vapor outlet connection. If this is not practical, a drip leg with heater should be installed. The piping can be heated to prevent the low temperature by electric tracer heater cable wrapped around such piping. Unless ambient temperatures are very low, insulating the piping may be sufficient to prevent too low a temperature of the gas vapor.

Usually venturi mixers employing atmospheric air must have an output mixture pressure less than 9 psig (62 kPa gauge) in order to produce a viable mixed gas. For mixer pressures above 10 psig (69 kPa gauge), air under pressure must be supplied to the venturi mixer from an external source (i.e., plant air system or air compressor). For this external air supply system to the venturi mixer some method must be provided on the venturi mixer to prevent 100 percent air or a flammable LP-Gas/air mixture from being supplied to the user's gas distribution

Figure 2.24 High Pressure Orifice Gas-Air Mixer. A high pressure [above 9 psig (62 kPa gauge)] variable orifice type gas-air mixer with inlet air and gas control trains located in a room complying with Chapter 7. Both the air and gas control trains include pressure regulators, check valves, shutoff valves, safety shutoff valves, temperature and pressure indicators and high-low pressure switches. Process air for the mixer may be furnished from an air compressor in a separate room or the plant air system. The air and gas pressures are both regulated to equal pressures on the inlet to the mixer. The variable orifice in the mixing valve will open and close depending on flow to provide a constant Btu valve. The propane-air is then distributed within the plant process piping.

piping. This is usually accomplished through special controls and low air pressure interlocks that will shut the mixer down in the event of malfunction.

Consideration should also be given to the method of installation of the downstream piping for LP-Gas/air from the mixer to the plant utilization point relative to possible recondensation of the LP-Gas/air mix. For most cases when commercial propane is the LP-Gas/air feedstock and the mixed LP-Gas/air mix is at or below 100 psig (690 kPa gauge), there would be no possible recondensation of propane in the mix until the mix gas temperature became 20°F (-6.7°C) or colder. But, if commercial butane is the LP-Gas feedstock, considerable care must be taken in the mixed gas piping design. At a 10 psig (69 kPa gauge) mixed gas pressure, butane will start recondensing at about 20°F (-6.7°C).

Thus, for higher mixed gas sendout pressures, recondensation of butane will occur at higher and higher temperatures [e.g., at 50 psig (345 kPa gauge) the recondensation temperature is about 60°F (15.6°C)]. Heat for the butane/air mix gas line must thus be provided in the form of heat tracing and/or insulation to guard against possible butane recondensation.

2-5.5 Strainers.

2-5.5.1 Strainers shall be designed to minimize the possibility of particulate materials clogging lines and damaging pumps, compressors, meters, or regulators. The strainer element shall be accessible for cleaning.

2-5.6 Meters.

2-5.6.1 Vapor meters of the tin or brass case type of soldered construction shall not be used at pressures in excess of 1 psig (7 kPa gauge).

2-5.6.2 Vapor meters of the die cast or iron case type may be used at any pressure equal to or less than the working pressure for which they are designed and marked.

2-5.7 Dispensing Devices.

2-5.7.1 Components of dispensing devices, such as meters, vapor separators, valves and fittings within the dispenser, shall comply with 2-5.1.2(b) and 2-5.1.3.

2-5.7.2 Pumps of dispensers used to transfer LP-Gas shall comply with 2-5.1.2(b), 2-5.1.3 and with 2-5.2. Such pumps shall be equipped to permit control of the flow and to minimize the possibility of leakage or accidental discharge. Means shall be provided on the outside of the dispenser to readily shut off the power in the event of fire or accident. This means may be integral with the dispenser or provided externally when the dispenser is installed (*see 3-2.10.6*).

2-5.7.3 Dispensing hose shall comply with 2-4.6.1 through 2-4.6.3. An excess-flow check valve or an automatic shutoff valve complying with 2-3.3.3(a), (b), and (c) and 2-4.5.3, or 2-3.3.3(d) and 2-4.5.4, shall be installed in or on the dispenser at the point at which the dispenser hose is connected to the liquid piping. A differential back pressure valve shall be considered as meeting these provisions.

A differential back pressure valve is normally supplied with an LP-Gas meter. The valve senses the difference in pressure before and after the meter and closes if they are significantly different. This insures that only liquid (not vapor) passes through the meter for accurate reading. If the dispensing hose should break or become disconnected the pressure downstream of the meter will drop and the valve will close. This provision is essentially an exception to the requirement for an excess-flow valve or an emergency shutoff valve in liquid transfer lines in paragraph 3-3.3.

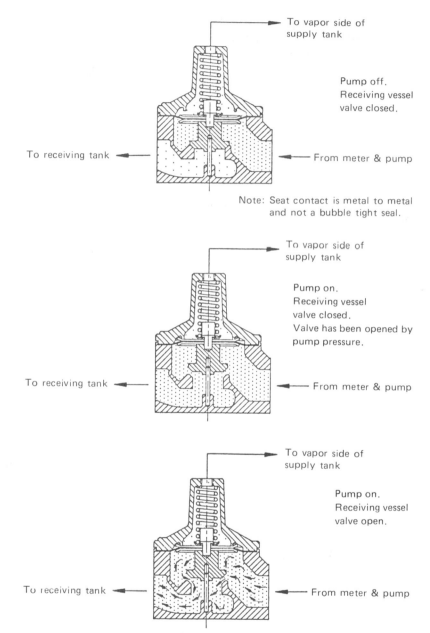

Figure 2.25 Differential back pressure valve.

2-5.8 Regulators.

This requirement was clarified in the 1989 Edition of the standard by changing "with one or both of the following" to "with one of the following" to clarify that an option is presented.

Table 2-5.8.1

Regulator Delivery Pressure in psig (kPa gauge)	Relief Valve Start-to-Leak Pressure Setting, % of Regulator Delivery Pressure	
	Minimum	Maximum
1 (7) or less	170%	300%
Above 1 (7), not over 3 (21)	140%	250%
Above 3 (21)	125%	250%

2-5.8.1 Final stage regulators (excluding appliance regulators) shall be equipped with one of the following [*see 3-2.5.2(b) for required protection from the elements which may be integral with the regulator*]:

(a) A pressure relief valve on the low pressure side having a start-to-leak pressure setting within limits specified in Table 2-5.8.1.

(b) A shutoff device that shuts the gas off at the regulator inlet when the downstream pressure reaches the overpressure limits specified in Table 2-5.8.1. Such a device shall not open to permit flow of gas until it has been manually reset.

The minimum relief valve setting prevents operation at too low a pressure and interference with operations. The maximum setting is to prevent excessive pressures building up downstream of the regulator, which could cause leakage of the equipment downstream.

2-5.9 Sight Flow Glasses.

2-5.9.1 Flow indicators, either of the simple observation type or combined with a backflow check valve, may be used in applications in which the observation of liquid flow through the piping is desirable or necessary.

2-6 Appliances.

2-6.1 General.

2-6.1.1 This section includes basic construction and performance provisions for LP-Gas consuming appliances.

In the U.S., Canada, and other economically developed countries, it has become customary to subject the more common LP-Gas and natural gas appliances to coverage by specific national consensus or governmental standards. In many instances, private or governmental listing agencies have been active in achieving a level of safety for these appliances by

evaluating them against the standards or through the exercise of their judgment. Because of this activity, it has not only been unnecessary to provide detailed technical coverage in NFPA 58, but such coverage could create difficult correlation problems. Therefore, NFPA 58 coverage has been limited to the following situations:

1.   Equipment not addressed in separate standards. This may be due to either a small number or limited use of appliances or to the use being in an environment under the control of specialists. Neither situation warrants the level of activity necessary to develop a separate standard.

2.   Coverage needed as an interim measure. It takes considerable time to develop a separate standard and frequently such a standard cannot be promulgated unless the use of the appliance is at least permitted by NFPA 58 and basic technical features included in it.

In such instances, while NFPA 58 serves a very useful function, the Technical Committee must be careful to amend the standard as necessary when a separate specific standard is developed. Such amendment could include deletion of provisions in NFPA 58.

In general, a specific appliance standard should govern in

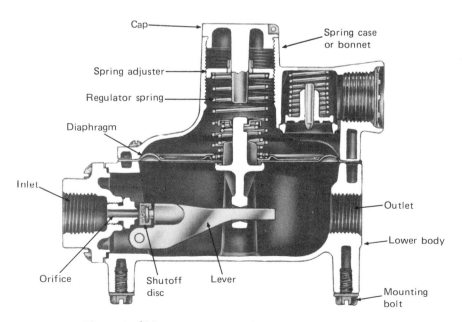

Figure 2.26(a) Lever Type Regulator for Vapor Service.

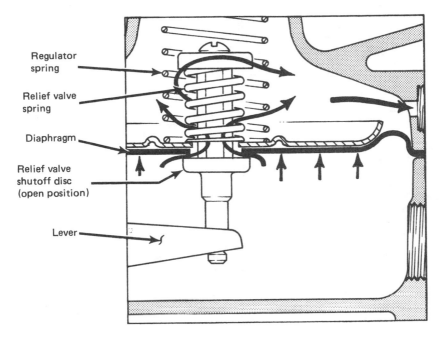

Regulator spring

Relief valve spring

Diaphragm

Relief valve shutoff disc (open position)

Lever

Figure 2.26(b) Operation of Regulator Relief Valve.

recognition of its more complete and intensive coverage, provided the standard has suitable status.

It is reemphasized that the requirement for approval in 2-6.2.1 does not include listing. The decision is that of the authority having jurisdiction. Such authorities have, however, demonstrated a pronounced disposition to accept, or even require, listed appliances and equipment whenever they are available.

It should also be noted that many of the appliances addressed in Section 2-6 are actually installed under circumstances that make them subject to NFPA 54, *National Fuel Gas Code* (ANSI Z223.1) rather than to NFPA 58, in accordance with 1-2.3.1(f). [*See commentary on 1-2.3.1(f).*]

2-6.2 Approved Appliances.

2-6.2.1 New residential, commercial, and industrial LP-Gas consuming appliances, except for those covered in 2-6.2.2 and 2-6.3.1, shall be approved.

Most residential and commercial LP-Gas appliances are covered by specific national consensus standards. The largest body of these is the ANSI Z21 series of standards promulgated by the ANSI-accredited Z21 Committee on Performance and Installa-

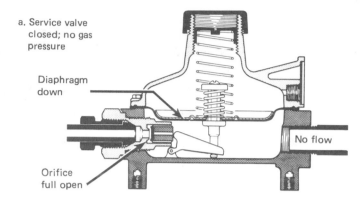

a. Service valve
   closed; no gas
   pressure

Diaphragm
down

No flow

Orifice
full open

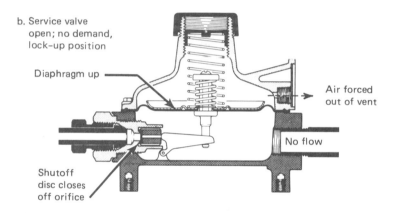

b. Service valve
   open; no demand,
   lock-up position

Diaphragm up

Air forced
out of vent

No flow

Shutoff
disc closes
off orifice

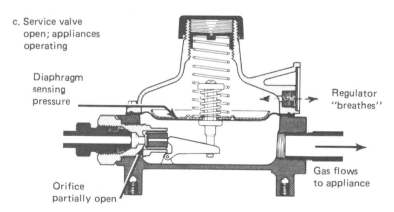

c. Service valve
   open; appliances
   operating

Diaphragm
sensing
pressure

Regulator
"breathes"

Gas flows
to appliance

Orifice
partially open

Figure 2.26(c) Regulator Operation.

tion of Gas-Burning Appliances and Related Accessories, sponsored by the American Gas Association (AGA). These appliances
are also certified ("listed," in NFPA 58 terminology) by the AGA
Laboratories. An analogous situation exists in Canada under the

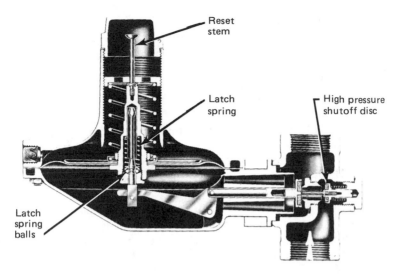

Figure 2.26(d) Regulator with High Pressure Shutoff.

auspices of the Canadian Standards Association and the Canadian Gas Association.

Many industrial appliances are also covered by specific national consensus standards. Included in these are the ANSI Z83 series of standards promulgated by the ANSI-accredited Z83 Committee on Industrial Gas Equipment Installation and Utilization, sponsored by the American Gas Association. These are also certified (listed) by the AGA Laboratories. Similar conditions exist in Canada. A list of current ANSI Z21 and Z83 standards is included at the end of this chapter.

In addition to the preceding, many LP-Gas appliances and equipment are listed by Underwriters Laboratories Inc. (U.S. and Canada). In many instances, these listings use national consensus standards adopted by ANSI under its canvass method.

In the industrial area, the listing activity (called "approval") of the Factory Mutual Engineering Corporation is notable with respect to industrial heat processing equipment.

2-6.2.2 For an appliance, class of appliance, or appliance accessory for which no applicable standard has been developed, approval of the authority having jurisdiction may be required before installation is made.

The intent here is to encourage such approval. In many instances where NFPA 58 is adopted as a regulatory instrument and often as a contractual matter, such approval is mandatory. For example, it is common for a purchaser to specify that an appliance be approved by its insurance carrier.

However, there are situations where such approval is not feasible—e.g., the authority does not have the needed expertise—and this prevents the standard from requiring such approval.

2-6.3 Provisions for Appliances.

2-6.3.1 Any appliance, originally manufactured for operation with a gaseous fuel other than LP-Gas, and in good condition, may be used with LP-Gas provided it is properly converted, adapted, and tested for performance with LP-Gas before being placed into use.

The conversion of, for example, a natural gas appliance to LP-Gas is not a complex operation. However, if the original appliance was listed or approved, the conversion may void the original listing or approval. Prudence is indicated in these situations.

2-6.3.2 Unattended heaters used inside buildings for animal or poultry production or care shall be equipped with approved automatic devices to shut off the flow of gas to the main burners, and pilots if used, in the event of flame extinguishment or combustion failure. (*See 3-5.1.3 for exception to this provision when such heaters are used in buildings without enclosing walls.*)

This provision is an example of the use of NFPA 58 to fill a gap due to the absence of a specific standard (*see commentary on 2-6.1.1*). It is noted that this provision addresses only one hazard of such a heater. Although it is a major hazard, others need to be considered. (*See commentary on 3-5.1.3.*)

2-6.3.3 Appliances using vaporizing-burners shall comply with 2-5.4.7.

Figure 2.27(a) Sight Flow Indicator with Backflow Check Valve.

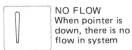 NO FLOW
When pointer is
down, there is no
flow in system

 ERRATIC FLOW
Unstable position
indicates vapor or
cavitation

 FULL FLOW
When pointer is in
position shown, there
is flow in the system

Figure 2.27(b) Flow Indicator with Backflow Check Valve.

2-6.3.4   Appliances used in mobile homes and recreational vehicles shall be approved for such service.

This provision is a carry-over from earlier editions that applied to mobile homes and recreational vehicles. While of questionable appropriateness in the current edition, the provision is consistent with the ANSI A119 and NFPA 501 series of standards and federal regulations covering these structures specifically.

2-6.3.5   LP-Gas appliances used on commercial vehicles (*see Section 3-9*) shall be approved for the service (*see 2-6.2*) and shall comply with 2-6.3.5(a) through (c).

This provision also fills gaps due to the absence of specific standards. (*See commentary on 2-6.1.1.*) However, there remains a decided dearth of either standards or listed appliances for commercial vehicles.

(a) Gas-fired heating appliances and water heaters shall be equipped with automatic devices designed to shut off the flow of gas to the main burner and the pilot in the event the pilot flame is extinguished.

(b) Catalytic heating appliances shall be equipped with an approved automatic device to shut off the flow of gas in the event of combustion failure.

(c) Gas-fired heating appliances and water heaters to be used in vehicles intended for human occupancy shall make provisions for complete separation of the combustion system and the living space. If this separation is not integral with the appliance, it shall be provided otherwise by the method of installation (*see 3-9.4.2*).

REFERENCES CITED IN COMMENTARY

The following publication is available from the National Fire Protection Association, Batterymarch Park, Quincy, MA 02269.

NFPA 54, *National Fuel Gas Code.*

The following publication is available from the Underwriters Laboratories Inc., 333 Pfingsten Rd., Northbrook, IL 60062.

UL 132, *Safety Relief Valves for Anhydrous Ammonia and LP-Gas.*

The following publications are available from the Compressed Gas Association, Inc., 1235 Jefferson Davis Highway, Arlington, VA 22202.

CGA Pamphlet C-2, *Recommendations for the Disposition of Unserviceable Compressed Gas Cylinders.*
CGA Pamphlet C-14, *Procedures for Fire Testing of DOT Cylinder Safety Relief Device Systems.*
CGA Standard V-1/ANSI B57.1, *Compressed Gas Cylinder Valve Outlet and Inlet Connections.*
CGA Pamphlet S-1-1, *Pressure Relief Device Standards.*

The following publications are available from the American Gas Association Laboratories, 8501 East Pleasant Valley Road, Cleveland, Ohio 44131.

ANSI B95.1, *Standard Terminology for Pressure Relief Devices.*

ANSI Z21 and Z83 Series Standards:

ANSI Z21.1, *Household Cooking Gas Appliances.*
ANSI Z21.5.1, *Gas Clothes Dryers, Volume I, Type 1 Clothes Dryers.*
ANSI Z21.5.2, *Gas Clothes Dryers, Volume II, Type 2 Clothes Dryers.*
ANSI Z21.10.1, *Gas Water Heaters, Volume I, Automatic Storage Water Heaters With Inputs of 75,000 Btu Per Hour or Less.*

ANSI Z21.10.3, *Gas Water Heaters, Volume III, Circulating Tank, Instantaneous and Large Automatic Storage Water Heaters.*

ANSI Z21.11.1, *Gas-Fired Room Heaters, Volume I, Vented Room Heaters.*

ANSI Z21.11.2, *Gas-Fired Room Heaters, Volume II, Unvented Room Heaters.*

ANSI Z21.13, and Addenda, Z21.13a, *Gas-Fired Low-Pressure Steam and Hot Water Boilers.*

ANSI Z21.19, *Refrigerators Using Gas Fuel.*

ANSI Z21.40.1, and Addenda, Z21.40.1a, *Gas-Fired Absorption Summer Air Conditioning Appliances.*

ANSI Z21.42, *Gas-Fired Illuminating Appliances.*

ANSI Z21.44, *Gas-Fired Gravity and Fan Type Direct Vent Wall Furnaces.*

ANSI Z21.47, *Gas-Fired Central Furnaces (Except Direct Vent and Separated Combustion System Central Furnaces).*

ANSI Z21.48, *Gas-Fired Gravity and Fan Type Floor Furnaces.*

ANSI Z21.49, *Gas-Fired Gravity and Fan Type Floor Furnaces.*

ANSI Z21.50, *Vented Decorative Gas Appliances.*

ANSI Z21.56, *Gas-Fired Pool Heaters.*

ANSI Z21.57, *Recreational Vehicle Cooking Gas Appliances.*

ANSI Z21.58, *Outdoor Cooking Gas Appliances.*

ANSI Z21.60, *Decorative Gas Appliances for Installation in Vented Fireplaces.*

ANSI Z21.61, *Gas-Fired Toilets.*

ANSI Z21.64, *Direct Vent Central Furnaces.*

ANSI Z83.3, *Standard for Gas Utilization Equipment in Large Boilers.*

ANSI Z83.4, *Direct Gas-Fired Make-Up Air Heaters.*

ANSI Z83.6, *Gas-Fired Infrared Heaters.*

ANSI Z83.7, *Gas-Fired Construction Heaters.*

ANSI Z83.8, *Gas Unit Heaters.*

ANSI Z83.9, *Gas-Fired Duct Furnaces.*

ANSI Z83.10, *Separated Combustion System Central Furnaces.*

ANSI Z83.11, *Gas Food Service Equipment — Ranges and Unit Broilers.*

ANSI Z83.12, *Gas Food Service Equipment — Baking and Roasting Ovens.*

ANSI Z83.13, *Gas Food Service Equipment — Deep Fat Fryers.*

ANSI Z83.14, *Gas Food Service Equipment — Counter Appliances.*

ANSI Z83.15, *Gas Food Service Equipment — Kettles, Steam Cookers and Steam Generators.*

ANSI Z83.16, *Unvented Commercial and Industrial Heaters.*

3 INSTALLATION OF LP-GAS SYSTEMS

3-1 Scope.

3-1.1 Application.

3-1.1.1 This chapter applies to the field installation of LP-Gas systems utilizing components, subassemblies, container assemblies and container systems fabricated in accordance with Chapter 2.

The intent of Chapter 3 is to include all information needed by the installer of LP-Gas systems with the exception of cargo tank vehicles, which are covered by Chapter 6. Prior to the 1972 Edition, this material was scattered throughout the standard.

Chapter 3 is the largest chapter in the standard. It actually comprises about three-fourths of the entire 1969 Edition in its scope. Its size is indicative of the importance of the installer's role in LP-Gas system safety.

The fire protection systems covered by Section 3-10 generally are not installed by the LP-Gas system installer and, as such, are somewhat of an anomaly in Chapter 3. Section 3-10, however, can be important from a facility location standpoint and, as such, must be considered with Section 3-2.

3-1.1.2 Section 3-2 includes general provisions applicable to most stationary systems. Sections 3-3 to 3-9 extend and modify Section 3-2 for systems installed for specific purposes.

3-1.1.3 Installation of systems used in the highway transportation of LP-Gas is covered in Chapter 6.

3-1.1.4 LP-Gas systems shall be installed in accordance with this standard and other national standards or regulations which may apply. These include:

The purpose of this provision is both to call attention to the existence of other standards covering specific LP-Gas system applications and to encourage their adoption and use on the same basis as NFPA 58. In fact, unless handled otherwise in enabling legislation, the adoption of 3-1.1.4 as part of NFPA 58 also adopts the standards cited in 3-1.1.4(a) through (j).

(a) NFPA 54, *National Fuel Gas Code* (ANSI Z223.1).

(b) NFPA 37, *Standard for the Installation and Use of Stationary Combustion Engines and Gas Turbines.*

(c) NFPA 501A, *Standard for Firesafety Criteria for Mobile Home Installations, Sites, and Communities.*

(d) NFPA 501C, *Standard on Firesafety Criteria for Recreational Vehicles.*

(e) NFPA 96, *Standard for the Installation of Equipment for the Removal of Smoke and Grease-Laden Vapors from Commercial Cooking Equipment.*

(f) NFPA 86, *Standard for Ovens and Furnaces.*

(g) NFPA 82, *Standard on Incinerators, Waste and Linen Handling Systems and Equipment.*

(h) NFPA 302, *Fire Protection Standard for Pleasure and Commercial Motor Craft.*

(i) NFPA 61B, *Standard for the Prevention of Fires and Explosions in Grain Elevators and Facilities Handling Bulk Raw Agricultural Commodities.*

(j) U.S. DOT Regulations, 49 CFR 191 and 192, for LP-Gas pipeline systems subject to DOT.

3-2 General Provisions.

3-2.1 Application.

3-2.1.1 This section includes location and installation criteria for containers; the installation of container appurtenances and regulators; piping service limitations; the installation of piping (including flexible connectors and hose); hydrostatic relief valves and equipment (other than vaporizers, *see Section 3-7*); and the testing of piping systems.

3-2.1.2 The provisions of this section are subject to modification for systems used for certain specific purposes (*see 3-1.1.2*).

3-2.1.3 For container appurtenances and gaskets installed on containers in excess of 3,500 gal (13 m^3) w.c., see 2-3.1.2(a) and 2-3.1.4.

3-2.2 Location of Containers.

3-2.2.1 LP-Gas containers shall be located outside of buildings except as follows:

From its inception, NFPA 58 has taken a basic position that LP-Gas containers should be located outdoors. (*See commentary on Section 3-4 for history.*)

The container holds the LP-Gas in both the liquid and vapor states to an extent dependent upon how much of its contents have been withdrawn in service—a "full" container being approximately 80 percent full of liquid. The pressure inside the container will vary with the temperature of the liquid but is substantial—about 130 psi (0.9 MPa) at 70°F (21°C) for propane. (*See Appendix B and texts covering thermodynamic properties of LP-Gases.*) If the LP-Gas escapes in the vapor state, the pressure can result in a substantial leakage rate. If the LP-Gas escapes in the liquid state, it will rapidly vaporize and produce large quantities of vapor. In the case of propane, one volume of liquid will vaporize to about 270 volumes of vapor. Either of these circumstances has the potential to lead to a substantial quantity of a flammable vapor/air mixture which, if ignited in an enclosed space, could result in a combustion explosion.

In this context it must also be noted that NFPA 58 requires that all LP-Gas containers be equipped with a pressure relief device. As explained earlier, a principal purpose of this device is to release gas and/or liquid to prevent container failure from overpressurization. This device will operate at normal atmospheric or room temperatures if the container has been overfilled. Therefore, it is not necessary to have a leak as a result of some sort of container or appurtenance failure in order to have vapor or liquid release.

The other major hazard is that of the Boiling Liquid-Expanding Vapor-Explosion (BLEVE). If an LP-Gas container fails in such a way that it suddenly comes apart in two or more pieces at a time when it contains liquid, the result is nearly instantaneous vaporization and dispersal of the contents (and consequent very rapid formation of a flammable mixture), propulsion of the container pieces, and creation of a shock wave. Whether the failure is the result of fire exposure (the usual case), weakening of the container (e.g., by impact or corrosion), or failure of a pressure relief device to operate (the latter two being infrequent), in an enclosed space the results can be similar to a combustion explosion.

Considering either type of explosion hazard, it is apparent that the presence of a room, building, or other enclosure is a key element in both the potential for and severity of the explosion. It follows that the absence of an enclosure is a highly desirable safeguard which should be achieved whenever it is feasible to do so.

In the more than 50 years since the first edition of NFPA 58, however, the convenience of "gas-in-a-bottle" has led society to demand applications where the container must be present inside a structure. These have evolved slowly but more or less in pace with the increasing wealth and technological complexity of the industrialized nations. The NFPA Technical Committee has approached these applications with great caution. Where considered acceptable, they have been accompanied by safeguards designed to minimize the chances of an explosion occurring and its severity. These include limiting the size of the container, controlling the presence of ignition sources, relating operation of systems to the presence or absence of persons in the building, and even designing the room or building so as to control its reaction to explosion or fire.

All of these safeguards and more are represented by the seven exceptions to the outdoor location cited in 3-2.2.1 and are addressed more fully in text provisions on the specific exceptions.

(a) Portable containers as specifically provided for in Section 3-4.

(b) Containers of less than 125 gal (0.5 m³) water capacity for the purposes of being filled in buildings or structures complying with Chapter 7.

(c) Containers on LP-Gas vehicles complying with, and parked or garaged in accordance with, Chapter 6.

(d) Containers used with LP-Gas stationary or portable engine fuel systems complying with Section 3-6.

(e) Containers used with LP-Gas fueled industrial trucks complying with 3-6.3.6.

(f) Containers on LP-Gas fueled vehicles garaged in accordance with 3-6.7.

(g) Portable containers awaiting use or resale when stored in accordance with Chapter 5.

3-2.2.2   Containers installed outside of buildings, whether of the portable type replaced on a cylinder exchange basis, or permanently installed and refilled at the installation, shall be located with respect to the nearest container, important building, group of buildings, or line of adjoining property which may be built upon, in accordance with Table 3-2.2.2, 3-2.2.3 and 3-2.2.5.

The provisions of 3-2.2.2 with respect to container location, together with those of Section 4-3 with respect to location of transfer operations—and, for larger installations, those of Section 3-10—comprise the basic outdoor container siting criteria in the standard. They should be studied together any time that a container is to be filled and the point of transfer is not at a container appurtenance, in order to avoid costly mistakes in allocation of land use.

The siting criteria in 3-2.2.2 provide for a container to be located certain distances from another container, an important building or group of buildings, and a line of adjoining property which may be built upon—and only from these. They reflect the hazard of the container to the items cited, and vice versa.

It is important to recognize that these distances are not sufficient, in and of themselves, to eliminate mutual exposures. A large, sustained liquid leak from a container connection or unignited liquid release during a BLEVE can produce a vapor cloud that can travel much farther than even the largest distance specified [400 ft (122 m)]. The thermal radiation, flying missiles, and shock wave from a BLEVE can also cause damage and injury at distances greater than any of those specified.

The distances specified are intended, then, to buy time and provide space for implementation of emergency activity. The great majority of leaks are of a size such that these distances are adequate to prevent ignition (most ignition sources are in buildings or beyond the property line). If ignition does occur, these distances provide for separation which facilitates the application of water and reduces thermal radiation exposure. For more information on conducting a firesafety analysis, refer to the supplement *Guidelines for Conducting a Firesafety Analysis.*

Experience has indicated some confusion with respect to the terms "important buildings" and "line of adjoining property which may be built upon." A building can be "important" for a number of reasons. Its replacement value is customarily recognized by all parties to a decision. Its importance by virtue of human occupancy or the value of its contents is also well recognized. Its importance by virtue of the vital role of its production equipment or business records to maintenance of production is not so often recognized.

A frequently overlooked reason for designating building importance is the building's effect upon leak and fire control activities by fire fighters and other emergency handling groups. As discussed in other sections, the application of water from hoses is a fundamental fire fighting technique. A poorly sited building having little monetary, structural, or content value could be quite important as an impediment to these control activities.

The bottom line is that the importance or non-importance of a building should be determined by consultation of all parties likely to be involved in the handling of an incident. In most cases, the facility owner, the fire department, and the insurance company have an interest.

In 1957, the NFPA Technical Committee issued Formal Interpretation LPG-5 concerned with "line of adjoining property which may be built upon." This interpretation remains valid and is reproduced below.

The following interpretation of the *Standard for the Storage and Handling of Liquefied Petroleum Gases*, NFPA Standard No. 58, has been released by the Interpretation Subcommittee of the NFPA Committee on Gases following concurrence by the entire Committee on Gases.

Question: Does the quoted language in Section B.6(b) of NFPA Standard No. 58, 1957 Edition, when referring to the location of a domestic tank mean the lot line of the property on which the tank is located or the center line of the street or alley, or is it intended that this language indicate the lot line of an adjoining property that may be built upon?

Answer: It is the Committee's opinion that the "line of adjoining property which may be built upon" refers to the property boundaries of the plot adjacent to the one upon which the tank is located. This is illustrated in the sketch below, taking into consideration a condition that involves property on the other side of a street, highway, or other right of way. It is the Committee's opinion that the minimum distance limitation is from the tank to the property line where that property line is common to plots of ground of different ownership and would also apply between the tank and the property line on the far side of a street or other public right of way as illustrated below.

In spite of editorial amendments to clarify, Note (a) to Table 3-2.2.2 still is subject to misinterpretation in the field. This provision was added to the standard as Note 1 to Table 1 in the 1969 Edition.

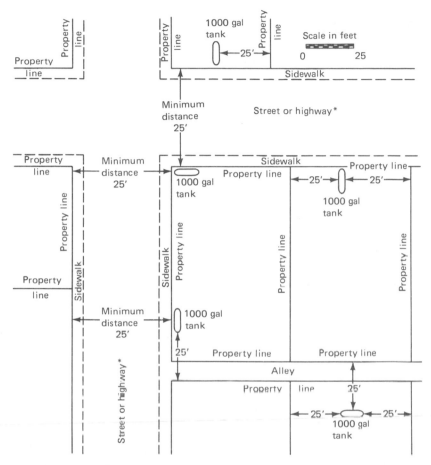

* Railroad right of ways, bodies of water, or property that cannot be built upon should receive the same consideration as streets, highways, or public right of ways.

Over its entire history, the standard has permitted containers of less than 125 gal (0.5 m³) water capacity to be installed alongside a building [but some distance away from building openings per Note (b) to Table 3-2.2.2] with no limit to the number of such containers. The presumption was that this number would not exceed 10-12 containers because normal industry practice would result in a large single container being used under such service conditions. While this is generally the case today, in the middle 1960s the Technical Committee became aware of several installations of a large number of manifolded containers to take advantage of limited space. An example of one such installation is shown in Figure 3.2.

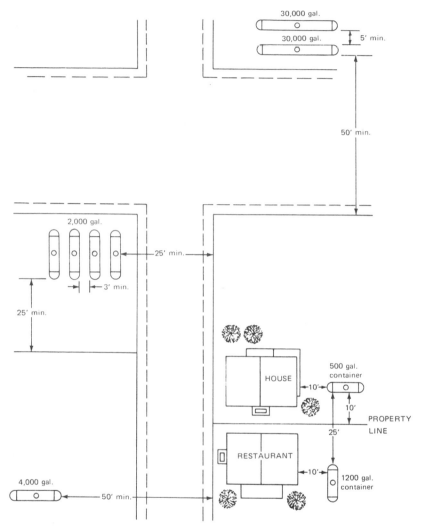

Note: 1. The requirements of table 4-3.3.2 must also be considered.

2. Installations over 4,000 gal. aggregate must comply with section 3-10.

Figure 3.1 Example of Multiple Container Installations and Various Size Containers. [Illustration of Table 3-2.2.2 Note (c).]

Note (a) to Table 3-2.2.2 corrects such abuses by placing an aggregate capacity on such multicontainer installations. However, Note (a) is applicable only where the water capacity of each individual container is less than 125 gal (0.5 m³). It is not applicable where larger containers are involved.

Figure 3.2 Fifty 100 lb (45 kg) LP-Gas Capacity [about 29 gal (110 L) water capacity each] DOT Specification Containers in Two Manifolds Installed Alongside a Multistory Office Building. [The installation is equivalent to about a single 1,200 gal (4.5 m³) container which would require a 10 ft (3 m) distance from the building per Note (c) to Table 3-2.2.2.]

Note (b) to Table 3-2.2.2 is applicable wherever containers are installed alongside a building [a DOT specification container cannot exceed 1,000 lb (454 kg) water capacity — about 120 gal (0.45 m³) of water — per DOT regulations]. The criteria are intended to minimize the chances of LP-Gas entering a building, fire exposure to a container during the early stages of a fire originating inside the building, and ignition by some ignition sources commonly found in the immediate vicinity of such installations.

Parts (1) and (2) of Note (b) are concerned with discharge from container pressure relief devices — primarily as a result of overfilling. In both instances noted, the relief device discharge is normally on the container or its shutoff valve (not piped away). A lesser distance [3 ft (1 m)] is permitted for all DOT containers than is permitted for an ASME container [5 ft (1.5 m)] because the start-to-discharge setting for a DOT container is higher than that for an ASME container [nominally 375 psi (2.6 MPa) and 250 psi (1.7 MPa) respectively], resulting in less opportunity for discharge from a DOT container. For this same reason, the relief device discharge from a DOT container can be beneath certain buildings but the discharge from an ASME container cannot. (These certain buildings, as represented by the "not enclosed for more than 50 percent of their perimeter" qualification, are most often found constructed on the sides of hills or on waterfronts).

The "building openings" cited are normally doors and windows which can be either closed or open at any particular time and which, when open, can have airflow through them in either direction. Where the airflow is normally into the building — e.g., a mechanical ventilation air intake — a 5 ft (1.5 m) distance is specified, which could affect the installation of a DOT container especially.

While direct vent appliances do not represent a pathway for LP-Gas into the building interior, they draw in outside air for combustion. If this air contains LP-Gas, in ignitable proportions, the appliance ignitor or burner flame can constitute an ignition source.

Mechanical ventilation system air intakes are on an obvious pathway for LP-Gas into a building interior. A mechanical ventilation system air outlet, e.g., a kitchen exhaust fan, is considered a building opening like a door or window.

Part (3) of Note (b) requires a 10 ft (3 m) [rather than 5 ft (1.5 m)] separation because it reflects conditions where LP-Gas vapor and liquid are released intentionally and reasonably often as a result of normal operations. This distance was derived from experience and tests.

The results of one such test are shown in Figure 3.3. Discharge of liquid from such a gauge for as long as three minutes is abnormal (normally the gauge discharge valve is closed the instant liquid appears). A lack of wind is conducive to the formation of high concentrations of LP-Gas and calm conditions are seldom found. In any event, in spite of the severity of this test, the concentration of LP-Gas at 10 ft (3 m) did not exceed 20 percent (one-fifth of the lower flammable limit) — or a safety factor of 5.

Note (c) of Table 3-2.2.2 relaxes the 25 ft (7.6 m) requirement basically applicable to a 1,200 gal (4.5 m³) or less water capacity container where only one such container is installed. This is a

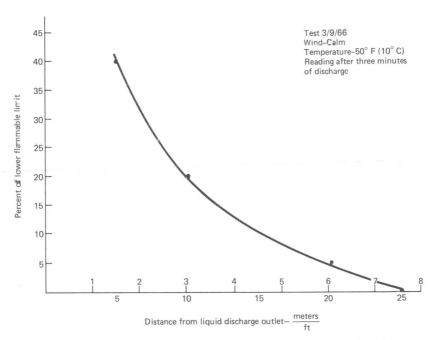

Figure 3.3 Concentration of Propane/Air Mixture Resulting from Discharge of Liquid Propane from Fixed Liquid Level Gauge on DOT/ICC/CTC Cylinder.

concession to the limited space often found in commercial areas and has proven to be adequate over many years of experience. *(See Figure 3.1.)*

The lesser distances for underground or mounded containers and the criteria in Note (d) to Table 3-2.2.2 reflect the fact that these containers are not subject to BLEVEs from fire exposure.

Note (f) reflects the exposure to larger overhanging eaves of buildings becoming more common in consideration of aesthetics and energy conservation. A torching container pressure relief device could lead to rapid penetration of fire into the structure.

In the 1989 Edition Table 3-2.2.2 was modified by extending coverage to containers having a capacity over 120,000 gal (454 m³). This was done in conjunction with the change in the scope of the standard to include refrigerated storage, marine and pipeline terminals, and portions of natural gas processing plants, refineries, and petrochemical plants in certain cases (*see 1-2.3.1*).

3-2.2.3 Where storage containers having an aggregate water capacity of more than 4,000 gal (15.1 m³) are located in heavily populated or congested areas, the siting provisions of 3-2.2.2 and Table 3-2.2.2 may be modified as indicated by the firesafety analysis described in 3-10.2.3.

Historically, the character of LP-Gas applications has been such that they have been concentrated in rural or suburban areas, or in commercial or industrial areas where the hazards are commensurate with other operations and exposure to the public has been limited accordingly. Inevitably, however, exceptions occur and the versatility of LP-Gas has increased the number of them. In recent years, for example, the popularity of LP-Gas powered vehicles has resulted in an increase in the number of refueling stations in heavily populated or congested areas. In addition to the use of the containers themselves, filling them means that LP-Gas transports are often located in such areas or may travel through such areas.

Paragraph 3-2.2.3 acknowledges that such circumstances may warrant further attention to the general siting criteria in 3-2.2.2. While the implication is that the distances should be increased in such locations, this is not necessarily the case (in fact, this may be impossible in a heavily populated or congested area). What 3-2.2.3 does require is a firesafety analysis to be conducted in such areas. The character of these areas is such that emergency handling measures are usually much more available than in more remote areas. A firesafety analysis is much more likely to result in distances no greater than those in Table 3-2.2.2 (or even less), and in the provisions of leak and fire control provisions exceeding the minimums prescribed — e.g., special protection. In extreme cases, such an analysis may result in a decision that a proposed site is simply unacceptable. (*For additional information see the supplement on conducting a firesafety analysis.*)

Table 3-2.2.2

Water Capacity Per Container Gallons (m³)	Minimum Distances		
	Mounded or Underground Containers [Note (d)]	Aboveground Containers [Note (f)]	Between Containers [Note (e)]
Less than 125 (0.5) [Note (a)]	10 ft (3 m)	None [Note (b)]	None
125 to 250 (0.5 to 1.0)	10 ft (3 m)	10 ft (3 m)	None
251 to 500 (1.0 + to 1.9)	10 ft (3 m)	10 ft (3 m)	3 ft (1 m)
501 to 2,000 (1.9 + to 7.6)	10 ft (3 m)	25 ft (7.6 m) [Note (c)]	3 ft (1 m)
2,001 to 30,000 (7.6 + to 114)	50 ft (15 m)	50 ft (15 m)	5 ft (1.5 m)
30,001 to 70,000 (114 + to 265)	50 ft (15 m)	75 ft (23 m)	
70,001 to 90,000 (265 + to 341)	50 ft (15 m)	100 ft (30 m)	(¼ of sum of
90,001 to 120,000 (341 + to 454)	50 ft (15 m)	125 ft (38 m)	diameters of adja-
120,001 to 200,000 (454 to 757)		200 ft (61 m)	cent containers)
200,001 to 1,000,000 (757 to 3 785)		300 ft (91 m)	
Over 1,000,000 (3 785)		400 ft (122 m)	

Notes to Table 3-2.2.2

Note (a): At a consumer site, if the aggregate water capacity of a multi-container installation comprised of individual containers having a water capacity of less than 125 gal (0.5 m³) is 501 gal (1.9 + m³) or more, the minimum distance shall comply with the appropriate portion of this table, applying the aggregate capacity rather than the capacity per container. If more than one such installation is made, each installation shall be separated from any other installation by at least 25 ft (7.6 m). Do not apply the MINIMUM DISTANCES BETWEEN CONTAINERS to such installations.

Note (b): The following shall apply to aboveground containers installed alongside of buildings:

(1) DOT specification containers shall be located and installed so that the discharge from the container pressure relief device is at least 3 ft (1 m) horizontally away from any building opening below the level of such discharge, and shall not be beneath any building unless this space is well ventilated to the outside and is not enclosed for more than 50 percent of its perimeter. The discharge from container pressure relief devices shall be located not less than 5 ft (1.5 m) in any direction away from any exterior source of ignition, openings into direct-vent (sealed combustion system) appliances, or mechanical ventilation air intakes.

(2) ASME containers of less than 125 gal (0.5 m³) water capacity shall be located and installed so that the discharge from pressure relief devices shall not terminate in or beneath any building and shall be located at least 5 ft (1.5 m) horizontally away from any building opening below the level of such discharge, and not less than 5 ft (1.5 m) in any direction away from any exterior source of ignition, openings into direct-vent (sealed combustion system) appliances, or mechanical ventilation air intakes.

(3) The filling connection and the vent from liquid level gauges on either DOT or ASME containers filled at the point of installation shall be not less than 10 ft (3 m) in any direction away from any exterior source of ignition, openings into direct-vent (sealed combustion system) appliances, or mechanical ventilation air intakes.

Note (c): This distance may be reduced to not less than 10 ft (3 m) for a single container of 1,200 gal (4.5 m³) water capacity or less provided such container is at least 25 ft (7.6 m) from any other LP-Gas container of more than 125 gal (0.5 m³) water capacity.

Note (d): Minimum distances for underground containers shall be measured from the pressure relief device and filling or liquid level gauge vent connection at the container, except that no part of an underground container shall be less than 10 ft (3 m) from a building or line of adjoining property which may be built upon.

Note (e): When underground multi-container installations are made of individual containers having a water capacity of 125 gal (0.5 m³) or more, such containers shall be installed so as to permit access at their ends or sides to facilitate working with cranes or hoists.

Note (f): In applying the distance between buildings and ASME containers of 125 gal (0.5 m³) or more water capacity, a minimum of 50 percent of this horizontal distance shall also apply to all portions of the building which project more than 5 ft (1.5 m) from the building wall and which are higher than the relief valve discharge outlet. This horizontal distance shall be measured from a point determined by projecting the outside edge of such overhanging structure vertically downward to grade or other level upon which the container is installed. Under no conditions shall distances to the building wall be less than those specified in Table 3-2.2.2.

Exception to Note (f): Not applicable to installations in which overhanging structure is 50 ft (15 m) or more above the relief valve discharge outlet.

The expression "heavily populated or congested area" is not defined quantitatively. In most instances it is subject to local determination, and often zoning criteria come into play.

3-2.2.4 Aboveground multi-container installations comprised of containers having an individual water capacity of 12,000 gal (45 m³) or more installed for use in a single location shall be limited to the number of containers in one group and with each group separated from the next group in accordance with the degree of fire protection provided in accordance with Table 3-2.2.4.

Table 3-2.2.4

Fire Protection Provided by	Maximum Number of Containers in One Group	Minimum Separation Between Groups—feet
Hose streams only— see 3-10.2.3	6	50 (15 m)
Fixed monitor nozzles per 3-10.3.5*	6	25 (7.6 m)
Fixed water spray per 3-10.3.4*	9	25 (7.6 m)
Insulation per 3-10.3.1	9	25 (7.6 m)

*In the design of fixed water spray and fixed monitor nozzle systems, the area of container surface to be protected may reflect portion of containers not likely to be subject to fire exposure as determined by good fire protection engineering practices.

Installations of greater numbers of larger containers were increasing and containers were being added to existing installations without adequate consideration of the fire control problems introduced. These provisions were added in the 1986 Edition.

3-2.2.5 In the case of buildings of other than wood-frame construction devoted exclusively to gas manufacturing and distribution operations, including LP-Gas service stations, the distances in Table 3-2.2.2 may be reduced provided that in no case shall containers having a water capacity exceeding 500 gal (1.9 m³) be located closer than 10 ft (3 m) to such gas manufacturing and distributing buildings.

3-2.2.6 The following provisions shall also apply:

The nine items in this provision are siting or operational controls designed primarily to protect the integrity of an installed LP-Gas container. However, some items are applicable to non-installed containers and may be more appropriate in other parts of the standard.

(a) Containers shall not be stacked one above the other.

Subparagraph (a) is applicable to both installed and non-installed containers. Its purpose is to avoid "bonfire" type incidents and, in the case of installed containers, opportunities for overfilling the lower containers due to migration of liquid under the influence of gravity.

(b) Loose or piled combustible material and weeds and long dry grass shall not be permitted within 10 ft (3 m) of any container.

Subparagraph (b) is applicable to both installed and non-installed containers.

(c) Suitable means shall be used to prevent the accumulation or flow of liquids having flash points below 200°F (93.4°C) under adjacent LP-Gas containers such as by dikes, diversion curbs or grading. Determination of flash points shall be in accordance with NFPA 321, *Standard on Basic Classification of Flammable and Combustible Liquids*.

(d) When tanks containing flammable or combustible liquids (*see NFPA 321 for definitions of these liquids*) are within a diked area, LP-Gas containers shall be outside the diked area and at least 10 ft (3 m) away from the centerline of the wall of the diked area.

(e) The minimum horizontal separation between aboveground LP-Gas containers and aboveground tanks containing liquids having flash points below 200°F (93.4°C) shall be 20 ft (6 m). This provision shall not apply when LP-Gas containers of 125 gal (0.5 m^3) or less water capacity are installed adjacent to fuel oil supply tanks of 660 gal (2.5 m^3) or less capacity. No horizontal separation is required between aboveground LP-Gas containers and underground tanks containing flammable or combustible liquids installed in accordance with NFPA 30, *Flammable and Combustible Liquids Code*. See 3-2.2.6(c) for flash point determinations.

Subparagraphs (c), (d), and (e) are applicable to both installed and non-installed containers. The provisions of (c) and (d) are aimed at keeping leaking flammable or combustible liquids from accumulating under or around LP-Gas containers. Subparagraph 3-2.2.6(e) reflects the mutual exposure between tanks or containers themselves, and was derived through a cooperative effort between the NFPA Technical Committees responsible for NFPA 30, *Flammable and Combustible Liquids*

Code, and NFPA 58. The separation distances are intended to facilitate cooling and fire extinguishing activities by fire departments. The exception in the second sentence reflects a fairly common farm or residential situation. It is a concession to limited space available and experience has shown it not to be a problem.

(f) The minimum separation between LP-Gas containers and oxygen or gaseous hydrogen containers shall be in accordance with Table 3-2.2.6(f) except that lesser distances are permitted where protective structures having a minimum fire resistance rating of two hours interrupt the line of sight between uninsulated portions of the oxygen or hydrogen containers and the LP-Gas containers. The location and arrangement of such structures shall minimize the problems cited in the Note to 3-2.2.8. Also, see NFPA 50 and 51 for oxygen systems and NFPA 50A on gaseous hydrogen systems. The minimum separation between LP-Gas containers and liquefied hydrogen containers shall be in accordance with NFPA 50B.

Subparagraph (f) is derived from several NFPA standards covering oxygen and hydrogen container installations. It is included in NFPA 58, along with Table 3-2.2.6(f) as a convenience. The caution stated in the Note to 3-2.2.8 is not in the other standards because the problems cited are not pertinent to oxygen or hydrogen.

(g) Where necessary to prevent flotation due to possible high flood waters around aboveground containers, or high water table for those underground, containers shall be securely anchored.

Subparagraph (g) is applicable to installed containers, as anchorage would be impractical for non-installed containers. In practice, anchorage has been applied only to ASME containers—the smaller DOT containers not being anchored because of their size. As a result, flood experience has revealed numerous examples of DOT containers being torn from their connections and being found miles away. Even though their valves may not have been closed in anticipation of the flood, DOT containers have not contributed significantly to the overall flood damage.
Anchorage usually consists of concrete pads or foundations as the basic element and utilizes basic civil engineering criteria.

(h) When LP-Gas containers are to be stored or used in the same area with other compressed gases, the containers shall be marked to identify their content in accordance with ANSI Standard 748.1, *Method of Marking Portable Compressed Gas Containers to Identify the Material Contained* (CGA C-4).

Table 3-2.2.6(f)

LP-Gas Containers Having An Aggregate water capacity of	Separation From Oxygen Containers Having An			Separation From Gaseous Hydrogen Containers Having An		
	Aggregate capacity of 400 CF (11 m³)* or less	Aggregate capacity of more than 400 CF (11 m³)* to 20,000 CF (566 m³),* including unconnected reserves.	Aggregate capacity of more than 20,000 CF (566 m³),* including unconnected reserves.	Aggregate capacity of less than 400 CF (11 m³)*	Aggregate capacity of 400 CF (11 m³)* to 3000 CF (85 m³)*	Aggregate capacity of more than 3000 CF (85 m³)*
1200 Gal (4.5 m³) or less	None	20 ft (6 m)	25 ft (7.6 m)			
Over 1200 Gal (4.5 m³)	None	20 ft (6 m)	50 ft (15 m)			
500 Gal (1.9 m³) or less				None	10 ft (3 m)	25 ft (7.6 m)
Over 500 Gal (1.9 m³)				None	25 ft (7.6 m)	50 ft (15 m)

*Cubic feet measured at 70°F and atmospheric pressure.

Subparagraph (h) is applicable to both installed and non-installed containers. It is noted that the identification is by marking with the name of the gas. Identification by color or other means is acceptable only if it is in addition to the name of the gas.

Steel LP-Gas cylinders and ASME containers are customarily painted for corrosion protection. While NFPA 58 does not stipulate paint colors, the color affects the rate of heat absorption from solar radiation and the consequent pressure in the container. This can result in operation of a pressure relief device.

Tests have shown a white paint with a titanium oxide pigment to reflect 90-95 percent of the solar light. Yellow (medium yellow chrome pigment) reflects about 80 percent. Aluminum reflects about 70 percent. Other colors reflect less than 15 percent and black reflects none.

(i) No part of an aboveground LP-Gas container shall be located in the area 6 ft (1.8 m) horizontally from a vertical plane beneath overhead electric power lines that are over 600 volts, nominal.

This reflects both the possibility of container failure through arc penetration and the hazard to fire fighters applying water. This subparagraph was added in the 1986 Edition.

3-2.2.7 Because of the anticipated "flash" of nonrefrigerated LP-Gas when released to the atmosphere dikes normally serve no useful purpose for nonrefrigerated installations.

This provision was changed in the 1989 Edition with the addition of LP-Gas refrigerated storage to the standard. The need to dike refrigerated LP-Gas storage containers has long been recognized because a spill can puddle or run as a liquid on the ground.

While the volatility of the LP-Gases covered by NFPA 58 varies substantially (see vapor pressures of propane and butane in Table B-1.2.1), in all cases where they are to be used as burner fuels (which is the usual case) it is highly desirable that the liquid in storage be above its atmospheric boiling point at the vapor pressure caused by ambient temperature.

Under temperate atmospheric conditions, liquid LP-Gas vaporizes rapidly when it escapes from a pressurized container. Much of the heat necessary for vaporization is contained in the liquid itself and the remainder needed is readily available from the air, ground or other material contacted by the liquid. Therefore, the establishment of a pool or flowing stream of LP-Gas over the ground or water is unlikely and a dike has no function to perform.

There have been incidents of LP-Gas liquid flow on the ground—usually of butane but occasionally of propane—under extreme conditions of very low atmospheric temperatures and ice and snow covered terrain. These have not proven to be serious, however, and the more commonly encountered conditions [in which a dike introduces hazards (*see 3-2.2.8*)] lead to the conclusion that a dike is generally to be avoided.

3-2.2.8 Structures such as fire walls, fences, earth or concrete barriers and other similar structures shall be avoided around or over installed nonrefrigerated containers.

Exception No. 1: Such structures partially enclosing containers are permissible if designed in accordance with a sound fire protection analysis.

Exception No. 2: Structures used to prevent flammable or combustible liquid accumulation or flow are permissible in accordance with 3-2.2.6(c).

Exception No. 3: Structures between LP-Gas containers and gaseous hydrogen containers are permissible in accordance with 3-2.2.6(f).

Exception No. 4: Fences are permissible in accordance with 3-3.6.1.

NOTE: The presence of such structures can create significant hazards, e.g., pocketing of escaping gas, interference with application of cooling water by fire departments, redirection of flames against containers, and impeding egress of personnel in an emergency.

This provision was inspired by a serious BLEVE of an above-ground propane container which had been enclosed in a roofed-over enclosure for aesthetic reasons. The enclosure not only contributed to ignition but made it difficult for the fire department to apply cooling water to the container. However, the Technical Committee was also aware of an increasing use of such methods to hide LP-Gas containers or to limit the travel of container pieces in the event of a BLEVE.

Exception No. 1 to 3-2.2.8 recognizes that the problems associated with such structures can be obviated by design, which eliminates the problems cited in the Note. The other exceptions were added in the 1989 Edition along with a corresponding deletion of the introduction to the paragraph which contained the exception. This was done to conform with the NFPA style manual, to make the provision clearer, and to eliminate needless page turning.

3-2.3 Installation of Containers.

3-2.3.1 Containers shall be installed in accordance with 3-2.3.1(a) through (f):

> This section provides basic general provisions common to the installation of all containers. More specific details for the installation of horizontal aboveground, vertical, portable storage, vehicular tanks, mounded, partially underground unmounded, and underground containers are given in subsequent sections.

(a) DOT cylinder specification containers shall be installed only aboveground, and shall be set upon a firm foundation, or otherwise firmly secured. Flexibility shall be provided in the connecting piping. (*See 3-2.7.5 and 3-2.7.8.*)

> DOT specification containers are designed primarily for transportation purposes but with the concept that they will be used as fuel supply storage. They are not designed for belowground installation. In an earlier edition of NFPA 58, an explanation was given that they could be installed in a niche in a slope or terrace wall as long as the container and regulator did not contact the ground and the compartment or recess was ventilated and drained. Flexibility in the connecting piping is generally obtained by the use of a pigtail (short length of copper tubing with POL* connectors).

(b) All containers shall be positioned so that the pressure relief valve is in direct communication with the vapor space of the container.

> Positioning the container so that the relief valve is in communication with the vapor space assures a minimal release of LP-Gas should abnormal conditions exist. Tilting or laying a cylinder on its side so that liquid LP-Gas may be withdrawn is not to be done. There are liquid withdrawal arrangements available without doing this that allow the proper position of the relief valve to be retained.

---

* The letters "POL" stand for "Prest-o-Lite" — a trademark of a pioneer LP-Gas operator (Pyrofax-Union Carbide). Because these fittings have a left-hand thread, they have also come to stand for "Put-on-Left."

(c) Where physical damage to LP-Gas containers, or systems of which they are a part, from vehicles is a possibility, precautions against such damage shall be taken.

This is intentionally written as a performance provision, rather than providing specific guidelines for when protection is needed or for the nature of the protection. The Technical Committee believes that it cannot anticipate all the ways that a vehicle may potentially threaten a container and does not intend to limit the types of protection. These are left to the user of the standard or the authority having jurisdiction.

Questions about this provision usually are the result of an incident and question the intent. For example, an incident occurred in which a tractor trailer transporting 63 head of cattle veered off a road and traveled almost 500 ft (152 m) before striking the piping associated with an 18,000 gal (68 m³) and a 30,000 gal (114 m³) LP-Gas container installed about 110 ft (34 m) from the edge of the road. Escaping LP-Gas ignited and the torch flame impinged on the 18,000 gal (68 m³) container resulting in a BLEVE after about ½ hour. In this case the installer could not have been reasonably expected to protect the containers from impact by a stray tractor trailer. Other incidents have occurred on construction sites where LP-Gas containers have been installed close to roadways or driveways used by delivery trucks and construction vehicles. In these cases the installer would be expected to provide protection for the anticipated vehicles which pass close to the container.

(d) The installation position of ASME containers shall make all container appurtenances accessible for their normally intended use.

Occasionally a storage container must be evacuated, before it is moved or for other reasons. There are fittings for this evacuation to eliminate the need to roll a container on its side to pump it out. (*See commentary on 2-2.3.3.*) Also, there are many other installation situations in which the container appurtenances may not be accessible unless attention is given to the container position before installation.

(e) Field welding on containers shall be limited to attachments to nonpressure parts, such as saddle plates, wear plates or brackets applied by the container manufacturer. Welding to container proper shall comply with 2-2.1.6.

(f) *Aboveground containers shall be kept properly painted.

A-3-2.3.1(f) Aboveground Storage Tank Paint Color. Generally, a light reflecting color paint is preferred unless the system is installed in an extremely cold climate.

See commentary on 3-2.2.6(h) for further information on paint colors.

3-2.3.2 Horizontal ASME containers designed for permanent installation in stationary service aboveground shall be placed on substantial masonry or noncombustible structural supports on concrete or firm masonry foundations, and supported as follows:

The primary consideration in the installation of horizontal ASME containers is to assure that the integrity of the supports is retained, or that damage under fire conditions is minimized. For large containers this is accomplished through masonry foundations and supports. This method may also be used for smaller containers [2,000 gal (7.6 m³) water capacity or less] or, as an alternate, when nonfireproofed structural supports are provided, they may be placed on fire-resistive foundations as long as certain heights from the container bottom to the ground are maintained. These minimum heights, together with a provision that flexibility be provided in the piping, assure that the container will not settle to an unsafe level. Two exceptions are given for these smaller containers. In a temporary installation, as described in 3-2.3.2(a)(2)(b), a 5 ft (1.5 m) height is permitted and fire resistive foundations and saddles are not required. The other exception is described in 3-2.3.2(b) for a permanent installation in an isolated location with the approval of the authority having jurisdiction. These higher elevation tanks are generally used for filling other containers by gravity.

(a) Horizontal containers shall be mounted on saddles in such a manner as to permit expansion and contraction, and not to cause an excessive concentration of stresses. Structural steel supports may be used as follows, or if in compliance with 3-2.3.2(b).

Exception No. 1: Temporary use as provided in 3-2.3.2(a)(2)b.

Exception No. 2: Isolated locations as provided in 3-2.3.2(b).

(1) Containers of more than 2,000 gal (7.6 m³) water capacity shall be provided with concrete or masonry foundations formed to fit the container contour, or if furnished with saddles in compliance with 2-2.5.1, may be placed on flat-topped foundations.

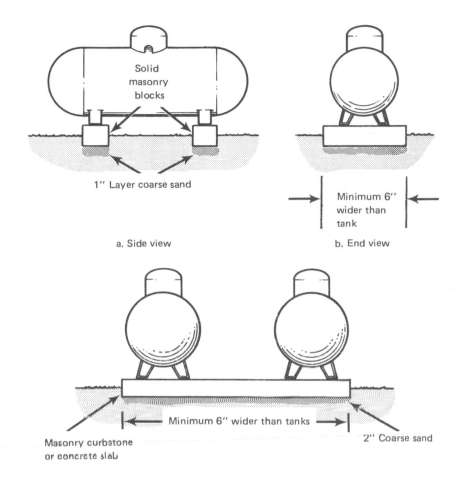

Figure 3.4(a) Typical Smaller ASME Container Aboveground Installation.

(2)   Containers of 2,000 gal (7.6 m³) water capacity or less may be installed on concrete or masonry foundations formed to fit the container contour, or if equipped with attached supports complying with 2-2.5.2(a), may be installed as follows:

a. If the bottoms of the horizontal members of the container saddles, runners or skids are to be more than 12 in. (305 mm) above grade, fire-resistive foundations shall be provided. A container shall not be mounted with the outside bottom of the container shell more than 5 ft (1.5 m) above the surface of the ground.

b. For temporary use at a given location, not to exceed 6 months, fire-resistive foundations or saddles are not required provided the outside bottom of the container shell is not more than 5 ft (1.5 m) above the ground and that flexibility in the connecting piping is provided. (*See 2-4.6.3.*)

Figure 3.4(b) ASME Storage Container with Concrete Saddles. A 30,000 gal (114 m³) container supported 3 ft (1 m) above grade. This installation utilized a concrete saddle with a ½ in. (13 mm) thick bitumastic felt pad installed between the steel container and concrete foundation.

(3)   Containers or container-pump assemblies mounted on a common base complying with 2-2.5.2(b) may be placed on paved surfaces or on concrete pads at ground level within 4 in. (102 mm) of ground level.

(b) With the approval of the authority having jurisdiction, single containers complying with 2-2.5.1 or 2-2.5.2 may be installed in isolated locations, with nonfireproofed steel supports resting on concrete pads or footings, provided the outside bottom of the container shell is not more than 5 ft (1.5 m) above the ground level.

(c) Suitable means of preventing corrosion shall be provided on that part of the container in contact with the saddles or foundations or on that part of the container in contact with masonry.

3-2.3.3 Vertical ASME containers over 125 gal (0.5 m³) water capacity designed for permanent installation in stationary service aboveground shall be installed on reinforced concrete or steel structural supports on reinforced concrete foundations which are designed to meet the loading provisions established in 2-2.2.3.

(a) Steel supports shall be protected against fire exposure with a material having a fire resistance rating of at least two hours. Continuous steel skirts having only one opening 18 in. (457 mm) or less in diameter need such fire protection applied only to the outside of the skirts.

Although their use is not as extensive as horizontal containers, vertical containers are useful for certain applications, such as dispensers in service stations.

This provision was changed in the 1989 Edition to reduce the minimum size of vertical containers covered from 2,000 gal (7.6 m³) to 125 gal (0.5 m³) to correct an omission of containers of 125 (0.5 m³) to 2000 gal (7.6 m³) made when the standard was reorganized in 1972.

3-2.3.4 Single containers constructed as portable storage containers (*see definition*) for temporary stationary service in accordance with 2-2.5.4(a) shall be placed on concrete pads, paved surfaces or firm earth for such temporary service (normally not more than 12 months at a given location) and the following shall apply:

Temporary single portable storage container installations find applications as fuel supplies for grain drying, road construction, etc.

(a) The surface on which they are placed shall be substantially level and, if not paved, shall be cleared (and kept cleared) of dry grass and weeds, and other combustible material within 10 ft (3 m) of the container.

(b) Flexibility shall be provided in the connecting piping.

(c) If such containers are to be set with the bottoms of the skids or runners above the ground, nonfireproofed structural supports may be used for isolated locations with the approval of the authority having jurisdiction, and provided the height of the outside bottom of the container shell above the ground does not exceed 5 ft (1.5 m). Otherwise, fire-resistive supports shall be provided.

3-2.3.5 If the container is mounted on, or is part of, a vehicle as provided in 2-2.5.4(b), the unit shall be parked in compliance with the provisions of 3-2.2.2 as to the location of a container of that capacity for normal stationary service, and in accordance with the following:

If the container is mounted as part of a vehicle, it is actually a cargo container and is not considered a permanent storage container.

(a) The surface shall be substantially level and if not paved shall be suitable for heavy vehicular use, and shall be cleared (and kept cleared) of dry grass and weeds, and other combustible material within 10 ft (3 m) of the container.

(b) Flexibility shall be provided in the connecting piping.

3-2.3.6 Portable containers of 2,000 gal (7.6 m³) water capacity or less complying with 2-2.5.5 may be installed for stationary service as provided in 3-2.3.2(a)(2) for stationary containers.

This permits these particular DOT (CTC) portable containers to be considered permanent storage containers.

3-2.3.7 Mounded containers shall be installed as follows:

A mounded container may be above or partially buried below the ground surface. Many times it is not desirable to bury a tank completely because of the water table, rock formations, and other reasons. Sometimes mounding is for aesthetic purposes and is done at the request of the customer. Mounding also provides insulation from low or high temperatures and is, therefore, a means of "special protection" (*see definition in Section 1-7*). Mounded containers should be protected from corrosion in the same manner required for underground containers.

Figure 3.4(c) Mounded Storage Containers. Installation of three 80,000 gal (302 m³) storage containers mounded for special protection. These containers have an epoxy coal tar coating for corrosion protection in addition to cathodic protection. A 12 in. (305 mm) covering of sand was provided around each container and a 6 in. (152 mm) layer of gravel installed over the three containers. The containers were provided with extended manways and relief valve nozzles to permit installation of all tank trim above the top surface of the final layer of gravel.

(a) Mounding material shall be earth or sand and shall provide minimum thickness of cover for the container of at least 1 ft (305 mm).

(b) Unless inherently resistant to erosion, a suitable protective cover shall be provided.

3-2.3.8 ASME container assemblies listed for underground installation, including interchangeable aboveground-underground container assemblies may be installed underground as follows:

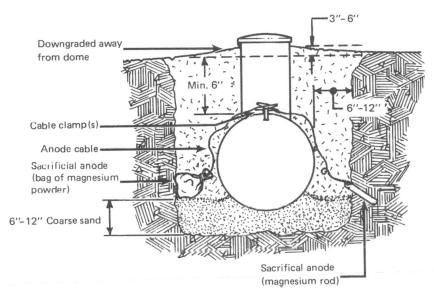

Figure 3.5 Typical Small ASME Container Underground Installation. Cathodic protection is not always needed.

Care should be used to see that the ASME containers for underground installation are marked by the manufacturer as suitable for underground or for the interchangeable aboveground/underground type of service. A container built for aboveground service should never be installed underground. Likewise, a container built for underground installation must not be installed aboveground, as the relief valve may be under-sized. Some people prefer underground or mounded tank installations for BLEVE protection. However, there are other procedures that have to be done differently with underground installations than with aboveground installations because the condition of the underground tank is not easily visible for checking.

It is very important that underground and mounded tanks be properly protected against corrosion by a suitable coating or

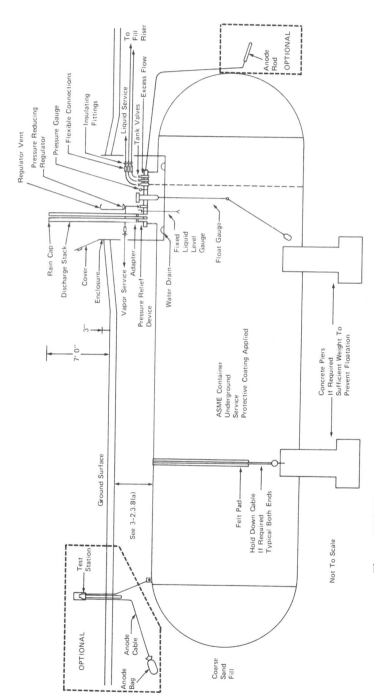

Figure 3.5(a) Typical Large ASME Container Underground Installation. Cathodic protection is not always needed.

cathodic protection. It is especially important not to allow any scarring of a protective coating while installing the tank, because just one small, unprotected spot can concentrate corrosion and lead to a leak. Tanks should never be moved by putting a chain or a cable around them without some cushioning material positioned between the chain or cable and the container coating. It is not necessary to have a concrete or metal saddle for an underground container installation, provided there is firm earth for it to rest on. But there can be conditions where it is advisable to have a proper foundation or slab on which to place the underground container. When the backfill is placed, it should be free of rocks and similar abrasive which can adversely affect the coating.

Backfill should be tamped so that there will not be settling. A major fire occurred when a vaporizer installed on untamped fill over a buried container settled and tipped, causing a piping break and liquid escape.

An underground tank should not be installed in a pit without its being properly filled in around the tank with sand or suitable earth. Such a container in a pit would provide a place for vapor to accumulate.

(a) The container shell shall be placed at least 6 in. (153 mm) below grade unless the container might be subject to abrasive action or physical damage from vehicular traffic within a parking lot area, driveway, or similar area. In this case, a noninterchangeable underground container shall be used and the container shell placed at least 18 in. (457 mm) below grade [see 3-2.3.8(c)] or equivalent protection shall be otherwise provided, such as the use of a concrete slab, to prevent imposing the weight of a vehicle directly on the container shell. Protection of the fitting housing, housing cover, tank connections, and piping shall be provided to protect against vehicular damage.

(b) Where containers are installed underground within 10 ft (3 m) where vehicular traffic may be reasonably expected, such as driveways and streets or within a utility easement subject to vehicular traffic, protection of the fitting housing, housing cover, tank connections, and piping shall be provided to protect against vehicular damage.

(c) Approved interchangeable aboveground-underground container assemblies installed underground shall not be placed with the container shell more than 12 in. (305 mm) below grade.

(d) Any party involved in construction and/or excavation in the vicinity of a buried container shall be responsible for determining the location of and providing protection for the container and piping against physical damage from vehicular traffic.

(e) The portion of the container to which the fitting cover or other connections are attached need not be covered. The discharge of the regulator vent shall be above the highest probable water level.

(f) Containers shall be protected against corrosion for the soil conditions at the container site by a method in accordance with good engineering practice. Precaution shall be taken to prevent damage to the coating during handling. Any damage to the coating shall be repaired before backfilling.

(g) Containers shall be set substantially level on a firm foundation (firm earth may be used) and surrounded by earth or sand firmly tamped in place. Backfill shall be free of rocks or similar abrasives.

(h) When a container is to be abandoned underground, the following procedure shall be followed:

(1)   Remove as much liquid LP-Gas as possible through the container liquid withdrawal connection.

(2)   Remove as much of the remaining LP-Gas vapor as possible by venting it through a vapor connection; either burning this vapor, or venting it to the open air at a safe location. The vapor shall not be vented at such a rapid rate as to exceed the vaporization rate of any residual liquid LP-Gas left after the liquid removal procedure of 3-2.3.8(h)(1).

NOTE: If vapor is vented too rapidly the pressure drop due to the refrigeration of the liquid may lead to the erroneous conclusion that no liquid remains in the container.

(3)   When only vapor LP-Gas at atmospheric pressure remains in the container, it shall be filled with water, sand or foamed plastic, or purged with an inert gas. The displaced vapor may be burned or vented to the open air at a safe location.

In the 1989 Edition, former (a) was rewritten and former (b) was rewritten and relocated to (e), and new (b), (c), and (d) were added to recognize incidents involving unprotected and unmarked underground containers caused by motor vehicle traffic in areas not normally subject to vehicle traffic. Most of these have been related to construction in the area of underground containers.

Paragraph (a) has been revised to change burial depth to 18 in. (0.5 m) from 24 in. (0.6 m) based on tests conducted in the 1950's and improvements in protective coatings and installation procedures since then, and incorporates the former (b).

The new (b) provides performance requirements for containers buried within 10 ft (3 m) of traffic. New (c) limits the depth that an interchangeable abovegound-underground container may be placed below grade in order to reduce the possibility of flooding in the fitting enclosure and to provide ease of servicing the fitting enclosure.

3-2.3.9 Partially underground, unmounded ASME containers shall be installed as follows:

(a) The portion of the container below the surface, and for a vertical distance of at least 3 in. (75 mm) above the surface, shall be protected to resist corrosion as required for underground containers. [*See 3-2.3.8(f).*]

(b) Containers shall be set substantially level on a firm foundation, with backfilling to be as required for underground containers. [*See 3-2.3.8(g).*]

(c) Spacing provisions shall be as specified for aboveground containers in 3-2.2.2 and Table 3-2.2.2.

(d) The container shall be located so as not to be subject to vehicular damage, or shall be adequately protected against such damage.

3-2.4 Installation of Container Appurtenances.

3-2.4.1 Pressure relief devices shall be installed on containers in accordance with 3-2.4.2 through 3 2.4.5 and positioned so that the relief device is in direct communication with the vapor space of the container.

It is important that the pressure relief device be installed so that it is always in the vapor space in normal operation (*see commentary on 2-2.3.5*). In locating such devices, it is important to remember that what seems to be the normal filling point in a container is radically changed if the container is filled at a low temperature and then is subjected to a rising temperature which expands the liquid. If the pressure relief device is not properly located, the liquid expansion will be enough to then place the pressure relief device in the liquid phase.

3-2.4.2 Pressure relief devices on portable DOT cylinder specification containers, or their equivalent of ASME construction, of 1,000 lb (454 kg) [120 gal (0.5 m³)] water capacity or less, shall be installed to minimize the possibility of relief device(s) discharge(s) impingement on the container.

Burning gas being discharged from a pressure relief device and impinging upon the container or an adjacent container is the most common reason for spread of fire from container to container and their subsequent BLEVEs. Because the relief valves on most DOT containers discharge horizontally when the container is in its normal position, this provision is difficult to implement where more than one or two containers are

present. Where the container has a protective collar, the collar should have an opening opposite the relief valve discharge outlet for the gas to pass through.

3-2.4.3 Pressure relief devices on ASME containers of 125 gal (0.5 m³) water capacity or more permanently installed in stationary service, portable storage containers (*see definition*), portable containers (tanks) of nominal 120 gal (0.5 m³) water capacity or more, or cargo tanks shall be installed so that any gas released is vented away from the container upward and unobstructed to the open air. The following provisions shall also apply:

(a) Means shall be provided, such as rain caps, to minimize the possibility of the entrance of water or other extraneous matter (which might render the relief device inoperative or restrict its capacity) into the relief device or any discharge piping. If necessary, provision shall be made for drainage. The rain cap or other protector shall be designed to remain in place except when the relief device operates and shall permit the relief device to operate at sufficient relieving capacity.

Rain caps that direct gas discharge downward or to the side should not be used. Because any rain cap that remains in place during discharge will throttle the discharge to some extent, rain caps should be used only after determining their effect upon the relieving capacity of the relief device.

(b) On each aboveground container of more than 2,000 gal (7.6 m³) water capacity, the relief device discharge shall be vertically upward and unobstructed to the open air at a point at least 7 ft (2 m) above the top of the container. The following also shall apply:

(1) Relief device discharge piping shall comply with 3-2.4.3(f).

(2) In providing for drainage in accordance with 3-2.4.3(a), the design of relief device discharge(s) and attached piping shall:

a. Be such as to protect the container against flame impingement which might result from ignited product escaping from the drain opening.

b. Be directed so that a container(s), piping or equipment which might be installed adjacent to container on which the relief device is installed is not subjected to flame impingement.

(c) On underground containers of 2,000 gal (7.6 m³) or less water capacity, except those installed in LP-Gas service stations covered in 3-2.4.3(e), the relief device may discharge into the manhole or housing, provided such manhole or housing is equipped with ventilated louvers, or their equivalent, of adequate area as specified in 3-2.4.6(d).

(d) On underground containers of more than 2,000 gal (7.6 m³) water capacity, except those installed in LP-Gas service stations, the discharge from relief devices shall be piped vertically and directly upward to a point at least 7 ft (2 m) above the ground. Relief device discharge piping shall comply with 3-2.4.3(f).

(e) On underground containers in LP-Gas service stations, the relief device discharge shall be piped vertically and directly upward to a point at least 10 ft (3 m) above the ground. Discharge piping shall comply with 3-2.4.3(f) and shall be adequately supported and protected against physical damage.

(f) The discharge terminals from relief devices shall be located so as to provide protection against physical damage. Discharge piping used shall be adequate in size to permit sufficient relief device relieving capacity. Such piping shall be metallic and have a melting point over 1500°F (816°C). Discharge piping shall be designed so that excessive force applied to the discharge piping will result in breakage on the discharge side of the valve rather than on the inlet side without imparing the function of the valve. Return bends and restrictive pipe or tubing fittings shall not be used.

The 7 ft (2 m) provision cited in 3-2.4.3(b) and (d) is intended to assure that the discharge will be above the head of anyone near or on a container. This is increased to 10 ft (3 m) in service stations [3-2.4.3(e)] because individuals may be on a vehicle. These distances also facilitate vapor disposal at a height above most building openings and ignition sources.

The variation permitted by 3-2.4.3(c) recognizes that the relief devices seldom operate on underground containers and that the discharge rate on underground containers 2,000 gal (7.6 m³) and less water capacity is reduced. This variation, however, is not permitted in service stations because, even though operation may be unlikely, such locations can present considerable public exposure.

(g) Shutoff valves shall not be installed between relief devices and the container, or between the relief devices and the discharge piping, except for specially designed relief device-shutoff valve combinations covered by 2-3.2.4(c), or where two or more separate relief devices are installed, each with its individual shutoff valve, and the shutoff valve stems are mechanically interconnected in a manner which will allow the rated relieving capacity required for the container from the relief device or devices which remain in communication with the container.

Shutoff valves between containers and relief devices greatly facilitate maintenance and testing because it is not necessary to empty and purge the container. However, experience revealed that these valves were closed all too often. The arrangements

cited will permit a shutoff valve to be installed while maintaining the desired protection at all times.

See Figure 2.12 for an illustration of equipment used for this purpose.

3-2.4.4   Pressure relief devices on portable storage containers (constructed and installed in accordance with 2-2.5.4 and 3-2.3.4 respectively) used temporarily in stationary type service shall be installed in accordance with the applicable provisions of 3-2.4.3.

3-2.4.5   Additional provisions (over and above the applicable provision in 3-2.4.2 and 3-2.4.3) apply to the installation of pressure relief devices in containers used in connection with vehicles as follows:

(a)   For containers installed on vehicles in accordance with Sections 3-6 and 3-9.

(b)   For cargo containers (tanks) installed on cargo vehicles in accordance with Section 6-3, see 6-3.2.1(a).

3-2.4.6   Container appurtenances other than pressure relief devices shall be installed and protected as follows:

(a)   All container openings except those used for pressure relief devices (*see 2-3.2*), liquid level gauging devices (*see 2-3.4*), pressure gauges (*see 2-3.5*), those equipped with double check valves as allowed in Table 2-3.3.2, and plugged openings shall be equipped with internal valves [*see 2-3.3.3(d)*] or with positive shutoff valves and either excess-flow or backflow check valves (*also see 2-3.3 for specific application*) as follows:

> **An excess-flow valve must never be installed in a pressure relief device connection because the high pressure condition that will cause the pressure operation of this safety device will result in a very high flow through the pressure relief device and the excess-flow valve, possibly closing the excess-flow valve and defeating the safety device.**
>
> **The exception of container openings used for level gauging devices and pressure gauges recognizes that these openings are restricted to a size where it is not a significant hazard and an excess-flow or backflow check valve is not only unnecessary but, if provided, could close and negate the gauging device or pressure gauge.**
>
> **On all other connections, either an excess-flow valve or a double check valve must be used.**

(1)   Except for DOT cylinders, excess-flow or backflow check valves shall be located between the LP-Gas in the container and the shutoff valves,

either inside the container, or at a point immediately outside where the line enters or leaves the container. If outside, installation shall be made so that any undue strain beyond the excess-flow or backflow check valve will not cause breakage between the container and such valve. All connections, including couplings, nozzles, flanges, standpipes and manways, which are listed on the ASME Manufacturers' Data Report for the container, are considered part of the container. On DOT cylinders, the excess-flow valve where required may be located at the outlet of the cylinder shutoff valve.

(2) Shutoff valves shall be located as close to the container as practicable. The valves shall be readily accessible for operation and maintenance under normal and emergency conditions, either because of location or by means of permanently installed special provisions. Valves installed in an unobstructed location not more than 6 ft (1.8 m) above ground level shall be considered accessible. Special provisions include, but are not limited to, stairs, ladders, platforms, remote operators or extension handles.

(3) The connections, or line, leading to or from any individual opening shall have greater capacity than the rated flow of the excess-flow valve protecting the opening.

An excess-flow valve can be used where there might be flow in either direction through a given connection. A backflow check valve can be used where the flow would be only in one direction (into the container) and where flow in the opposite direction must be prevented.

It is preferable to have an excess-flow valve or backflow check valve inside of the container so that if there is a breakage of the piping outside of the container the safety device will not be adversely affected. Where necessary, however, they can be outside but should be upstream of the manual shutoff valve.

It should also be recognized that an excess-flow valve will operate only if the flow is in excess of the manufacturer's rated flow. If a partial break in a line occurs which will not allow a flow up to the rated capacity, the excess-flow valve will not operate. Many operators have felt that an excess-flow valve was 100 percent protection in case of a break, but this is true only if the flow from the break is of sufficient size.

It is also important to size the piping downstream of the excess-flow valve so that it does not restrict the flow such that the excess-flow valve will not work. In many cases it is necessary to put in additional excess-flow valves in the downstream line so as to be assured of a shutoff with a break some distance from the container or where there is a reduction in the piping size. It is also important that the operator, in opening a line or container valve, open it slowly, so that the excess-flow valve will not "slug" in to shut off the flow when it is not supposed to. In some cases where this has occurred, the operator has removed the working portion of the excess-flow valve which, of course, negates its function as a safety device. A number of accidents have occurred because of this improper practice.

(b) Valves, regulators, gauges and other container appurtenances shall be protected against physical damage.

(c) Valves in the assembly of portable multicontainer systems shall be arranged so that replacement of containers can be made without shutting off the flow of gas in the system. This provision shall not be construed as requiring an automatic changeover device.

(d) Connections to containers installed underground shall be located within a substantial dome, housing or manhole and with access thereto protected by a substantial cover. Underground systems shall be installed so that all terminals for connecting hose and any opening through which there can be a flow from pressure relief devices or pressure regulator vents are located above the normal maximum water table. Terminals for connecting hoses, openings for flow from pressure relief devices, and the interior of domes, housing and manholes shall be kept clean of debris. Such manholes or housings shall be provided with ventilated louvers or their equivalent. The area of such openings shall equal or exceed the combined discharge areas of the pressure relief devices and other vent lines which discharge into the manhole or housing.

The housing around appurtenances on underground containers is subject to accumulations of water and debris and should be checked often.

(e) Container inlet and outlet connections, except pressure relief devices, liquid level gauging devices and pressure gauges, on containers of 2,000 gal (7.6 m³) water capacity or more, or on containers of any capacity used in LP-Gas service stations, shall be labeled to designate whether they communicate with the vapor or liquid space. Labels may be on valves. (*See Sections 3-6 and 3-9 for requirements for labeling smaller containers used for vehicular installations.*)

(f) Every storage container of more than 2,000 gal (7.6 m³) water capacity shall be provided with a suitable pressure gauge (*see 2-3.5*).

This paragraph was changed in the 1989 Edition by deletion of a provision that containers of any capacity used in LP-Gas service stations be equipped with a pressure gauge. The Technical Committee could find no reason to continue the requirement which apparently came from a former UL requirement.

a. Die-stamped on cylinder

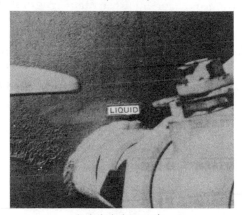

b. Labeled on tank

c. Tagged to service valve

Figure 3.6 Typical Container Inlet and Outlet Connection Markings.

3-2.5 Regulator Installation.

3-2.5.1 Regulators used to control distribution or utilization pressure shall be as close to the container or vaporizer outlets as is practicable. First stage regulating equipment shall be outside of buildings except as used with containers and liquid piping systems covered by 3-2.2.1(a), (b), (d), (e) and (f), and 3-2.6.1(d).

A serious fire hazard could occur if LP-Gas vapor were allowed to recondense in long lengths of piping under cold weather conditions. Liquid LP-Gas would be present at the point of use where vapor usage was intended. To prevent this from happening, first stage regulators must be located as close to the supply container or outlet of the vaporizer as possible.

The provision for locating first stage regulating equipment outside of buildings dates from the 1927 Edition of the first NFPA standard on this subject and is intended to keep LP-Gas at container pressure from entering a building. This rule may not apply in certain cases, such as for special applications where first stage pressure regulators must be used indoors (*see 3-4*), or for engine fuel uses (*see 3-6*), industrial applications, and some gas distribution facilities.

3-2.5.2 Regulators shall be securely attached to container valves, containers, supporting standards or building walls.

(a) First stage regulators shall be either directly connected to the container shutoff valve or outlet of vaporizer where used, unless attached thereto with flexibility provided in the connecting piping or the interconnecting piping of manifolded containers or vaporizers.

In a single container or single vaporizer installation, the first stage regulator can be connected directly to the container service valve or vaporizer outlet (where used). As an alternate, a flexible connector such as a pigtail (short length of copper tubing with POL connectors) may be used.

In the case of manifolded containers or vaporizers, it is impractical to install a regulator at each container service valve or vaporizer. Pressure imbalances would occur causing regulators to counteract each other resulting in an unstable pressure condition. Thus, a single regulator may be used with this type of system with interconnecting piping in the manifolded system.

Note that the requirement that the regulator be connected directly to the container shutoff valve can be met in a manifolded container installation by connecting the regulator

Figure 3.7(a) and (b) Some Regulator Installation Arrangements.

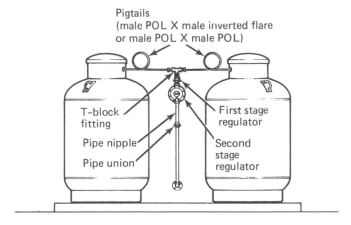

Pigtails
(male POL X male inverted flare
or male POL X male POL)

T-block fitting

Pipe nipple

Pipe union

First stage regulator

Second stage regulator

3.7(a) Dual Cylinder Installation (Stationary Service).

Figure 3.7(b) Automatic Changeover Regulator.

to a tee that is directly connected to the container shutoff valves with only connectors as required for spacing or flexibility. [*See Figure 3-7(a)*]. Flexibility in the piping system is necessary to provide for expansion or contraction.

(b) All regulators for outdoor installations, except regulators used for portable industrial applications, shall be designed, installed, or protected so their operation will not be affected by the elements (freezing rain, sleet, snow, ice, mud, or debris). This protection may be integral with the regulator.

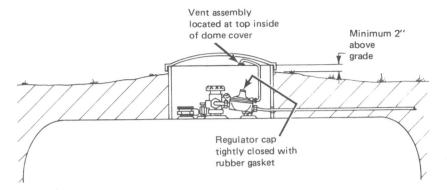

Figure 3.8 Use of Vent Extension to Protect Regulator Operation from Elements on Underground Container. Vent assembly must be above grade level.

If a regulator vent is blocked, such as with ice, and the regulator relief valve inside the regulator bonnet is activated, the relief valve would not be able to discharge to the atmosphere. High pressures in the gas distribution piping would result. It is, therefore, imperative that the vent on regulators used outdoors (except for portable industrial uses) be protected against the elements. This can be done by either enclosing the regulator in a housing or mounting the regulator with the vent opening vertically downward if the vent has a drip lip with an inside diameter of not less than $^{11}/_{16}$ in. (17 mm) and an outside diameter at least ¾ in. (19 mm). These dimensions evolved from a test project by Underwriters Laboratories Inc. where different designs were subjected to freezing rain and wind conditions. Current UL listed regulators are marked according to whether they must be installed under a protective cover or in the downward position or equivalent. Drip lip adapters with dimensions previously mentioned are available for those regulators without this protection. Also, it is important to assure that the protective housing, if used, is installed correctly. Plastic closures have been used with the regulator upside down and the housing became filled with water and froze. Sleet, freezing rain and glaze may cause havoc with LP-Gas distribution pressures, resulting in fires unless the regulator is properly protected.

3-2.5.3 On regulating equipment installed outside of buildings, the discharge from a pressure relief device shall be located not less than 3 ft (1 m) horizontally away from any building opening below the level of such discharge, and not beneath any building unless this space is well ventilated to the outside and is not enclosed for more than 50 percent of its perimeter.

This is intended to minimize the infiltration of LP-Gas into a building.

3-2.5.4   On regulators installed inside buildings, the discharge from the pressure relief device and from above the regulator and relief device diaphragms shall be vented to the outside air with the discharge outlet located not less than 3 ft (1 m) horizontally away from any building opening below the level of such discharge. This provision shall not apply to appliance regulators otherwise protected (*see NFPA 54*), or to regulators used in connection with containers in buildings as provided for in 3-2.2.1(a), (b), (d), (e) and (f).

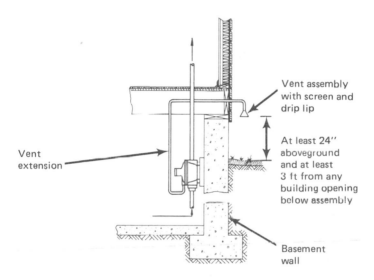

Vent assembly with screen and drip lip

Vent extension

At least 24" aboveground and at least 3 ft from any building opening below assembly

Basement wall

Figure 3.9 Venting Regulator Installed in a Building.

Whether or not a regulator is installed outdoors, or a second stage regulator (not an appliance regulator) is installed indoors with the vent and relief valve piped outdoors, the discharge of the vent and relief valve must be not less than 3 ft (1 m) horizontally from a building opening below the level of this discharge. In piping the vent and relief valve discharge from second stage regulators to the outside, a vent adapter of the dimensions given in the commentary on 3-2.5.2(b) should be used.

3-2.6 Piping System Service Limitations.

3-2.6.1   This subsection describes the physical state (vapor or liquid) and pressure at which LP-Gas may be transmitted through piping systems under various circumstances:

With the exception of 3-2.6.1(b), these provisions reflect the combustion explosion and fire hazard inside of structures, as determined by the quantity of flammable mixtures likely to be formed as a result of escape of LP-Gas from a system. This quantity, in turn, is expressed in terms of the physical state (liquid or gas) and, for gas, the pressure.

(a) LP-Gas liquid or vapor may be piped at all normal operating pressures outside of buildings.

(b) Polyethylene piping systems shall be limited to vapor service not exceeding 30 psig (208 kPa gauge).

This restriction reflects the unsuitability of this plastic where LP-Gas liquid may be present. The presence of liquid can lead to development of excessive pressure and also cause loss of strength from chemical incompatibility. Even though the use of plastic pipe is limited to underground service, ground temperatures could cause condensation of vapor into liquid if the pressure exceeds 30 psi (208 kPa).

(c) LP-Gas vapor at pressures not exceeding 20 psig (138 kPa gauge) may be piped into any building.

When piping vapor at elevated pressures condensation of LP-Gas vapor must be avoided or liquid can be fed directly to the gas utilization equipment with potentially dangerous results. If the ambient temperature can fall below -5°F (-21°C) at any time it may be necessary to heat trace and/or insulate the outdoor portion the piping. The condensation point of propane at different pressures is shown below. When piping at pressures exceeding 5 psig (34 kPa gauge) and not exceeding 20 psig (138 kPa gauge) in buildings covered by NFPA 54, *National Fuel Gas Code*, paragraph 2.5.1 places additional constraints on the piping system including approval by the authority having jurisdiction. (*See NFPA 54, paragraph 2.5.1.*)

Pressure (psig)	Temperature (°F)	Pressure (kPa)	Temperature (°C)
20	−5	140	−21
40	20	276	−7
63	40	434	4

(d) LP-Gas vapor at pressures exceeding 20 psig (138 kPa gauge) or LP-Gas liquid shall not be piped into any building except those meeting the following descriptions:

(1) Buildings, or separate areas of buildings, constructed in accordance with Chapter 7, and used exclusively to:

a. House equipment for vaporization, pressure reduction, gas mixing, gas manufacturing or distribution.

b. House internal combustion engines, industrial processes, research and experimental laboratories, or equipment or processing having a similar hazard.

Exception: Complete compliance with Chapter 7 for buildings, or separate areas of buildings, housing industrial processes and other occupancies cited in 3-2.6.1(d)(1)b may not be necessary depending upon the prevailing conditions. Construction of buildings or separate areas of buildings housing certain internal combustion engines is covered in NFPA 37.

(2) Buildings or structures under construction or undergoing major renovation, provided the temporary piping meets the provisions of 3-4.2 and 3-4.10.2.

The occupancies listed in 3-2.6.1(d) are those into which LP-Gas vapor above 20 psig may be piped. In recognition of the greater than normal potential hazard in these uses, provisions are made to significantly reduce the probability of ignition for ventilation and for structure construction to minimize explosion damage (compliance with Chapter 7). In addition, the occupancies are limited to those where LP-Gas is the product or major raw material [(a)], or installations where the operators will have training in the safe handling of LP-Gas [(b)].

The exception permits a flexible approach in industrial occupancies, and specifically recognizes the installation of stationary combustion engines as covered by NFPA 37, *Standard for the Installation and Use of Stationary Combustion Engine and Gas Turbines,* based on the favorable experience with these installations.

Subparagraph 3-2.6.1(d)(2) recognizes that these applications are temporary and subject to a large degree of supervision.

(3) In buildings or structures other than those covered by 3-2.6.1(d)(1) and (2) in which liquid feed systems are used, liquid piping may enter the building or structure to connect to a vaporizer provided heavy walled seamless brass or copper tubing not exceeding 3/32 in. (2.4 mm) internal diameter and with a wall thickness not less than 3/64 in. (1.2 mm) is used.

The quantity of liquid that can escape from such piping is limited by the small pipe diameter. The likelihood of leakage is small because of the pipe wall thickness specified.

3-2.7 Installation of Pipe, Tubing, Pipe and Tubing Fittings, Valves, and Hose.

3-2.7.1 LP-Gas normally is transferred into containers as a liquid, but may also be conveyed as a liquid or vapor under container or lower regulated pressure. Metallic piping except safety relief discharge piping (*see 3-2.4.3*) shall comply with the following:

(a) Piping used at pressures higher than container pressure, such as on the discharge side of liquid transfer pumps, shall be suitable for a working pressure of at least 350 psig (2.4 MPa gauge).

(b) Vapor LP-Gas piping with operating pressures in excess of 125 psig (0.9 MPa gauge), and liquid piping not covered by 3-2.7.1(a), shall be suitable for a working pressure of at least 250 psig (1.7 MPa gauge).

(c) Vapor LP-Gas piping, subject to pressures of not more than 125 psig (0.9 MPa gauge), shall be suitable for a working pressure of at least 125 psig (0.9 MPa gauge).

At a temperature of 70°F (21°C), the pressure in a container of propane (the most volatile LP-Gas covered by NFPA 58), is in the range of 125-130 psi (about 0.9 MPa).

3-2.7.2 Metallic pipe joints may be threaded, flanged, welded or brazed using pipe and fittings complying with 2-4.2 and 2-4.4 as follows:

(a) When joints are threaded or threaded and back welded:

(1) For LP-Gas vapor at pressures in excess of 125 psig (0.9 MPa gauge), or for LP-Gas liquid, the pipe and nipples shall be Schedule 80 or

(2) For LP-Gas vapor at pressures of 125 psig (0.9 MPa gauge) or less, the pipe and nipples shall be Schedule 40 or heavier.

(b) When joints are welded or brazed:

(1) The pipe shall be Schedule 40 or heavier.

(2) The fittings or flanges shall be suitable for the service in which they are to be used.

(3) Brazed joints shall be made with a brazing material having a melting point exceeding 1,000°F (538°C).

(c) Gaskets used to retain LP-Gas in flanged connections in piping shall be resistant to the action of LP-Gas. They shall be made of metal or other suitable material confined in metal having a melting point over 1,500°F (816°C) or shall be protected against fire exposure, except that aluminum

O-rings and spiral wound metal gaskets are acceptable. When a flange is opened, the gasket shall be replaced.

The provisions in 3-2.7.2 consider mechanical strength and fire resistance. The provisions concerned with mechanical strength are expressed simply and have proved adequate for those installing systems by NFPA 58 over many years. It is recognized that more elaborate criteria, e.g., ANSI B31, *Code for Pressure Piping*, can also be used to achieve a safe system.

3-2.7.3   Metallic tubing joints may be flared, or brazed using tubing and fittings, and brazing material complying with 2-4.3 and 2-4.4.

Refer to commentary following 2-4.4.1(a)(4).

3-2.7.4   Piping in systems shall be run as directly as is practicable from one point to another, and with as few restrictions, such as ells and bends, as conditions will permit, giving consideration to provisions of 3-2.7.5.

The number of fittings (ells,tees,etc.) is usually minimized in a piping system when it is designed for cost and ease of installation considerations. Frequently, however, piping is field run or modified resulting in excessive fittings. The intent of the requirement is to minimize the number of fittings as each fitting is a potential source of leakage, and the more fittings the more resistance to flow and slower transfer. This can be especially critical in pump suction piping.

Although the number of fittings must be minimized, the need for flexibility in the piping system cannot be ignored as the piping will shrink when liquid vaporizes, cooling to about -40 °F (-40°C).

(a) Where condensation of vapor may occur, metallic and nonmetallic piping shall be pitched back to the container or suitable means provided for revaporizing the condensate.

The temperature at which LP-Gas vapor will condense is dependent upon which LP-Gas is involved and its pressure. For example, at 10 psig (69 kPa) condensation of propane will occur at about -20°F (-29°C) and below; 20 psig (140 kPa) at about -5°F (-21°C) and below; and 60 psig (414 kPa) at about 30°F (-1°C) and below. (Additional data is available from LP-Gas regulator manufacturers.)

When installing piping between first and second stage regulators precautions must be taken to insure that any liquid that may condense does not reach the second stage regulator. This could result in liquid passing through the regulator and vaporizing, resulting in overfeeding the appliance and increasing the flame size which can be a significant fire hazard. This may be prevented by sloping the piping away from the second stage regulator or other means.

3-2.7.5 Provision shall be made in piping including interconnecting of permanently installed containers, to compensate for expansion, contraction, jarring and vibration, and for settling. Where necessary, flexible connectors complying with 2-4.6 may be used (*see 3-2.7.8*). The use of nonmetallic pipe, tubing or hose for permanently interconnecting such containers is prohibited.

3-2.7.6 Metallic piping outside buildings may be underground or aboveground or both. Aboveground piping shall be well supported and protected against physical damage. Where underground piping is beneath driveways, roads or streets, possible damage by vehicles shall be taken into account. Nonmetallic piping, including the nonmetallic portions of transition fittings, shall be installed outside, a minimum of 12 in. (305 mm) underground and in accordance with the piping manufacturer's instructions.

Aboveground piping should be well supported so that there will not be sags in the piping. In addition, the pipe supports must be substantial so there will not be side pressure which can put stresses on the pipe.

Encasing underground pipe with a larger pipe is a common method of providing protection for piping subject to vehicular loads.

3-2.7.7 Underground metallic piping shall be protected against corrosion as warranted by soil conditions. Corrosion protection shall comply with the following:

Most soils are corrosive to metallic pipe. While protection has traditionally been provided by coatings, cathodic protection is increasingly being used.

If coating is used, fittings such as collars, must be coated after installation. Coated piping should not be installed in soil where there are rocks which can scrape the coating and expose the pipe to the effect of corrosion. It is also important that, where dissimilar metals are connected together, an insulating fitting is installed so as to eliminate cathodic action on the piping. If this insulating fitting is not used, the piping system can be adversely affected rather rapidly.

(a) Underground piping shall be protected as needed with a suitable coating to retard the effects of the corrosion conditions existing in the local soil. Coated pipe shall extend at least 6 in. (152 mm) aboveground on all risers.

(b) When dissimilar metals are joined underground, an insulating fitting shall be installed to electrically isolate them from each other.

(c) If cathodic protection is used, insulating fittings shall be installed to electrically isolate the cathodically protected underground system from all aboveground piping and systems.

(d) LP-Gas piping shall not be used as a grounding electrode.

3-2.7.8 Flexible components used in piping systems shall comply with 2-4.6 for the service in which they are to be used, shall be installed in accordance with the manufacturer's instructions, and shall also comply with the following:

Flexibility is often necessary in LP-Gas piping to allow for thermal expansion and contraction of the piping and settling of containers and equipment. Flexible connections often fit in very well in an LP-Gas system but should not be used unless necessary because their expected life is generally not as long as permanent piping. Flexible connectors should be limited in length because the possibility of leakage increases with length. Flexibility can be designed into metallic piping systems, and an engineer experienced in piping design should be consulted if required.

(a) Flexible connectors in lengths up to 36 in. (1 m) (*see 2-4.6.3 and 2-4.6.4*) may be used for liquid or vapor piping, on portable or stationary tanks, to compensate for expansion, contraction, jarring, vibration and settling. This is not to be construed to mean that flexible connectors shall be used if provisions were incorporated in the design to compensate for these effects.

(b) Hoses may be installed if flexibility is required for liquid or vapor transfer. The use of wet hose (*see 4-2.3.4 for explanation of term "wet hose"*) is recommended for liquid.

3-2.7.9 On new installations, and by December 31, 1980, on existing installations, (1) stationary single container systems of over 4,000 gal (15.1 m³) water capacity, or (2) stationary multiple container systems with an aggregate water capacity of more than 4,000 gal (15.1 m³) utilizing a common or manifolded liquid transfer line, shall comply with 3-2.7.9(a) and (b).

(a) When a hose or swivel type piping 1 ½ in. (38 mm) or larger is used for liquid transfer or a 1 ¼ in. (32 mm) or larger vapor hose or swivel type piping is used in this service (excluding flexible connectors in such liquid and vapor piping), an emergency shutoff valve complying with 2-4.5.4 shall be installed in the fixed piping of the transfer system within 20 ft (6 m) of lineal pipe from the nearest end of the hose or swivel type piping to which the hose or swivel type piping is connected. The preceding sizes are nominal. Where the flow is only in one direction, a backflow check valve may be used in lieu of an emergency shutoff valve if installed in the fixed piping downstream of the hose or swivel type piping, provided the backflow check valve has a metal-to-metal seat or a primary resilient seat with a secondary metal seat not hinged with combustible material. When either a liquid or vapor line has two or more hoses or swivel type piping of the sizes designated, either an emergency shutoff valve or a backflow check valve shall be installed in each leg of the piping.

Formal Interpretation   79-1
Reference: 3-2.7.9(a)

Question: Is it the intent of 3-2.7.9(a) to require either an emergency shutoff valve or a backflow check valve in each leg of the piping when two or more hoses are used?

Answer: Yes.

Committee Comment: Unless these provisions are made, it would be possible for flow from one leg of the piping to escape through a leak in the other leg.

Issue Edition: 1979
Reference: 3168(a)
Date: November 1979

Formal Interpretation   79-2
Reference: 3-2.7.9(a)

Question: In an LP-Gas installation subject to the provisions of 3-2.7.9 of NFPA 58 by virtue of the container capacity qualifications, the vapor piping used in liquid transfer operations is 1¼-in. nominal size. However, a vapor hose permanently affixed to the delivery end of this piping (by the use of a 1¼-in.-to-1-in. reducing elbow) is 1-in. nominal size. No backflow check valve is installed in this piping.

Is it the intent of 3-2.7.9(a) of NFPA 58 to require that an emergency shutoff valve be installed in the fixed vapor piping?

Answer: No.

Committee Comment: The Committee notes that, in the absence of either an emergency shutoff valve or a backflow check valve, 3-3.3.4(a) or (b) would require an excess flow valve in the fixed vapor piping cited.

Issue Edition: 1979
Reference: 3168(a)
Date: November 1980

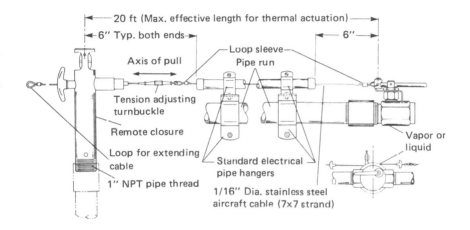

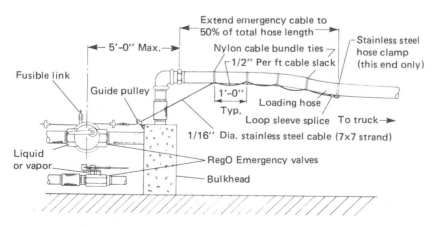

Figure 3.10(a) Installation of Mechanically Operated Emergency Shutoff Valve. (Top) Conventional. (Bottom) Incorporates hose pullaway protection.

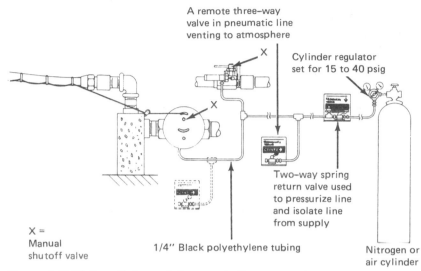

Figure 3.10(b) Installation of Pneumatically Operated Emergency Shutoff Valve.

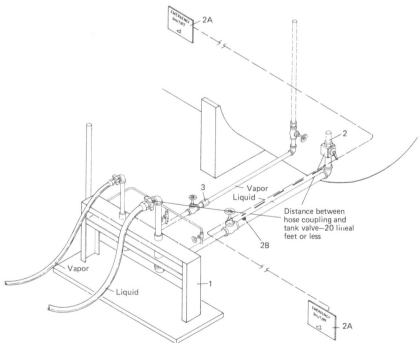

Figure 3.10(c) Installation of Emergency Shutoff Valves on ASME Storage Containers. (1) Concrete bulkhead or equivalent anchorage, or use of a predictable breakaway point. (2) Emergency valve with means for manual shutoff at the valve: (2A) Remote shutoff control; (2B) Thermal release within five feet from the nearest end of the hose, or swivel type piping. (3) Vapor line does not require an emergency valve because the hose diameter is less than 1¼ in. If pipe is larger, an emergency shutoff valve must also be installed in this line.

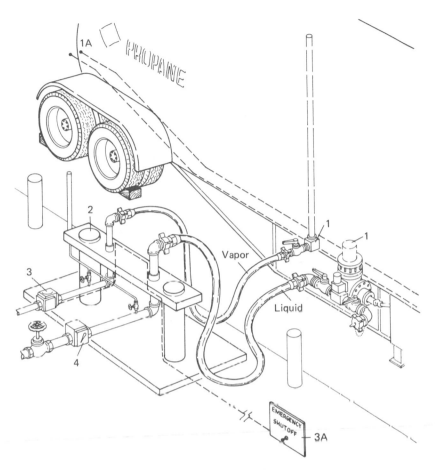

Figure 3.10(d) Installation of Emergency Shutoff Valves at Transport Unloading Station. Transports must have the following safety devices unless emergency valves are provided at transport end of hoses. (1) Internal valves, which must include a manual means for shutoff at internal valve and thermal release (temperature sensitive element) at internal valve, within five feet of hose. (1A) Remote control to close internal valves. Note: Drawing does not include all DOT requirements. Plant transport unloading area must have the following safety devices: (2) Concrete bulkhead or equivalent anchorage, with a predictable breakaway point to retain intact the valves and piping on the plant side of the connections. (3) Vapor line must have an emergency valve with means for manual shutoff, thermal release and remote control provisions. (3A) Remote shutoff control. (4) Liquid line must have either a backflow check valve (may be part of sight flow unit) or an emergency valve with means for manual shutoff, thermal release and remote shutoff control.

(1)   Emergency shutoff valves shall be installed so that the temperature sensitive element in the valve, or a supplemental temperature sensitive element [250°F (121°C) maximum] connected to actuate the valve, is not more than 5 ft (1.5 m) from the nearest end of the hose or swivel type piping connected to the line in which the valve is installed.

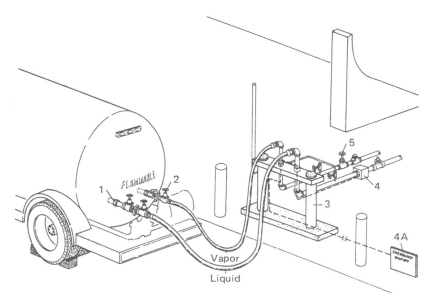

Figure 3.10(e) Installation of Emergency Shutoff Valves at Delivery Cargo Vehicle Loading Station. Cargo vehicle being filled directly into tank must have the following devices: (1) Liquid fill must have a backflow check valve mounted directly into tank, or an internal valve. (2) Vapor does not require emergency valve because hose diameter is less than 1¼ in. Plant loading riser must have the following safety devices: (3) Concrete bulkhead, or equivalent anchorage, or use of a predictable breakaway point to retain intact the valve(s) on the plant side of the connection(s). (4) Liquid line must have an emergency valve with means for manual shutoff and thermal release. (4A) Remote shutoff control. (5) Vapor line does not need an emergency valve because the hose diameter is less than 1¼ in.

(b) The emergency shutoff valve(s) or backflow check valve(s) specified in 3-2.7.9(a) shall be installed in the plant piping so that any break resulting from a pull will occur on the hose or swivel type piping side of the connection while retaining intact the valves and piping on the plant side of the connection. This may be accomplished by use of concrete bulkheads or equivalent anchorage or by the use of a weakness or shear fitting. Such anchorage is not required for tank car unloading.

This provision first appeared in the 1976 Edition of the standard and was a rare incidence of retroactive application. It resulted from an extensive study of accidents — especially of those that resulted in a BLEVE of a stationary container.

The accident experience was dominated by release of liquid and vapor during liquid transfer operations as a result of hose failures, hose coupling failures, and piping and component failures. In many instances, these failures resulted from a cargo vehicle being driven away before the transfer hose was discon-

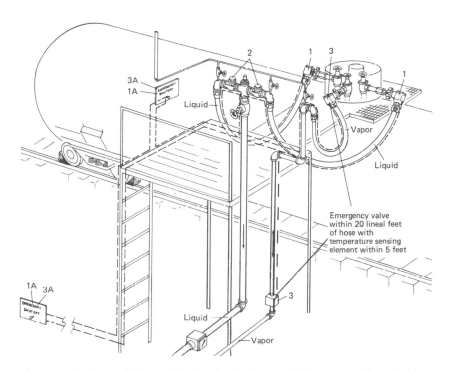

Figure 3.10(f) Installation of Mechanically Operated Emergency Shutoff Valves at Tank Car Loading or Unloading Station. As an alternate, emergency valves may be installed in the tank car unloading adapters [see Figure 3.10(g)]. Tank car unloading risers must have the following safety devices: (1) Liquid hoses at tank car end must have an emergency valve with means for manual shutoff and thermal release. (1A) Remote shutoff control. (2) Riser ends of liquid hose connections must have backflow check valves. (3) Vapor hose must have an emergency valve at each end with means for manual shutoff and thermal release. (3A) Remote shutoff control. (4) When two hoses or swivel type piping are used on tank car unloading riser, each leg of the piping should be protected by backflow check valve(s) or an emergency shutoff valve.

nected. In some cases, valves were torn out of the vehicle female connections — especially brass valves inserted into steel fittings.

This provision essentially requires an automatic and manual means of stopping the escape of LP-Gas (other than by an excess-flow check valve) from either side of a leak on installations where both the capacity of the container (or containers) and the diameter of the hoses or swivel type piping exceed those stipulated. There are installations where the container capacity qualifies, but not the size of the hoses or swivel type piping. (See previous Formal Interpretation.)

Where flow is in two directions, an emergency shutoff valve complying with 2-4.5.4 is used. In 1976, compact, simple, and

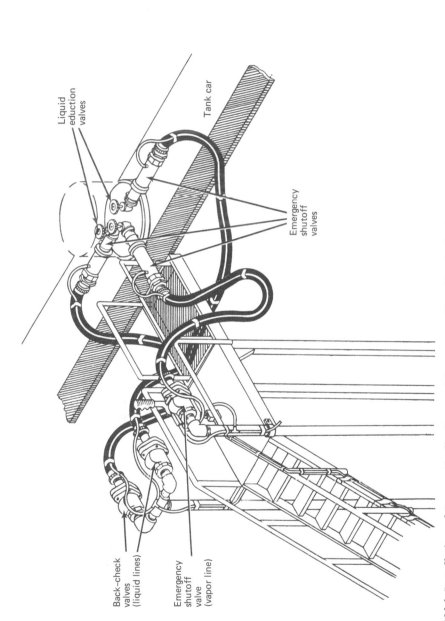

Liquid
eduction
valves

Tank car

Emergency
shutoff
valves

Back-check
valves
(liquid lines)

Emergency
shutoff
valve
(vapor line)

Figure 3.10(g) Installation of Pneumatically Operated Emergency Shutoff Valve/Unloading Adapter Combination at Tank Car Loading or Unloading Station. See Figure 2.18(b) for details of ESV/unloading adapter combination.

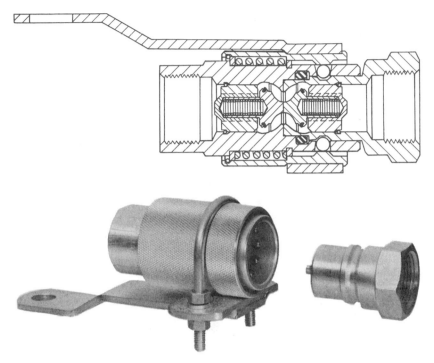

Figure 3.10(h) Pullaway Valve. Valve closes automatically if vehicle is moved with hose still connected. This is a way to incorporate a "predictable breakaway point."

economical valves having the desired characteristics were not available. As a result, the retroactive date had to be amended by Tentative Interim Amendments (TIAs) to provide time for these valves to be manufactured, tested, and listed (listing is not required but the manufacturers felt that it was highly desirable and this facilitated approval by the authorities having jurisdiction).

Where flow is only in one direction, a check valve can be used provided it has a degree of fire resistance.

Automatic actuation of an emergency shutoff valve occurs through sensing of heat from a fire [3-2.7.9(a)(1)] with the sensing element located near the source of leakage. In most instances, the sensing element is a point source fusible element or pressurized plastic tubing. The plastic tubing system is actually a line sensor capable of sensing fire at other locations (as could be a system using several fusible elements).

In addition to automatic actuation, an emergency shutoff valve must be installed so that it can be operated manually from a remote location and, also at its installed location.

Subparagraph 3-2.7.9(b) reflects the cargo vehicle "pull-away" experience.

3-2.7.10 Hose may be used on the low pressure side of regulators to connect to other than domestic and commercial appliances as follows:

(a) The appliance connected shall be of a portable type.

(b) For use inside buildings, the hose shall be of a minimum length, not exceeding 6 ft (1.8 m) [except as provided for in 3-4.2.3(b)], and shall not extend from one room to another, nor pass though any partitions, walls, ceilings or floors (except as provided for in 3-4.3.7). It shall not be concealed from view or used in concealed locations. For use outside buildings, hose length may exceed 6 ft (1.8 m), but shall be kept as short as practicable.

(c) Hose shall be securely connected to the appliance. The use of rubber slip ends is not permissible.

(d) A shutoff valve shall be provided in the piping immediately upstream of the inlet connection of the hose. When more than one such appliance shutoff is located near another, precautions shall be taken to prevent operation of the wrong valve.

(e) Hose used for connecting appliances to wall or other outlets shall be protected against physical damage.

The prohibition of hoses in domestic and commercial applications essentially restricts them to agricultural and industrial ones. Because of their vulnerability to mechanical and thermal damage, and their limited service life, they should be used only where really necessary and where they can be readily inspected and maintained.

3-2.8 Hydrostatic Relief Valve Installation.

Over a temperature range from 30°F to 90°F (-1°C to 32°C), liquid propane will expand an average of about 1.6 percent for each 10 Fahrenheit degrees (5.5 Celsius degrees) it is heated. If the liquid is trapped in a length of pipe by, for example, being between two closed valves, and there is no room for expansion, the pressure developed can be very high (thousands of psi) and pipe or valve failures can occur. Operation of a hydrostatic relief valve prevents this by discharging liquid. Unlike vapor pressure relief valves, the quantity that needs to be discharged is quite small. Being liquid, however, it represents a greater quantity of vapor and its low temperature presents a personnel hazard.

3-2.8.1 A hydrostatic relief valve complying with 2-4.7.1 or a device providing pressure relieving protection shall be installed in each section of piping (including hose) in which liquid LP-Gas can be isolated between shutoff valves so as to relieve the pressure which could develop from the trapped liquid to a safe atmosphere or product-retaining section.

3-2.9 Testing Piping Systems.

After assembly of piping (but before appliances are connected) the industry practice is to admit full container pressure to the system and check all connections for leaks with a soap or leak-testing solution. There are leak test solutions listed for this purpose. (Some soap solutions can be corrosive to piping). It is important, particularly with copper tubing or fittings, that the leak test solution contain no ammonia.

A widely used and very sensitive leakage test is that using a tee block fitting with a pressure gauge between the container service shutoff valve and the first stage regulator. This test is described in Appendix D of NFPA 54, *National Fuel Gas Code*. Essentially, it consists of admitting full container pressure to the system, closing the container shutoff valve and lowering the pressure reading on the gauge 10 psi by bleeding off a small amount of gas in the system. A very small leak will be accentuated in the small volume involved at the pressure gauge location. If leakage occurs, the source is detected by checking all fittings and connections with a leak testing solution.

Bearing in mind that a high proportion of LP-Gas systems from the outlet of the first stage regulator are covered by NFPA 54, *National Fuel Gas Code*, reference should be made to Part 4, "Gas Piping Inspection, Testing, and Purging," of that standard.

3-2.9.1 After assembly, piping systems (including hose) shall be tested and proven free of leaks at not less than the normal operating pressure. Piping within the scope of NFPA 54, *National Fuel Gas Code*, [see 1-2.3.1(f)], shall be pressure tested in accordance with that Code. Tests shall not be made with a flame.

3-2.10 Equipment Installation.

3-2.10.1 Pumps shall be installed as recommended by the manufacturer and in accordance with 3-2.10.1(a) through (c).

Because LP-Gas liquid under pressure in a container or pipeline will vaporize when the pressure is reduced, and the vapor thus produced can lead to cavitation in a pump, the types of pumps commonly used in installations covered by NFPA 58 are installed so that they hand a positive suctionhead. In practice, this means that their suction inlets are well below the lowest liquid level in a container, and the suction piping is large

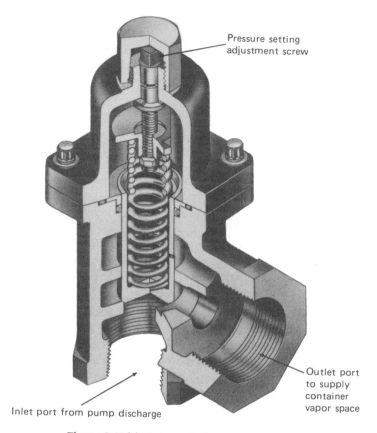

Pressure setting
adjustment screw

Outlet port
to supply
container
vapor space

Inlet port from pump discharge

Figure 3.11(a) Automatic Pump Bypass Valve.

enough to minimize friction loss.

The suction inlet pressures are substantial—e.g., up to 250-312.5 psi (1.7-2.2 MPa) for propane containers—and, therefore, discharge pressures can also be substantial. In order to permit the use of readily available and economical piping system components, the standard limits discharge pressures to a normal maximum of 350 psi (2.4 MPa) with transient maximums up to 400 psi (2.8 MPa).

These pressures, combined with the totally off or totally on operation in the typical installation, require provisions to prevent excessive vibration and strain from inlet and discharge connections.

Many pumps are started and stopped from a location remote from the pump—e.g., cylinder filling areas and dispensers. For the safety of anyone who might be working on a pump, 3-2.10.1(c) requires a means for preventing startup to be located near the pump.

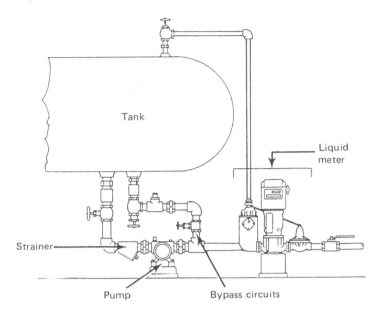

Figure 3.11(b) Typical Pump and Meter Installation.

(a) Installation shall be made so that the pump casing shall not be subjected to excessive strains transmitted to it by the suction and discharge piping. This shall be accomplished by piping design, the use of flexible connectors or expansion loops, or by other effective methods, in accordance with good engineering practice.

(b) Positive displacement pumps shall be installed in accordance with 2-5.2.2.

(1) The bypass valve or recirculating device to limit the normal operating discharge pressure to not more than 350 psig (2.4 MPa gauge) shall discharge either into a storage container (preferably the supply container from which the product is being pumped) or into the pump suction.

(2) If this primary device is equipped with a shutoff valve, an adequate secondary device designed to operate at not more than 400 psig (2.8 MPa gauge) shall, if not integral with the pump, be incorporated in the pump piping. This secondary device shall be designed or installed so that it cannot be rendered inoperative, and shall discharge either into the supply container or into the pump suction.

(c) A pump operating control or disconnect switch shall be located near the pump. Remote control points shall be provided as necessary for other plant operations such as container filling, loading or unloading of cargo vehicles and tank cars, or operation of motor fuel dispensers.

3-2.10.2 Compressors shall be installed as recommended by the manufacturer and in accordance with 3-2.10.2 (a) and (b).

(a) Installation shall be made so that the compressor housing shall not be subjected to excessive strains transmitted to it by the suction and discharge piping. Flexible connectors may be used where necessary to accomplish this.

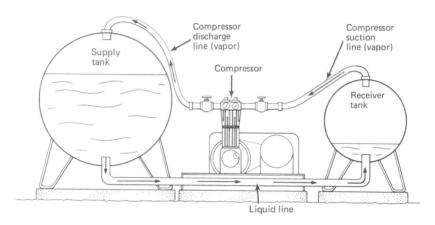

Figure 3.12 Transferring with Compressors.

(b) If the compressor is not equipped with an integral means to prevent the LP-Gas liquid entering the suction (*see 2-5.3.3*), a suitable liquid trap shall be installed in the suction piping as close to the compressor as practicable. Portable compressors used with temporary connections are excluded from this requirement.

(c) Engines used to drive portable compressors shall be equipped with exhaust system spark arrestors and shielded ignition systems.

See commentary on 2-5.3.4.

3-2.10.3 The installation of vaporizers of the types covered by 2-5.4 is covered in Section 3-7 and of engine fuel vaporizers in Section 3-6. Integral vaporizing-burners, such as are used for weed burners or tar kettles, are considered to be part of these units (or "appliances"). For appliance installation standards, see Section 3-5.

3-2.10.4 Strainers shall be installed so that the strainer element can be serviced.

3-2.10.5 Liquid or vapor meters shall be installed as recommended by the manufacturer, and in compliance with the applicable provisions of 3-2.10.5(a) and (b).

(a) Liquid meters shall be securely mounted and shall be installed so that the meter housing is not subjected to excessive strains from the connecting piping. If not provided in the piping design, flexible connectors may be used where necessay to accomplish this.

(b) Vapor meters shall be securely mounted and installed so as to minimize the possibility of physical damage.

3-2.10.6 LP-Gas engine fuel dispensing devices installed in service stations shall be installed as recommended by the manufacturer and in accordance with 3-2.10.6(a) through (h).

While 3-2.10.6 covers a liquid transfer operation and the intent of Chapter 4 was to cover all liquid transfer operations, provisions for the LP-Gas service station were placed in Chapter 3 because of the presence of the general public in the routine operation of the service station. However, these provisions remain essentially as originally developed in the late 1930s, when transfer was almost always only into containers mounted on vehicles for vehicle propulsion engine supply, and only LP-Gas was dispensed in the station.

The number of LP-Gas powered vehicles has increased greatly since the late 1970s to the point where originally gasoline-only service stations are also dispensing LP-Gas. In addition, the use of LP-Gas containers in recreational activities — e.g., recreational vehicles and gas grills — has increased and these service stations often are arranged for filling these vehicles and portable cylinders. Finally, following the pattern of gasoline service stations, there is considerable pressure for self-service LP-Gas dispensing in these service stations.

The Technical Committee feels that neither 3-2.10.6 nor Chapter 4 addresses these installations completely enough for current conditions and has had the matter under study. In the meantime, it will be necessary to apply judgment to installations where both gasoline and LP-Gas are handled and portable containers are also filled. It is suggested that LP-Gas containers not be located on gasoline dispensing islands and that self-service dispensing of LP-Gas not be permitted in service stations open to the general public.

(a) Installation shall not be within a building, but may be under weather shelter or canopy, provided this area is adequately ventilated and is not enclosed for more than 50 percent of its perimeter.

(b) Dispensing devices shall be located as follows:

(1) Not less than 10 ft (3 m) from aboveground storage containers of more than 2,000 gal (7.6 m³) water capacity.

(2) Not less than 20 ft (6 m) from any building [not including canopies covered in 3-2.10.6(a)], basement, cellar, pit or line of adjoining property which may be built upon.

(3) Not less than 10 ft (3 m) from sidewalks, streets or thoroughfares.

Formal Interpretation
Reference: 3-2.10.6(b)(3), 4-3.4.1, 3-8.4.1(a), 3-8.4.1(b)

Question: A 1600-gallon LP-Gas storage container and associated liquid transfer equipment are centrally installed in a parking area serving a suburban shopping center for the purpose of filling containers mounted on vehicles and portable containers. The vehicles and portable containers are in the immediate vicinity of the storage container while they are being filled. The lane for vehicles being serviced is also a driveway for parking of vehicles and for movement within the parking area. Is such an installation and operation specifically covered by the provisions of NFPA 58?

Answer: Yes. Specifically, such an installation is a "Distributing Point" as defined in Section 1-7 of NFPA 58 and is subject to all provisions applicable to Distributing Points.

Applicable provisions include 3-2.10.6(b)(3), 4-3.4.1, 3-8.4.1(a) and 3-8.4.1(b), and Parts 6 and 7 of Table 4-3.3.2. In the context of an installation in a parking area open to the public, the Committee is of the opinion that portions of such an area accessible to the public for activities normally associated with the movement and parking of vehicles should be considered as public ways, streets, sidewalks, and thoroughfares as these terms are used in 3-2.10.6(b)(3), 4-3.4.1, and Parts 6 and 7 of Table 4-3.3.2.

The Committee noted that while it would be possible to install such a distributing point "centrally" in such a parking area, provisions for restricting movement of the public and vehicles when not associated with the functioning of the installation [especially to assure compliance with 3-8.4.1(b)] could make a less central location more practical.

Issue Edition: 1974
Reference: 3195(b)(3), 4012, 4060(a), 4060(b)
Date: October-November 1976

(c) Dispensing devices shall either be installed on a concrete foundation or be part of a complete storage and dispensing unit mounted on a common base [to be mounted as provided in 3-2.3.1(b) and (d)]. In either case, they shall be adequately protected against physical damage.

(d) Control for the pump used to transfer LP-Gas through the dispensing device into motor vehicle tanks shall be provided at the device in order to minimize the possibility of leakage or accidental discharge. The following also shall apply:

(1) Means shall be provided at some point outside the dispensing device, such as a remote switch [see 3-2.10.1(c)], to shut off the power in the event of fire or accident.

(2) A manual shutoff valve and an excess-flow check valve of suitable capacity shall be located in the liquid line between the pump and dispenser inlet only when the dispensing device is installed at a remote location and not part of a complete storage and dispensing unit mounted on a common base.

(e) Provision shall be made for venting the LP-Gas contained in the dispenser to a safe location.

(f) The dispensing hose shall comply with 2-4.6. An excess-flow check valve, or an automatic shutoff valve [see 2-3.3.3(d) and 2-4.5.4] shall be installed at the terminus of the liquid piping at the point of attachment of the dispensing hose. A differential back pressure valve shall be considered as meeting this provision.

See commentary on 2-5.7.3.

(g) Piping leading to, and within the dispenser, and the dispensing hose shall be provided with hydrostatic relief valves as specified in 3-2.8.1 (see also 2-4.7.1).

(h) No drains or blowoffs from the dispensing device shall be directed toward, or be in close proximity to sewer systems.

3-3 Distributing and Industrial LP-Gas Systems.

3-3.1 Application.

3-3.1.1 This section includes provisions for LP-Gas systems installed at distributing plants, industrial plants and distributing points (see definitions). These provisions extend and modify the provisions of Section 3-2 for these applications.

Section 3-3 is primarily a compilation of provisions scattered elsewhere in the standard that are applicable to facilities where

larger quantities of LP-Gas are used and liquid transfer operations are frequent. To this extent, its purpose is one of convenience to those concerned with such facilities.

Much of Section 3-3 consists of references to other provisions
which are commented upon therein.

3-3.2 General.

3-3.2.1   The location and installation of storage containers and the installation of container appurtenances, piping, and equipment shall comply with
Section 3-2.

3-3.3 Installation of Liquid Transfer Facilities.

3-3.3.1   Points of transfer (*see definition*) or the nearest part of a structure
housing transfer operations shall be located in accordance with 4-3.2 and
4-3.3.

3-3.3.2   Separate buildings, and attachments to or rooms within other
buildings, housing points of transfer or transfer pumps and compressors,
constructed or converted to such use after December 31, 1972, shall
comply with Chapter 7.

3-3.3.3   The track of the railroad siding or the roadway surface at the
transfer points shall be relatively level. Adequate clearances from buildings,
structures, or stationary containers shall be provided for the siding or
roadway approaches to the unloading or loading points. Substantial bumpers shall be provided at the ends of sidings, and as necessary to protect
storage containers and points of transfer.

**The primary reason for the track of railroad sidings and the
roadway surface at tank truck transfer points to be level is so
that the gauging device in the unit will be as accurate as
possible and not be affected by the transport unit being on a
slant. It also helps prevent a "run-away" situation from occurring in the event wheel chocks become dislodged (or are not
used).**

3-3.3.4   Safeguards shall be provided to prevent the uncontrolled discharge of LP-Gas in the event of failure in the hose or swivel type piping.
The provisions of 3-2.7.9 shall apply. For all other LP-Gas systems, the
following shall apply:

(a) The connection, or connecting piping, larger than ½ in. (13 mm)
internal diameter into which the liquid or vapor is being transferred shall be
equipped with:

(1)   A backflow check valve, or

(2)   An emergency shutoff valve complying with 2-4.5.4, or

(3)   An excess-flow valve properly sized in accordance with 3-2.4.6(a)(3).

(b) The connection, or connecting piping, larger than ½ in. (13 mm) internal diameter from which the liquid or vapor is being withdrawn shall be equipped with:

(1)   An emergency shutoff valve complying with 2-4.5.4, or

(2)   An excess-flow valve properly sized in accordance with 3-2.4.6(a)(3).

3-3.3.5   See 4-2.3.6 for railroad tank car transfer operations.

3-3.3.6   If gas is to be discharged from containers inside a building, the installation provisions of 4-4.2.1 shall apply.

3-3.4 Installation of Gas Distribution Facilities.

3-3.4.1   This subsection applies to the installation of facilities used for gas manufacturing, gas storage, gas-air mixing and vaporization, and compressors not associated with liquid transfer.

3-3.4.2   Separate buildings and attachments to or rooms within other buildings housing gas distribution facilities, constructed or converted to such use after December 31, 1972, shall comply with Chapter 7.

Exception No. 1: Facilities for vaporizing LP-Gas and gas-air mixing shall be designed, located and installed in accordance with Section 3-7.

Exception No. 2:   Facilities for storing LP-Gas in portable containers at industrial plants and distributing points shall comply with Chapter 5.

3-3.4.3   Buildings housing vapor compressors shall be located in accordance with 4-3.3.2 considering the building as one housing a point of transfer.

3-3.4.4   The use of pits to house gas distribution facilities shall be avoided unless automatic flammable vapor detecting systems are installed in the pit. Drains or blowoff lines shall not be directed into or in proximity of sewer systems.

3-3.4.5   If gas is to be discharged from containers inside a building, the installation provisions of 4-4.2.1 shall apply.

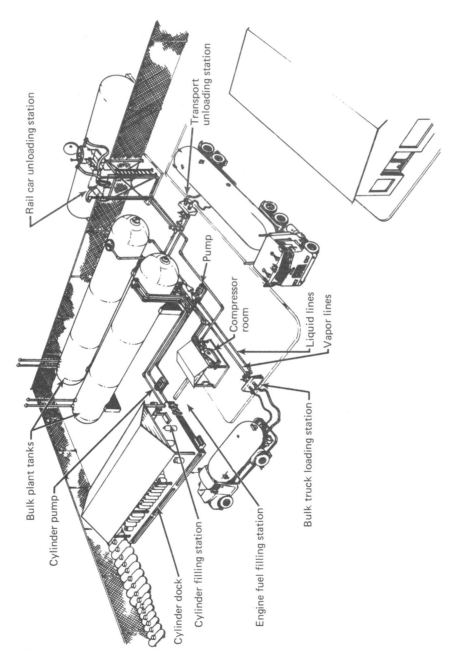

Rail car unloading station

Transport unloading station

Pump

Compressor room

Liquid lines

Vapor lines

Bulk plant tanks

Cylinder pump

Cylinder dock

Cylinder filling station

Engine fuel filling station

Bulk truck loading station

Figure 3.13 LP-Gas Bulk Plant.

3-3.5 Installation of Electrical Equipment.

3-3.5.1 Installation of electrical equipment shall comply with Section 3-8.

3-3.6 Protection Against Tampering for Section 3-3 Systems.

3-3.6.1 To minimize the possibilities for trespassing and tampering, the area which includes container appurtenances, pumping equipment, loading and unloading facilities and container filling facilities shall be protected by one of the following methods:

(a) Enclosure with at least a 6-ft (1.8-m) high industrial-type fence, unless otherwise adequately protected. There shall be at least two means of emergency access from the fenced or other enclosure. Clearance shall be provided to permit maintenance to be performed and a clearance of at least 3 ft (1 m) shall be provided to allow emergency access to the required means of egress. If guard service is provided, it shall be extended to the LP-Gas installation. Guard personnel shall be properly trained. (*See 1-6.1.1.*)

Exception: If a fenced or otherwise enclosed area is not over 100 sq ft (9 m²) in area, the point of transfer is within 3 ft (1 m) of a gate and containers being filled are not located within the enclosure, a second gate need not be provided.

(b) As an alternate to fencing the operating area, suitable devices which can be locked in place shall be provided. Such devices, when in place, shall effectively prevent unauthorized operation of any of the container appurtenances, system valves or equipment.

These security provisions are appropriate for most circumstances but are not intended to deter those intent upon doing damage. Tests and experience have shown that an LP-Gas container can withstand rifle and pistol projectiles.

Of the two methods described in 3-3.6.1, fencing is most common. Locking devices are used mostly in distributing points (*see definitions in Section 1-7*).

While an industrial-type fence is not described in further detail, the chain-link type is the most common. A solid fence should not be used for the reasons cited in the commentary on 3-2.2.9.

Access to and from a fenced enclosure is for both operators/ maintainers and emergency personnel, and two exits are required for safety in most cases. The Exception in 3-3.6.1 recognizes that both the need to evacuate and the difficulty in doing so is lessened under the circumstances described. As an operational matter, all gates should be open when anyone is inside.

3-3.7 Lighting.

3-3.7.1   If operations are normally conducted during other than daylight hours, adequate lighting shall be provided to illuminate storage containers, containers being loaded, control valves and other equipment.

3-3.8 Ignition Source Control.

3-3.8.1   Ignition source control shall comply with Section 3-8.

3-4 LP-Gas Systems in Buildings or on Building Roofs or Exterior Balconies.

3-4.1 Application.

3-4.1.1   This section includes installation and operating provisions for LP-Gas systems containing liquid LP-Gas located inside of, or on the roofs or exterior balconies of, buildings or structures. Systems covered include those utilizing portable containers inside of or on the roofs or exterior balconies of buildings, and those in which the liquid is piped from outside containers into buildings or onto the roof. These systems are permitted only under the conditions specified in 3-4.1.1(a) through (d) and in accordance with 3-4.1 and 3-4.2. Containers in use shall mean connected for use.

> Paragraph 3-2.2 summarizes the only situations in which LP-Gas containers can be located inside of buildings. Section 3-4 covers one of these situations — namely, the use of portable container systems inside of or on the roofs of buildings. In addition, this section also covers the piping of liquid LP-Gas into buildings or onto the roofs of buildings from containers located outside. This important requirement of the standard dates back to the first NFPA standard on LP-Gas in 1927 and has been part of the requirements ever since.
>
> Prior to considering any installation of an LP-Gas container in a building, this Section (3-4.1, Application) must be read in its entirety.

(a)   The portable use of containers indoors shall be only for the purposes specified in 3-4.3 through 3-4.8. Such use shall be limited to those conditions where operational requirements make portable use of containers necessary and location outside is impractical.

> The use of portable containers indoors (including balconies of buildings) is permitted only for the purposes specified in this section, and all other uses are prohibited. The use of portable cylinder systems having capacities larger than 1 lb (0.45 kg) of LP-Gas and the associated storage of such cylinders

indoors are limited only to uses in construction and renovation of buildings, industrial applications, education, research, training, and temporary heating in the case of emergencies. These limited applications acknowledge the good experience and presence of trained personnel in industrial uses and the temporary nature and lack of alternate fuel sources for certain appliances needed at certain times in buildings.

No other uses, including that for normal, routine, comfort heating, are permitted. This philosophy was affirmed by the Technical Committee in preparing the 1986 Edition of the standard by their rejection of a proposal to permit the use of portable nonvented indoor heating appliances fueled by integral 20 lb (9 kg) LP-Gas containers (cabinet heaters).

Where indoor use of LP-Gas cylinders is permitted in the section, the installer must first attempt to locate the container outdoors and may then install the container indoors only if it is impractical to locate them outdoors. In determining if an outdoor location is impractical, the authority having jurisdiction may have to be consulted.

(b) Installations using portable containers on roofs shall be as specified in 3-4.9.1. Such use shall be limited to those conditions where operational requirements make portable use of containers necessary and location not on roofs of buildings or structures is impractical.

Provisions for the installation of portable LP-Gas containers on roofs were incorporated into the standard to provide fuel for emergency generators and microwave relay stations. Again, such systems are to be used only when the use of portable containers is necessary, and other outdoor locations are impractical.

(c) Installations using portable containers on exterior balconies shall be as specified in 3-4.9.2.

This requirement was added in the 1989 Edition to clarify that the use of portable containers on balconies and the transportation of containers within buildings were covered by the standard.

(d) Liquid LP-Gas shall be piped into buildings or structures only for the purposes specified in 3-2.6.1(d).

While LP-Gas vapor at pressures up to and including 20 psig (138 kPa gauge) is permitted to be piped into buildings, there are certain limitations for piping liquid LP-Gas at pressures exceeding 20 psig (138 kPa gauge) into buildings. These are covered in 3-2.6.1(d) and 3-4.10.

3-4.1.2 Storage of containers awaiting use shall be in accordance with Chapter 5.

> This reference was added in the 1989 Edition to clarify that stored cylinders, including cylinders that are part of an appliance but not connected to the appliance, are covered by the standard and that Chapter 5 applies.

3-4.1.3 Transportation of containers within a building shall be in accordance with 3-4.2.7.

3-4.1.4 These provisions are in addition to those specified in Section 3-2.

3-4.1.5 Liquid transfer systems are covered in Chapter 4.

3-4.1.6 Engine fuel systems used inside buildings are covered in Section 3-6.

3-4.1.7 LP-Gas transport or cargo vehicles stored, serviced or repaired in buildings are covered in Chapter 6.

3-4.2 General Provisions for Containers, Equipment, Piping, and Appliances.

> Certain conditions are set out for specific applications of portable container systems in buildings in 3-4.3 through 3-4.8 and for permanent systems under 3-4.9. Subsection 3-4.2 specifies certain general provisions that are applicable to all of these specific applications.

3-4.2.1 Containers shall comply with DOT cylinder specifications (*see 2-2.1.3 and 2-2.2.1*), shall not exceed 245 lb (111 kg) water capacity [nominal 100 lb (45 kg) LP-Gas capacity] each, shall comply with other applicable provisions of Section 2-2 and be equipped as provided in Section 2-3 (*see 2-3.3 and Table 2-3.3.2*). They shall also comply with the following:

> Basic considerations are given for the type of container to be used. They must be DOT cylinders with a maximum capacity of 100 lb (45 kg) of LP-Gas. This is the largest size that can be moved by personnel from a practical standpoint. In the past there have been some 500 gal (1.9 m³) ASME tanks used at different floor levels on large construction sites but this is in variance with NFPA 58 and must receive special approval. The

NFPA 58 approach with respect to the use of ASME bulk tanks in this instance would be to locate them at ground level on the outside and pipe the LP-Gas into the building in accordance with 3-4.10.

The importance of compliance with this requirement was highlighted by an incident. Employees in a building housing a radiator repair business were moving a 500 gal (1.9 m³) ASME propane container on the third floor with an industrial lift truck. The container was secured on the truck with a length of 2 × 4 in. (51 × 102 mm) wood. The container fell, shearing off a valve leaving an opening about ¹¹⁄₁₆ in. (17 mm) through which liquid LP-Gas was released. After unsuccessfully attempting repairs, the workers turned off the electrical supply to the building and left, and called the fire department. Just as the fire engine arrived the propane ignited and the explosion killed 5 firemen and 2 civilians, and injured about 50 others. (For a full report of the incident see the March, 1984 Edition of *Fire Command.*)

(a) Containers shall be marked as provided in 2-2.6.

(b) Containers with water capacities greater than 2 ½ lb (1 kg) [nominal 1 lb (0.45 kg) LP-Gas capacity] shall be equipped with shutoff and excess-flow valves as provided in 2-3.3.2 (Column 3, Table 2-3.3.2). The installation of excess-flow valves shall take into account the type of valve protection provided for the container in accordance with 2-2.4.1.

Column 3 of Table 2-3.3.2 covers DOT cylinders of water capacity 2.5 to 245 lbs (1 to 110 kg) when used in buildings. All connections (filling, withdrawal, and exchange) must have a positive (manual) shutoff and excess-flow check valve (internal or external). If an external excess-flow valve is uses it must be installed so that it is protected from breakage. This excludes the very prevalent 20 lb (9 kg) gas grill cylinders from all indoor uses unless they are equipped with an external excess-flow check valve. Other provisions in this section limit their use to temporary heating in buildings, in case of emergency, construction, industrial occupancies, etc.

(c) Valves on containers shall be protected in accordance with 2-2.4.1.

Paragraph 2-2.4.1 requires that all container valves be protected against physical damage by either recessing the valve into the container or protecting the valve with a ventilated cap or collar. Note that when a removable cap is used it must be in place when the container is not in use.

(d) Containers having water capacities greater than 2 ½ lb (1 kg) [nominal 1 lb (0.45 kg) LP-Gas capacity] connected for use shall stand on a firm and substantially level surface. If necessary, they shall be secured in an upright position.

(e) Containers and the valve protecting devices used with them shall be oriented so as to minimize the possibility of impingement of the pressure relief device discharge on the container and adjacent containers.

> It is important here that the container pressure relief discharge is directed through a hole in the cap or collar. Also, in positioning a group of cylinders, attention should be given to ensure that pressure relief valves are not directed at adjacent containers.

3-4.2.2 Regulators, if used, shall be suitable for use with LP-Gas. Manifolds and fittings connecting containers to pressure regulator inlets shall be designed for at least 250 psig (1.7 MPa gauge) service pressure.

3-4.2.3 Piping, including pipe, tubing, fittings, valves, and hose, shall comply with Section 2-4, except that a minimum working pressure of 250 psig (1.7 MPa gauge) shall apply to all components. The following also shall apply:

(a) Piping shall be installed in accordance with the provisions of 3-2.7 for liquid piping or for vapor piping for pressures above 125 psig (0.9 MPa gauge). [*See 3-2.7.1(b).*]

(b) Hose, hose connections, and flexible connectors used shall be designed for a working pressure of at least 350 psig (2.4 MPa gauge), shall comply with 2-4.6, and be installed in accordance with 3-2.7.10. Hose length may exceed that specified by 3-2.7.10(b), but shall be as short as practicable, although long enough to permit compliance with the spacing requirements (*see 3-4.3.3 and 3-4.3.4*) without kinking or straining hose or causing it to be close enough to a burner to be damaged by heat. See 3-4.9 for permanent roof installations.

> General piping provisions are referred to in 3-2.7. However, special attention is given here to the hose used with portable container systems inside buildings. While basic provisions for hose are given through reference to 3-2.7.10 and 2-4.6, two exceptions are noteworthy. First, the hose must be designed for a 350 psig (2.4 MPa gauge) working pressure. This is not so much for the pressures involved, but to assure a stronger type hose is used in this service, particularly at construction sites where rough usage may be encountered. Secondly, a 6 ft (2 m) length is the general maximum size permitted but longer lengths are allowed to permit some of the spacing requirements

set out in this section. Too short a length could possibly decrease safety. With regard to permanent installations on roofs, the use of hose connections to containers is prohibited. See Formal Interpretation of 3-4.5.

3-4.2.4 Containers, regulating equipment, manifolds, pipe, tubing, and hose shall be located so as to minimize exposure to abnormally high temperatures (such as might result from exposure to convection and radiation from heating equipment or installation in confined spaces), physical damage or tampering by unauthorized persons.

When containers are exposed to abnormally high temperatures, liquid LP-Gas expands and the container could become liquid-full, causing high hydrostatic pressures resulting in the discharge of the container pressure relief valve. Heating equipment should be positioned so that infrared heaters do not focus on containers and convection heat is not directed at them. Also, particular attention should be given to installing cylinders in confined spaces where temperatures may build up, such as in pits, tunnels, etc. at construction areas. Although indirectly related, appliance location in confined areas is equally important from the standpoint of providing sufficient air for combustion.

3-4.2.5 Heat producing equipment shall be located and used so as to minimize the possibility of the ignition of combustibles.

Tarpaulins, plastic sheeting, wood scaffolding, etc. should not be positioned in such a way that they may be ignited by heaters.

3-4.2.6 When containers are located on a floor, roof, or balcony, provisions shall be made to minimize the possibility of containers falling over the edge.

The addition of balconies in the 1989 Edition was to clarify the Technical Committee's intent that containers over 1 lb (0.45 kg) propane capacity must not be brought into any part of buildings, including balconies.

(a) Filling containers on roofs or balconies is prohibited. See 4-3.1.1(b).

3-4.2.7 Transportation (movements) of containers within a building shall comply with 3-4.2.7(a) through (d).

This 1989 amendment applies provisions formerly applicable to movement associated with containers installed on roofs, to all movement inside buildings.

(a) Movement of containers having water capacities greater than 2 ½ lb (1 kg) [nominal 1 lb (0.45 kg) LP-Gas capacity] within a building shall be restricted to movement directly associated with the uses covered by Sections 3-4.3 through 3-4.9 and be conducted in accordance with these provisions and 3-4.2.7(b) through (d).

(b) Valve outlets on containers having water capacities greater than 2 ½ lb (1 kg) [nominal 1 lb (0.45 kg) LP-Gas capacity] shall be tightly plugged and the provisions of 2-2.4.1 shall be complied with.

(c) Only emergency stairways not generally used by the public shall be used and reasonable precautions shall be taken to prevent the container from falling down the stairs.

(d) Freight or passenger elevators may be used when occupied only by those engaged in moving the container.

3-4.2.8 Portable heaters, including salamanders, shall be equipped with an approved automatic device to shut off the flow of gas to the main burner, and pilot if used, in the event of flame extinguishment or combustion failure. Such portable heaters shall be self-supporting unless designed for container mounting (*see 3-4.3.4*). Container valves, connectors, regulators, manifolds, piping, or tubing shall not be used as structural supports. The following shall also apply:

Although standards exist for the listing of portable heaters by nationally recognized testing laboratories [e.g., ANSI Z83.7 (*see 2-6.2.1*)], not all heaters are tested and listed. This part of Section 3-4 is intended to provide basic requirements for the authorities to utilize in extending approvals.

(a) Portable heaters manufactured on or after May 17, 1967, having an input of more than 50,000 Btuh (53 MJ/h), and those manufactured prior to May 17, 1967, with inputs of more than 100,000 Btuh (105 MJ/h), shall be equipped with either:

    (1) A pilot which must be lighted and proved before the main burner can be turned on, or

    (2) An approved electric ignition system.

(b) The provisions of 3-4.2.8 are not applicable to the following:

Except for tar kettle burners, hand torches, or melting pots, which are attended, and smaller heaters less than 7,500 Btuh (8 MJ/h) connected to a 1 lb (0.45 kg) LP-Gas container, all portable heaters used indoors must have flame failure or, in the case of catalytic heaters, combustion failure protection. This

protection is an approved automatic device to shut off the flow of gas to the main burner — and pilot, if used — in the event of such failures. Additionally, these portable heaters designed for container mounting must not use valves, piping, regulators, etc. as structural supports for the heater.

The possibility of operators being burned during ignition of heaters may exist with larger heaters designed to operate on higher inlet pressures if the provisions of this part are not followed. Unless proved pilot lights or electric ignition are used, a large flame rollout may occur. Pilots on these larger heaters may encounter problems of thermocouple premature failure or outages due to wind, etc. This can result in operators bypassing controls. Therefore, for older heaters (which, for the most part, have subsequently been replaced) and newer heaters of certain sizes, assurance must be made that delayed ignition of large volumes of gas will not occur.

(1)   Tar kettle burners, hand torches or melting pots.

(2)   Portable heaters with less than 7,500 Btuh (8 MJ/h) input if used with containers having a maximum water capacity of 2 ½ lb (1 kg).

3-4.3 Buildings Under Construction or Undergoing Major Renovation.

These are the basic provisions for the transportation and use of portable container systems for construction or major renovation of buildings not occupied by the public. Prior approval is particularly noted when the building is partially occupied by the public. Sometimes occupancy is started before all construction is completed. Paragraph 3-4.4 covers those instances where minor renovation is done when the building is frequented by the public.

Note that paragraph 3-4.3.1 was changed in the 1989 Edition to specifically extend coverage to transportation and use of containers.

3-4.3.1   Containers may be used and transported in buildings or structures under construction or undergoing major renovation when such buildings are not occupied by the public or, if partially occupied by the public, containers may be used and transported in the unoccupied portions with the prior approval of the authority having jurisdiction. Such use shall be in accordance with 3-4.3.1 through 3-4.3.8.

3-4.3.2   Containers, equipment, piping, and appliances shall comply with 3-4.2.

3-4.3.3   For temporary heating, such as curing concrete, drying plaster, and similar applications, heaters (other than integral heater-container units covered in 3-4.3.4) shall be located at least 6 ft (1.8 m) from any LP-Gas container.

3-4.3.4 Integral heater-container units specifically designed for the attachment of the heater to the container, or to a supporting standard attached to the container, may be used, provided they are designed and installed so as to prevent direct or radiant heat application to the container. Blower and radiant type units shall not be directed toward any LP-Gas container within 20 ft (6 m).

3-4.3.5 If two or more heater-container units of either the integral or nonintegral type are located in an unpartitioned area on the same floor, the container(s) of each such unit shall be separated from the container(s) of any other such unit by at least 20 ft (6 m).

> Paragraphs 3-4.3.3, 3-4.3.4, and 3-4.3.5 reflect the basic requirements set out in 3-4.2, but incorporate specific distances to accomplish them. Six ft (1.8 m) separation between nonintegral heaters and other LP-Gas containers is given, while 20 ft (6 m) is required for integral heater-container units. The latter are infrared or larger blower type units which have a more pronounced effect on heat transmission to nearby containers.

3-4.3.6 If heaters are connected to containers manifolded together for use in an unpartitioned area on the same floor, the total water capacity of containers manifolded together serving any one heater shall not be greater than 735 lb (333 kg) [nominal 300 lb (136 kg) LP-Gas capacity], and if there is more than one such manifold it shall be separated from any other by at least 20 ft (6 m).

> The 300 lb (136 kg) LP-Gas maximum for manifolded systems has a long and successful history in NFPA 58. A 20 ft (6 m) distance is set out for the separation of separate manifolded systems in the same unpartitioned floor area.

3-4.3.7 On floors on which no heaters are connected for use, containers may be manifolded together for connection to a heater or heaters on another floor, provided:

(a) The total water capacity of the containers connected to any one manifold is not greater than 2,450 lb (1111 kg) [nominal 1,000 lb (454 kg) LP-Gas capacity], and

(b) Manifolds of more than 735 lb (333 kg) water capacity [nominal 300 lb (136 kg) LP-Gas capacity], if located in the same unpartitioned area, shall be separated from each other by at least 50 ft (15 m).

3-4.3.8 The provisions of 3-4.3.5, 3-4.3.6, and 3-4.3.7 may be altered by the authority having jurisdiction if compliance is impractical.

3-4.4 Buildings Undergoing Minor Renovation when Frequented by the Public.

Renovation of buildings during the hours when the public is present requires special considerations. The maximum size of container permitted is 20 lb (9 kg) LP-Gas capacity. The number of cylinders is not to exceed the number of workers assigned to them. The cylinders are not to be left unattended at any time. At other times, the provisions of 3-4.3 apply.

3-4.4.1 Containers may be used and transported for repair or minor renovation in buildings frequented by the public as follows:

(a) During the hours of the day the public normally is in the building the following shall apply:

(1) The maximum water capacity of individual containers shall be 50 lb (23 kg) [nominal 20 lb (9 kg) LP-Gas capacity] and the number of containers in the building shall not exceed the number of workers assigned to using the LP-Gas.

(2) Containers having a water capacity greater than 2 ½ lb (1 kg) [nominal 1 lb (0.45 kg) LP-Gas capacity] shall not be left unattended.

(b) During the hours of the day when the building is not open to the public, containers may be used and transported in the building for repair or minor renovation in accordance with 3-4.2 and 3-4.3, provided, however, that containers with a greater water capacity than 2 ½ lb (1 kg) [nominal 1 lb (0.45 kg) LP-Gas capacity] shall not be left unattended.

3-4.5 Buildings Housing Industrial Occupancies.

Formal Interpretation 79-3
Reference: 3-4.5

Question: An LP-Gas fired infrared space heater and an LP-Gas cylinder are located and used inside of a foundry. They are connected, through a regulator, by means of a hose. The pressure in the hose is less than 1 psi.

What provisions in NFPA 58 characterize the hose that should be used?

Answer: This application is covered under 3-4.5, "Buildings Housing Industrial Occupancies," of NFPA 58. It is, therefore, also subject

206   LIQUEFIED PETROLEUM GASES HANDBOOK

to the provisions of 3-4.1 and 3-4.2.3 of Section 3-4, which, through references, characterize the type of hose to be used as follows:

Paragraph 3-4.2.3 provides: "Piping, including pipe, tubing, fittings, valves and hose, shall comply with Section 2-4, except that a minimum working pressure of 250 psig shall apply to all components. The following shall also apply:

(b)   Hose, hose connections and flexible connectors used shall be designed for a working pressure of at least 350 psig (2.4 MPa gauge), shall comply with 2-4.6 . . ."

Paragraph 2-4.6.1 requires that the hose be fabricated of materials resistant to the action of LP-Gas both as liquid and vapor and, if wire braid is used for reinforcement, it shall be corrosive resistant material such as stainless steel.

Paragraph 2-4.6.2 provides: "Hose and quick connectors shall be approved (*see Section 1-7, Approved*)."

Paragraph 2-4.6.3(a) reiterates the requirement in 3-4.2.3(b) for a 350 psi (2.4 MPa) working pressure regardless of the actual pressure, and stipulates hose marking and other pressure criteria applicable to the assembly of hose and hose connections.

Issue Edition: 1979
Reference: 334
Date: August 1982

3-4.5.1   Containers may be used in buildings housing industrial occupancies for processing, research, or experimental purposes as follows:

(a) Containers, equipment, and piping used shall comply with 3-4.2.

(b) If containers are manifolded together, the total water capacity of the connected containers shall be not more than 735 lb (333 kg) [nominal 300 lb (136 kg) LP-Gas capacity]. If there is more than one such manifold in a room, it shall be separated from any other by at least 20 ft (6 m).

This provision of a maximum of 300 lb (136 kg) of LP-Gas manifolded together has a long history in NFPA 58. See commentary on 3-4.3.6. If there is more than one manifold, a separation of 20 ft (6 m) is required.

(c) The amount of LP-Gas in containers for research and experimental use in the building shall be limited to the smallest practical quantity.

3-4.5.2 Containers may be used to supply fuel for temporary heating in buildings housing industrial occupancies with essentially noncombustible contents, if portable equipment for space heating is essential and a permanent heating installation is not practicable, provided containers and heaters comply with and are used in accordance with 3-4.3.

3-4.6 Buildings Housing Educational and Institutional Occupancies.

3-4.6.1 Containers may be used in buildings housing educational and institutional laboratory occupancies for research and experimental purposes, but not in classrooms, as follows:

(a) The maximum water capacity of individual containers used shall be:

(1)   50 lb (23 kg) [nominal 20 lb (9 kg) LP-Gas capacity] if used in educational occupancies.

(2)   12 lb (5.4 kg) [nominal 5 lb (2 kg) LP-Gas capacity] if used in institutional occupancies.

(b) If more than one such container is located in the same room, the containers shall be separated by at least 20 ft (6 m).

(c) Containers not connected for use shall be stored in accordance with Chapter 5, except that they shall not be stored in a laboratory room.

In these occupancies, the maximum size container is 20 lb (9 kg) of LP-Gas for educational buildings and 12 lb (5.4 kg) LP-Gas for institutional occupancies, with a separation of 20 ft (6 m) if more than one container is located in the same room. Containers are not to be used in classrooms (there have been some portable demonstration cabinets proposed, but these are not recognized in NFPA 58). Storage must be in accordance with Chapter 5.

3-4.7 Temporary Heating in Buildings in Emergencies.

3-4.7.1 Containers may be used in buildings for temporary emergency heating purposes if necessary to prevent damage to the buildings or contents, and if the permanent heating system is temporarily out of service, provided the containers and heaters comply with and are used and transported in accordance with 3-4.2 and 3-4.3, and the temporary heating equipment is not left unattended.

This provision is strictly an emergency measure if the permanent heating system is temporarily out of service. It is not intended to apply to supplemental or zone heating. Also, the emergency heating equipment must be attended at all times. If

someone has to be hired to provide attendance, the problem with the permanent heating system is more likely to be solved promptly.

3-4.8 Use in Buildings for Demonstrations or Training, or in Small Containers.

3-4.8.1 Containers having a maximum water capacity of 12 lb (5.4 kg) [nominal 5 lb (2 kg) LP-Gas capacity] may be used temporarily inside buildings for public exhibitions or demonstrations, including use in classroom demonstrations. If more than one such container is located in the same room, the containers shall be separated by at least 20 ft (6 m).

> For temporary use in buildings for exhibitions or demonstrations, containers up to 5 lb (2 kg) LP-Gas capacity may be used. The use of a 20 lb (9 kg) container (e.g., for a barbeque grill) filled with 5 lb (2 kg) of LP-Gas is not permitted as there is no way to easily verify that only 5 lb (2 kg) of LP-Gas are in the container. This provision (and 3-4.8.3) permits the demonstration of a portable cooking device with a 5 lb (2 kg) LP-Gas container at an indoor trade show, but does not permit an identical device to be used at the show to prepare food for sale.

3-4.8.2 Containers may be used temporarily in buildings for training purposes related to the installation and use of LP-Gas systems, provided:

(a) The maximum water capacity of individual containers shall be 245 lb (111 kg) [nominal 100 lb (45 kg) LP-Gas capacity], but not more than 20 lb (9 kg) of LP-Gas may be placed in a single container.

(b) If more than one such container is located in the same room, the containers shall be separated by at least 20 ft (6 m).

(c) The training location shall be acceptable to the authority having jurisdiction.

(d) Containers shall be promptly removed from the building when the training class has terminated.

> For training in buildings LP-Gas 100 lb (45 kg) containers may be used, but they may only be filled with 20 lbs (9 kg) of LP-Gas and approval of the authority having jurisdiction is required. Note that this differs from paragraph 3-4.8.1 (covering public exhibitions or demonstrations) where only small containers [12 lb (5 kg) maximum] are permitted.

3-4.8.3* Except as stipulated in 3-4.8.3(a) containers having a maximum water capacity of 2 ½ lb (1 kg) [nominal 1 lb (0.45 kg) LP-Gas capacity] may be used in buildings as part of approved self-contained torch assemblies or similar appliances.

This provision relates to the use of 1 lb (0.45 kg) LP-Gas containers; it is important to note that their use with mobile cooking units is prohibited. Many requests are received for LP-Gas fueled mobile cooking units for restaurant use, not only with the 1 lb (0.45 kg) LP-Gas size but also 20 lb (9 kg) cylinders. Subparagraph (a) was added in the 1986 Edition, and modified in the 1989 Edition, to clarify that none of this is permitted.

(a) Containers of any capacity used to supply appliances for residential or commercial food service shall not be used in buildings except as provided in 3-4.8.1.

A-3-4.8.3 The weight will be affected by the specific gravity of the liquefied petroleum (LP) gas. Weights varying from 16.0 (454g) to 16.8 oz (476g) are recognized as being within the range of what is nominal.

The text in Appendix A was added in the 1989 Edition to clarify the term "nominal" which has been misinterpreted frequently.

3-4.9 Portable Containers on Roofs or Exterior Balconies.

Roof installations are generally used for microwave stations and emergency electric generating units. These systems can be installed only on roofs of buildings that are unlikely to sustain major structural failure from fire. Specific conditions for installation are outlined, but it is noteworthy that no container refilling can take place on roofs and that certain conditions are set forth for the movement of replacement cylinders to the roof location.

3-4.9.1 Containers may be permanently installed on roofs of buildings of fire-resistive construction, or noncombustible construction having essentially noncombustible contents, or of other construction or contents which are protected with automatic sprinklers (*see NFPA 220, Standard on Types of Building Construction*) in accordance with 3-4.2 and the following:

(a) The total water capacity of containers connected to any one manifold shall not be greater than 980 lb (445 kg) [nominal 400 lb (181 kg) LP-Gas capacity]. If more than one manifold is located on the roof, it shall be separated from any other by at least 50 ft (15 m).

(b) Containers shall be located in areas where there is free air circulation, at least 10 ft (3 m) from building openings (such as windows and doors) and at least 20 ft (6 m) from air intakes of air conditioning and ventilating systems.

(c) Containers shall not be located on roofs which are entirely enclosed by parapets more than 18 in. (457 mm) high unless either (1) the parapets are breached with low-level ventilation openings no more than 20 ft (6 m) apart, or (2) all openings communicating with the interior of the building are at or above the top of the parapets.

(d) Piping shall be in accordance with 3-4.2.3, provided, however, that hose shall not be used for connecting to containers.

(e) The fire department shall be advised of each such installation.

In the 1989 Edition two subparagraphs were deleted: the former (d) which prohibited container filling on roofs [which was redundant to 3-4.2.6(a)], and (e) which required valve plugs and limited use of stairs and elevators (which was incorporated into 3-4.2.7).

3-4.9.2   Containers having water capacities greater than 2 ½ lb (1 kg) [nominal 1 lb (0.5 kg) LP-Gas capacity] shall not be located on balconies above the first floor attached to a multiple family dwelling of three or more living units located one above the other.

Exception: Not applicable when such balconies are served by outside stairways and when only such stairways are used to transport the container.

This requirement was added in the 1989 Edition of the standard to clearly state that the use of LP-Gas cylinders over 1 lb (0.5 kg) on most balconies is prohibited.

3-4.10 Liquid Piped into Buildings or Structures.

Liquid LP-Gas may be piped into buildings at pressures higher than 20 psig (138 kPa gauge) for certain applications listed in 3-2.6.1(d). The piping system must comply with 3-2.7 and other specific provisions of 3-4.10. These include maximum size of piping, protection against breakage and against exposure to high ambient temperatures, accessible shutoff valves, excess-flow valves, and use of hydrostatic relief valves. Protection against release of fuel when disconnecting is obtained through either an automatic quick closing coupling or shutting off the system and allowing the appliance to burn off the fuel.

3-4.10.1 Liquid LP-Gas piped into buildings in accordance with 3-2.6.1(d)(1) shall comply with 3-2.7.

3-4.10.2 Liquid LP-Gas piped into buildings in accordance with 3-2.6.1(d)(2) from containers located and installed outside the building or structure in accordance with 3-2.2 and 3-2.3 shall comply with the following:

(a) Liquid piping shall not exceed ¾ in. I.P.S. and shall comply with 3-2.6 and 3-2.7. If approved by the authority having jurisdiction, copper tubing complying with 2-4.3.1(c)(1) and with a maximum outside diameter of ¾ in. may be used. Liquid piping in buildings shall be kept to a minimum, and shall be protected against construction hazards by:

(1) Securely fastening it to walls or other surfaces to provide adequate protection against breakage.

(2) Locating it so as to avoid exposure to high ambient temperatures.

(b) A readily accessible shutoff valve shall be located at each intermediate branch line where it leaves the main line. A second shutoff valve shall be located at the appliance end of the branch and upstream of any flexible appliance connector.

(c) Excess-flow valves complying with 2-3.3.3(b) and 2-4.5.3 shall be installed in the container outlet supply line, downstream of each shutoff valve, and at any point in the piping system where the pipe size is reduced. They shall be sized for the reduced size piping.

(d) Hose shall not be used to carry liquid between the container and the building, or at any point in the liquid line except as the appliance connector. Such connectors shall be as short as practicable and shall comply with 2-4.6, 3-2.7.8, and 3-2.7.10.

(e) Hydrostatic relief valves shall be installed in accordance with 3-2.8.

(f) Provision shall be made so that the release of fuel when any section of piping or appliances are disconnected shall be minimized by use of one of the following methods:

(1) An approved automatic quick-closing coupling which shuts off the gas on both sides when uncoupled.

(2) Closing the shutoff valve closest to the point to be disconnected and allowing the appliance or appliances on that line to operate until the fuel in the line is consumed.

For additional information involving the provisions of Section 3-4, refer to the following publications of the National LP-Gas Association:

NPGA 603, *How to Use LP-Gas Safely at Construction Sites;* NPGA 604, *Safe Use of LP-Gas for Heating Tar;* NPGA 605, *Safe Use of LP-Gas in Temporary Space Heating with Portable Containers;* NPGA 606, *Safe Use of LP-Gas with Portable Cylinders for Cutting, Brazing;* NPGA 800, *Recommended Procedures for the Temporary Use of LP-Gas in Places of Public Assembly Indoors.*

3-5 Installation of Appliances.

3-5.1 Application.

3-5.1.1 This section includes installation provisions for LP-Gas appliances fabricated in accordance with Section 2-6.

See commentary on Section 2-6.

3-5.1.2 Installation of appliances on commercial vehicles is covered in Section 3-9.

3-5.1.3 With the approval of the authority having jurisdiction, unattended heaters used for the purpose of animal or poultry production inside structures without enclosing walls need not be equipped with an automatic device designed to shut off the flow of gas to main burners and pilot, if used, in the event of flame extinguishment or combustion failure.

This exception to 2-6.3.2 acknowledges the difficulty in getting an accumulation of LP-Gas in a structure having no walls. This provision is of considerable vintage and it is doubtful if modern cost/benefit factors would result in the approval of the authority having jurisdiction.

3-5.2 Reference Standards.

3-5.2.1 LP-Gas appliances shall be installed in accordance with this standard and other national standards which may apply. These include:

This material is included for the same reasons given in the commentary on 3-1.1.4.

(a) NFPA 37, *Standard for the Installation and Use of Stationary Combustion Engines and Gas Turbines.*

(b) NFPA 54, *National Fuel Gas Code* (ANSI Z223.1).

(c) NFPA 61B, *Standard for the Prevention of Fires and Explosions in Grain Elevators and Facilities Handling Bulk Raw Agricultural Commodities.*

(d) NFPA 82, *Standard on Incinerators, Waste and Linen Handling Systems and Equipment.*

(e) NFPA 86, *Standard for Ovens and Furnaces.*

(f) NFPA 96, *Standard for the Installation of Equipment for the Removal of Smoke and Grease-Laden Vapors from Commercial Cooking Equipment.*

(g) NFPA 302, *Fire Protection Standard for Pleasure and Commercial Motor Craft.*

(h) NFPA 501A, *Standard for Firesafety Criteria for Mobile Home Installations, Sites, and Communities.*

(i) NFPA 501C, *Standard on Firesafety Criteria for Recreational Vehicles* (ANSI A119.2).

3-6 Engine Fuel Systems.

LP-Gas has been used as an engine fuel for stationary engines (ranch water pumps, pipeline compressors, etc.) since the 1920s and for vehicle propulsion in farm tractors, buses, etc. since shortly thereafter. It was the fuel used by the *Graf Zeppelin* in the world voyage of 1928. Standards for LP-Gas engine fuel installations first appeared in the 1937 Edition of NFPA 58 in a separate chapter.

All provisions relating to LP-Gas engine installations of any sort are now located in Section 3-6 of NFPA 58. There is little need to look elsewhere in the standard for information on this subject. This treatment of a specific application of LP-Gas is the only exception to the basic format of the standard instituted in the 1972 Edition. Coverage of specific applications can make the standard as a whole hard to understand and can lead to inconsistencies which are difficult to explain. However, engine fuel installations are a special field. It became readily apparent in the early 1980s that a different approach was needed, since it was confusing to search through the entire standard to find the requirements for a proper engine fuel installation.

The consolidation of Section 3-6 was made in the 1983 Edition. In addition, changes were made to remove inconsistencies, clarify the provisions, and provide for newer methods and equipment. Many of the changes reflected a study of worldwide regulations, practices, trends, and experience. To allow for flexibility and innovation, the provisions are generally in terms of performance rather than detailed product specifications.

Section 3-6 applies to fuel systems on vehicles for any purpose, stationary and portable engines, and the garaging of vehicles having LP-Gas fueled engines. Also, everyone making an installation, repairing, servicing, or refueling is required to be properly trained in these procedures.

3-6.1 Application.

3-6.1.1 This section applies to fuel systems using LP-Gas as a fuel for internal combustion engines. Included are provisions for containers, container appurtenances, carburetion equipment, piping, hose and fittings, and provisions for their installation. This section covers engine fuel systems for engines installed on vehicles for any purpose, as well as fuel systems for stationary and portable engines. It also includes provisions for garaging of vehicles upon which such systems are installed.

See Section 3-9 for systems on vehicles for purposes other than for engine fuel.

3-6.1.2 Containers supplying fuel to stationary engines, or to portable engines used in lieu of stationary engines, shall be installed in accordance with Section 3-2 (*see Section 3-4 for portable engines used in buildings, roofs, or exterior balconies under certain conditions*).

3-6.1.3 Containers supplying fuel to engines on vehicles, regardless of whether the engine is used to propel the vehicle or is mounted on it for other purposes, shall be constructed and installed in accordance with this section.

3-6.1.4 In the interest of safety, each person engaged in installing, repairing, filling, or otherwise servicing an LP-Gas engine fuel system shall be properly trained in the necessary procedures.

3-6.2 General Purpose Vehicle Engines Fueled by LP-Gas.

General purpose vehicles and industrial trucks are the two categories of vehicles using LP-Gas engines for their propulsion. General purpose vehicles covered in this section include practically any non-industrial truck vehicles using internal combustion engines. Most general purpose vehicles are used in fleet operations serviced from central sources of supply not available to the public. LP-Gas is not readily available in large enough quantities to accommodate masses of individual private vehicle owners. All over-the-road general purpose vehicles powered by LP-Gas have to be identified that they are so powered. (*See 3-6.2.10.*)

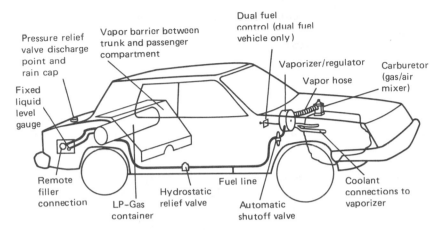

Figure 3.14 Typical LP-Gas (Propane) Fuel System on a Passenger Car.

3-6.2.1 This section covers the installation of fuel systems supplying engines used to propel vehicles such as passenger cars, taxicabs, multipurpose passenger vehicles, buses, recreational vehicles, vans, trucks (including tractors, tractor semi-trailer units, and truck trains), and farm tractors.

3-6.2.2 Containers.

The basic provisions for LP-Gas containers in Chapter 2 are repeated here for general purpose vehicle engine fuel containers. There are several special considerations for those constructed to the ASME Code to recognize environmental factors unique to vehicles.

(a) *Containers designed, fabricated, tested, and marked (or stamped) in accordance with the regulations of the U.S. Department of Transportation (DOT); or the "Rules for Construction of Unfired Pressure Vessels," Section VIII, Division I, ASME *Boiler and Pressure Vessel Code*, applicable at the date of manufacture shall be used as follows:

A-3-6.2.2(a) Prior to April 1, 1967, these regulations were promulgated by the Interstate Commerce Commission. In Canada, the regulations of the Canadian Transport Commission apply. Available from the Canadian Transport Commission, Union Station, Ottawa, Canada.

(1) Adherence to applicable ASME Code Case Interpretations and Addenda shall be considered as compliance with the ASME Code.

(2) Containers fabricated to earlier editions of regulations, rules or codes may be continued in use in accordance with 1-2.4.1. (*See Appendices C and D.*)

(3)   Containers which have been involved in a fire and showing no distortion shall be requalified for continued service in accordance with the Code under which they were constructed before being reused.

(4)   DOT containers shall be designed and constructed for at least 240 psig (1.6 MPa gauge) service pressure.

(5)   DOT specification containers shall be requalified in accordance with DOT regulations. The owner of the container shall be responsible for such requalification. (*See Appendix C.*)

(6)   ASME containers covered in this section shall be constructed for a minimum 250 psig (1.7 MPa gauge) design pressure except that containers installed in enclosed spaces on vehicles and all engine fuel containers for industrial trucks, buses (including school buses), and multipurpose passenger vehicles shall be constructed for at least a 312.5 psig (2.1 MPa gauge) design pressure.

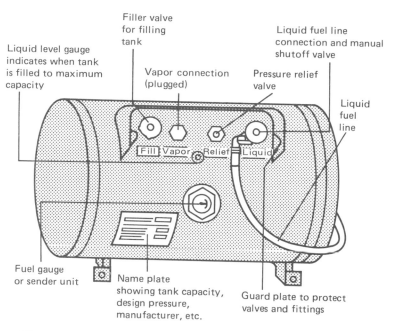

Filler valve
for filling
tank

Liquid fuel line
connection and manual
shutoff valve

Liquid level gauge
indicates when tank
is filled to maximum
capacity

Vapor connection
(plugged)

Pressure relief
valve

Liquid
fuel
line

Fill Vapor Relief Liquid

Fuel gauge
or sender unit

Name plate
showing tank capacity,
design pressure,
manufacturer, etc.

Guard plate to protect
valves and fittings

Figure 3.15 Typical Engine Fuel Container and Appurtenances.

ASME Code containers installed in enclosed spaces (i.e., trunks), buses (including school buses), multipurpose vehicles (10 persons or fewer), and industrial trucks must have a minimum 312.5 psig (2.1 MPa gauge) design pressure. All other containers must have a minimum 250 psig (1.7 MPa gauge) design pressure. Specifying a 312.5 psig (2.1 MPa gauge) design pressure is done primarily to provide a higher pressure relief valve setting for these particular containers. Containers in this type of service may be subjected to higher temperatures or be on vehicles containing large numbers of persons (especially children) and, by having a higher setting, unnecessary premature discharge of the relief valve is minimized.

(7) Repair or alterations of containers shall comply with the Regulations, Rules or Code under which the container was fabricated. Field welding on containers shall be limited to attachments to nonpressure parts, such as saddle pads, wear plates, lugs, or brackets applied by the container manufacturer.

(8) Containers showing serious denting, bulging, gouging, or excessive corrosion shall be removed from service.

(b) Containers shall comply with 3-6.2.2(a) or shall be designed, fabricated, tested, and marked using criteria which incorporate an investigation to determine that they are safe and suitable for the proposed service, are recommended for that service by the manufacturer, and are acceptable to the authority having jurisdiction.

(c) ASME containers shall be marked in accordance with 3-6.2.2(c)(1) through (12). The markings specified shall be on a stainless steel metal nameplate attached to the container so located as to remain visible after the container is installed. The nameplate shall be attached in such a way to minimize corrosion of the nameplate or its fastening means and not contribute to corrosion of the container.

The only difference in the marking on nameplates for ASME engine fuel containers is the obvious deletion of the reference to underground service. The nameplate markings must be visible after the container is installed. In some instances, a lamp and mirror may have to be used.

(1) Service for which the container is designed; i.e., aboveground.

(2) Name and address of container manufacturer or trade name of container.

(3) Water capacity of container in lb or U.S. Gallons.

(4) Design pressure in psig.

(5) The wording "This container shall not contain a product having a vapor pressure in excess of 215 psig at 100°F (37.8°C)."

(6) Tare weight of container fitted for service for containers to be filled by weight.

(7)   Outside surface area in sq ft.

(8)   Year of manufacture.

(9)   Shell thickness ____ head thickness ____.

(10)   OL ____ OD ____ HD ____.

(11)   Manufacturer's Serial Number.

(12)   ASME Code Symbol.

(d) LP-Gas fuel containers used on passenger carrying vehicles shall not exceed 200 gal (0.8 m³) aggregate water capacity.

(e) Individual LP-Gas containers used on other than passenger carrying vehicles normally operating on the highway shall not exceed 300 gal (1 m³) water capacity.

Total fuel capacity for passenger carrying vehicles is limited to 200 gal (0.8 m³) water capacity [approximately 160 gal (0.6 m³) of LP-Gas]. More than one tank may be used and manifolded together, but the total volume cannot exceed this limit. For nonpassenger vehicles the maximum size of individual containers that may be used is 300 gal (1.1 m³) water capacity [approximately 240 gal (0.9 m³) of LP-Gas].

(f) Containers covered in this section shall be equipped for filling into the vapor space.

Exception: Containers having a water capacity of 30 gal (0.1 m³) or less may be filled into the liquid space.

This exception, added in the 1989 Edition, brings the filling requirements for motor fuel containers into agreement with paragraph 2-2.3.2. Engine fuel containers smaller than 30 gal (0.1 m³) had previously not been allowed to be filled in the liquid space to insure equilibrium between liquid and vapor in the container, with corresponding pressure reduction on filling (*see commentary following 2-2.3.2*). This was changed because of problems experienced in the field with some automatic stop-fill valves [required by paragraph 3-6.2.3(a)(8)]. The most reliable stop-fill valves operate by filling into the liquid space.

(1)   The connections for pressure relief valves shall be located and installed in such a way as to have direct communication with the vapor space of the container and shall not reduce the relieving capacity of the relief device.

(2) If the connection is located in any position other than the uppermost point of the container, it shall be internally piped to the uppermost point practical in the vapor space of the container.

This requirement was modified in the 1989 Edition. See commentary following 2-2.3.5.

(g) The container openings, except those for pressure relief valves and gauging devices, shall be labeled to designate whether they communicate with the vapor or liquid space. Labels may be on valves.

3-6.2.3 Container Appurtenances.

Basic provisions for container appurtenances in Chapter 2 are also repeated here. Additional special considerations recognize factors unique to these applications.

(a) Container appurtenances (such as valves and fittings) shall comply with Section 2-3 and 3-6.2.3(a). Container appurtenances subject to working pressures in excess of 125 psig (0.9 MPa gauge) but not to exceed 250 psig (1.7 MPa gauge) shall be suitable for a working pressure of at least 250 psig (1.7 MPa gauge).

(1) Manual shutoff valves shall be designed to provide positive closure under service conditions and be equipped with an internal excess-flow check valve designed to close automatically at the rated flows of vapor or liquid specified by the manufacturers.

In the event the service manual valve is broken off at the container, the excess-flow check valve is designed to close and keep the gas from flowing freely. This excess-flow check valve is not intended to close in the event of partial breakage of the service manual valve or leakage in the system downstream of this valve.

(2) Double backflow check valves shall be of the spring loaded type and shall close when flow is either stopped or reversed. This valve shall be installed in the fill opening of the container for either remote or direct filling.

(3) Containers shall be fabricated so they can be equipped with a fixed liquid level gauge capable of indicating the maximum permitted filling level in accordance with 4-5.2.3. Fixed liquid level gauges in the container shall be designed so the bleeder valve maximum opening to the atmosphere is not larger than a No. 54 drill size. If the bleeder valve is installed at a remote location away from the container, the container fixed liquid level gauge opening and the remote bleeder valve shall be orificed to a No. 54 drill size.

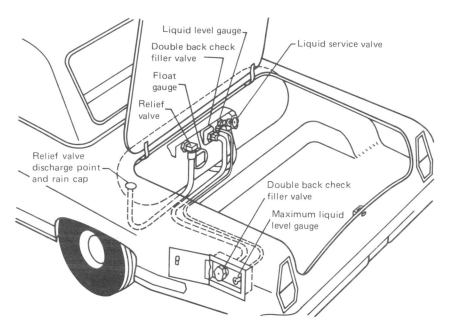

Figure 3.16 Container Shown in Figure 3.15 Installed and Arranged for Remote Filling and for External Pressure Relief Valve Discharge.

(4)   ASME containers shall be equipped with internal type spring loaded pressure relief valves conforming with applicable requirements of UL 132, *Safety Relief Valves for Anhydrous Ammonia and LP-Gas*, or other equivalent pressure relief valve standards. The start-to-leak setting of such pressure relief valve, with relation to the design pressure of the container, shall be in accordance with Table 2-3.2.3. These relief valves shall be plainly and permanently marked with (1) the pressure in psig (MPa gauge) at which the valve is set to start to leak; (2) the rated relieving capacity in cu ft per minute of air at 60°F (15.6°C) and 14.7 psia (0.1 MPa absolute); and (3) the manufacturer's name and catalog number. Fusible plugs shall not be used.

(5)   DOT containers shall be equipped with internal pressure relief valves in accordance with DOT regulations (*see Appendix E for additional information*). Fusible plugs shall not be used.

Pressure relief valves on ASME and DOT containers must be of the spring loaded internal type as the working elements must remain intact within the container so they can still function in the event of an accident. A shear section is generally employed. The prohibition of fusible plugs acknowledges that containers on these vehicles can be subjected to considerable heat in use and garaging.

(6)   A float gauge if used shall be designed and approved for use with LP-Gas.

(7)   A solid steel plug shall be installed in unused openings.

(8)   Containers fabricated after January 1, 1984, for use as engine fuel containers on vehicles shall be equipped or fitted with an automatic means to prevent filling in excess of the maximum permitted filling density.

The effective date was changed in the 1989 Edition from June 30, 1983 to January 1, 1984 in recognition of the fact that ASME containers are marked with the year of fabrication only.

a. An over-filling prevention device may be installed on the container or exterior of the compartment when remote filling is used, provided that a double back check valve is installed in the container fill valve opening.

Over-filled containers potentially can be hazardous through expansion of liquid LP-Gas to the point where the container becomes liquid full, causing the pressure relief valve to open and discharge LP-Gas. To minimize this hazard on general purpose vehicles, containers manufactured after January 1, 1984, must be fitted or equipped with an automatic means to prevent overfilling. Automatic "stop-fill" valves utilizing a float are listed for this purpose. There are other means available to accomplish this objective, e.g., externally operated solenoid valves.

3-6.2.4 Carburetion Equipment.

Carburetion equipment is specially designed and tested for this application. An important feature of LP-Gas carburetion systems is the automatic shutoff valve. This is either an approved electric solenoid valve controlled by vacuum or oil pressure, or a vacuum lockoff which will not permit fuel flow even if the ignition is in the "on" position (it must not provide fuel when the engine is not running). This valve is located as close to the regulator as possible to minimize the volume involved. A primer valve is used for starting.

(a) Carburetion equipment shall comply with 3-6.2.4(b) through (e) or shall be designed, fabricated, tested, and marked using criteria which incorporate an investigation to determine that they are safe and suitable for the proposed service, are recommended for that service by the manufacturer, and are acceptable to the authority having jurisdiction. Carburetion equipment subject to working pressures in excess of 125 psig (0.9 MPa

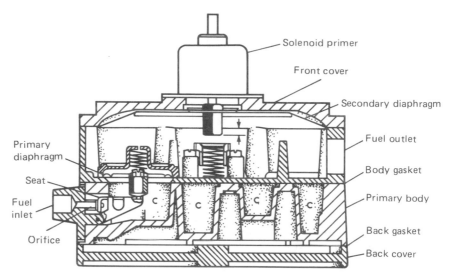

Figure 3.17 Vaporizer-Regulator. Engine coolant circulating through chambers (C) vaporizes liquid LP-Gas.

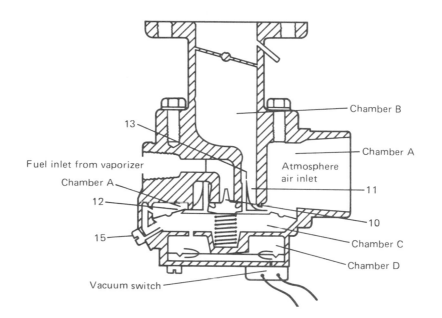

Figure 3.18 Updraft Type Gas/Air Mixer (Carburetor). The fuel metering valve (10) is connected to the air valve (11) and diaphragm assembly (12). Fuel enters the mixer from the vaporizer under slight pressure. It passes through the fuel metering valve (12), which also acts as the secondary regulator valve, and is reduced to subatmospheric pressure. Opening of the air valve is determined by engine vacuum in chamber B, which is sensed through passage (13). The power adjusting valve is (15). This unit also has a vacuum switch which serves as a control for an electric automatic shutoff valve [3-6.2.4(d)].

gauge) but not to exceed 250 psig (1.7 MPa gauge) shall be suitable for a working pressure of at least 250 psig (1.7 MPa gauge).

(b) *Vaporizer.*

(1) Vaporizers shall be fabricated of materials suitable for LP-Gas service and resistant to the action of LP-Gas under service conditions. Such vaporizers shall be designed and approved for engine fuel service and shall comply with the following:

a. The vaporizer proper, any part of it or any devices used with it which may be subjected to container pressure, shall have a design pressure of at least 250 psig (1.7 MPa gauge), where working pressures do not exceed 250 psig (1.7 MPa gauge), and shall be plainly and permanently marked at a readily visible point with a design pressure of the fuel containing portion in psig (MPa gauge).

(2) The vaporizer shall not be equipped with a fusible plug.

(3) Each vaporizer shall have a valve or suitable plug located at or near the lowest portion of the section occupied by the water or other heating liquid to permit substantially complete drainage. The engine cooling system drain or water hoses may serve this purpose, if effective.

(4) Engine exhaust gases may be used as a direct source of heat to vaporize the fuel if the materials of construction of those parts of the vaporizer in contact with the exhaust gases are resistant to corrosion from these gases and if the vaporizer system is designed to prevent pressure in excess of 200 psig (1.4 MPa gauge).

(5) Devices which supply heat directly to the fuel container shall be equipped with an automatic device to cut off the supply of heat before the pressure in the container reaches 200 psig (1.4 MPa gauge).

(c) *Regulator.* The regulator shall be approved and can either be part of the vaporizer unit or a separate unit.

(d) *Automatic Shutoff Valve.* An approved automatic shutoff valve shall be provided in the fuel system as close as practical to the inlet of the gas regulator. The valve shall prevent flow of fuel to the carburetor when the engine is not running even if the ignition switch is in the "on" position. Atmospheric type regulators (zero governors) shall not be considered as automatic shutoff valves for this purpose.

(e) Fuel Filter.   Fuel filters if used shall be approved and can be either a separate unit or part of a combination unit.

3-6.2.5 Piping, Hose, and Fittings.

Piping and hose requirements are identical to those specified in Chapter 2 with one exception with respect to hose. Subparagraph 3-6.2.5(d)(1) specifies that hose used for vapor or liquid in excess of 5 psig (34.5 kPa gauge) must meet certain requirements as specified in Chapter 2 and repeated here. This type of hose is tested and listed and used for remote filling and for providing fuel from the container to the carburetion equipment. Hose used for vapor service at 5 psig (34.5 kPa gauge) or less need be constructed only of materials resistant to the action of LP-Gas. This low pressure hose is used between the regulator or vaporizer-regulator, sometimes called "converter," and the carburetor, sometimes called "gas-air mixer." It is generally either under a vacuum or slightly positive pressure and there is no need to have hose in this service of the quality required for higher pressures.

Note that in the 1989 Edition soldering was deleted as a method of joining pipe and tubing. [*See commentary following 2-4.4.1(a)(4).*]

(a) Pipe.

(1)   Pipe shall be wrought iron or steel (black or galvanized), brass, or copper and shall comply with the following:

a. Wrought iron pipe; ANSI B36.10, *Wrought-Steel and Wrought Steel Pipe.*

b. Steel pipe; ANSI B125.1, *Specification for Pipe, Steel, Black and Hot-Dipped, Zinc-Coated Welded and Seamless Steel Pipe* (ASTM A 53).

c. Steel pipe; ANSI B125.30, *Specification for Seamless Carbon Steel Pipe for High-Temperature Service* (ASTM A 106).

d. Steel pipe; ANSI B125.2, *Specification for Pipe, Steel, Black and Hot-Dipped Zinc-Coated (Galvanized) Welded and Seamless, for Ordinary Uses* (ASTM A 120).

e. Brass pipe; ANSI H27.1, *Specification for Seamless Red Brass Pipe*, Standard Sizes (ASTM B 43).

f. Copper pipe; ANSI H26.1, *Specification for Seamless Copper Pipe, Standard Sizes* (ASTM B 42).

(2)   For LP-Gas vapor in excess of 125 psig (0.9 MPa gauge) or for LP-Gas liquid, the pipe shall be Schedule 80 or heavier. For LP-Gas vapor at pressures of 125 psig (0.9 MPa gauge) or less, the pipe shall be Schedule 40 or heavier.

(b) Tubing.

(1)   Tubing shall be steel, brass or copper and shall comply with the following:

a. Steel tubing; ASTM A 539, *Specification for Electric-Resistance-Welded Coiled Steel Tubing for Gas Fuel Oil Lines*, with a minimum wall thickness of 0.049 in.

b. Copper tubing; Type K or L, ANSI H23.1, *Specification for Seamless Copper Water Tube* (ASTM B 88).

c. Copper tubing; ANSI H23.5, *Specification for Seamless Copper Tube for Air Conditioning and Refrigeration Field Service* (ASTM B 280).

d. Brass tubing; ANSI H36.1, *Specification for Seamless Brass Tube* (ASTM B 135).

(c) Pipe and Tube Fittings.

(1)   Cast iron pipe fittings such as ells, tees, crosses, couplings, unions, flanges or plugs shall not be used. Fittings shall be steel, brass, copper, malleable iron or ductile iron and shall comply with the following:

a. Pipe joints in wrought iron, steel, brass, or copper pipe may be screwed, welded, or brazed. Tubing joints in steel, brass, or copper tubing shall be flared, brazed, or made up with approved gas tubing fittings.

See commentary following 2-4.4.1(a)(4).

(i)   Fittings used with liquid LP-Gas, or with vapor LP-Gas at operating pressures over 125 psig (0.9 MPa gauge), where working pressures do not exceed 250 psig (1.7 MPa gauge), shall be suitable for a working pressure of at least 250 psig (1.7 MPa gauge).

(ii)   Fittings for use with vapor LP-Gas at pressures in excess of 5 psig (34.5 kPa gauge) and not exceeding 125 psig (0.9 MPa gauge) shall be suitable for a working pressure of 125 psig (0.9 MPa gauge).

(iii)   Brazing filler material shall have a melting point exceeding 1,000°F (538°C).

(d) Hose, Hose Connections, and Flexible Connectors.

(1)   Hose, hose connections, and flexible connectors (*see definition*) used for conveying LP-Gas liquid or vapor at pressures in excess of 5 psig (34.5 kPa gauge) shall be fabricated of materials resistant to the action of LP-Gas both as liquid and vapor, and be of wire braid reinforced construction. The wire braid shall be stainless steel. The hose shall comply with the following:

a. Hose shall be designed for a working pressure of 350 psi (240 MPa) with a safety factor of 5 to 1 and be continuously marked "LP-GAS," "PROPANE," "350 PSI WORKING PRESSURE" and the manufacturer's name or trademark. Each installed piece of hose shall contain at least one such marking.

b. Hose assemblies after the application of connections shall have a design capability of withstanding a pressure of not less than 700 psig (4.8 MPa gauge). If a test is made, such assemblies shall not be leak tested at pressures higher than the working pressure [350 psig (2.4 MPa gauge) minimum] of the hose.

(2)   Hose used for vapor service at 5 psig (34.5 kPa gauge) or less shall be constructed of material resistant to the action of LP-Gas.

(3)   Hose in excess of 5 psig (34.5 kPa gauge) service pressure and quick connectors shall have the approval for this application of any of the authorities listed in 1-3.1.1.

3-6.2.6 Installation of Containers and Container Appurtenances.

Particular attention is given in this subsection to the protection of the container and its fittings from damage through collision, overturn, running over objects (such as curbs), or brushing against objects. Protection is also considered for road material thrown up from the ground and exposure to heat from the engine or exhaust. There are limitations as to where the fuel container may be located. It cannot be ahead of the front axle or beyond the rear bumper. Also, it cannot be mounted directly on vehicle roofs; the container or its fittings are not to protrude beyond the widest or highest point of the vehicle.

Specific provisions are given for clearances between the road and the container and its fittings. Figure 3-6.2.6(e) depicts shaded areas beneath the vehicle where the container must be installed. This procedure is primarily directed to those installations where a vehicle is being converted to LP-Gas from another fuel or to operate as a dual fuel vehicle. Original manufacturers can deviate from this procedure as they have the facilities for individual design and road testing to verify their design. Installers may install a substitute container within the space used by the original manufacturer.

(a) Containers shall be located in a place and in a manner to minimize the possibility of damage to the container and its fittings. Containers located in the rear of the vehicles, when protected by substantial bumpers, shall be considered in conformance with this requirement. In case the fuel container must be installed near the engine or exhaust system, it shall be shielded against direct heating.

(b) Container markings shall be readable after a container is permanently installed on a vehicle. A portable lamp and mirror may be used when reading markings.

(c) Container valves, appurtenances, and connections shall be adequately protected to prevent damage due to accidental contacts with stationary objects or from stones, mud, or ice thrown up from the ground, and from damage due to overturn or similar vehicular accident. Location on the container where parts of the vehicle furnish the necessary protection or a fitting guard furnished by the manufacturer of the container may meet these requirements.

(d) Containers shall not be mounted directly on roofs or ahead of the front axle or beyond the rear bumper of the vehicles. So as to minimize the possibility of physical damage, no part of a container or its appurtenances shall protrude beyond the sides or top of the vehicle.

This provision was changed in the 1989 Edition by the deletion of the words: "at the point where it is installed" at the end of the paragraph. The revised wording will permit a side mounted container as long as it does not protrude beyond the widest point of the vehicle.

(e) Containers shall be installed with as much road clearance as practicable. This clearance shall be measured to the bottom of the container or the lowest fitting, support, or attachment on the container or its housing, if any, whichever is lowest, as follows [*see Figure 3-6.2.6(e)*]:

(1) Containers installed between axles shall comply with 3-6.2.6(e)(3) or be not lower than the lowest point forward of the container on:

a. the lowest structural component of the body;

b. the lowest structural component of the frame or subframe if any;

c. the lowest point on the engine;

d. the lowest point of the transmission (including the clutch housing or torque converter housing as applicable) [Part 1, Figure 3-6.2.6(e)].

(2) Containers installed behind the rear axle and extending below the frame shall comply with 3-6.2.6(e)(3) or be not lower than the lowest of the following points and surfaces.

a. Not lower than the lowest point of a structural component of the body, engine, transmission (including clutch housing or torque convert-

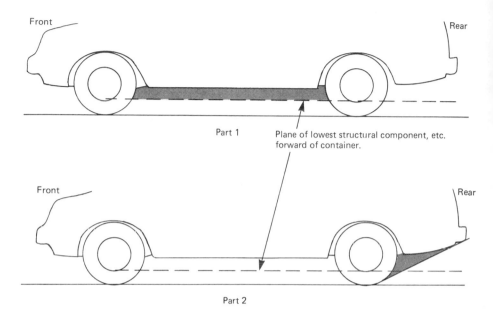

Figure 3-6.2.6(e) Container Installation Clearances.

er housing, as applicable), forward of the container. Also not lower than lines extending rearward from each wheel at the point where the wheels contact the ground directly below the center of the axle to the lowest and most rearward structural interference (i.e. bumper, frame, etc.). [Part 2, Figure 3-6.2.6(e).]

b. Where there are two or more rear axles the projections shall be made from the rearmost one of them.

(3)   Where an LP-Gas container is substituted for the fuel container installed by the original manufacturer of the vehicle (whether or not that fuel container was for LP-Gas), the LP-Gas container shall either fit within the space in which the original fuel container was installed or comply with 3-6.2.6(e)(1) or (2).

(f) Fuel containers shall be securely mounted to prevent jarring loose and slipping or rotating, and the fastenings shall be designed and constructed to withstand without permanent visible deformation static loading in any direction equal to four times the weight of the container filled with fuel.

Not only should the strength of fasteners for mounting the container be considered, but also the strength of that portion of the vehicle to which the container is mounted. For example, reinforcement must be used when mounting containers to thin metal decking. A preferable method is to mount the container on the chassis frame.

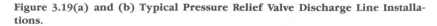

Figure 3.19(a) and (b) Typical Pressure Relief Valve Discharge Line Installations.

Figure 3.19(a) Pressure Relief Valve Discharge Line Extended above the Roof of the Cab within 15 Degrees of Vertical.

(g) Welding for the repair or alterations of containers shall comply with 3-6.2.2(a)(7).

(h) Main shutoff valves on a container for liquid and vapor shall be readily accessible without the use of tools, or other means shall be provided to shut off the container valves.

(i) Pressure relief valve installations shall comply with the following requirements:

Pressure relief valve discharge location must be considered in the installation of engine fuel containers as they should not be directed toward persons near the vehicle, adjacent vehicles, or upon the vehicle itself. Also, they should be piped to the

Figure 3.19(b) LP-Gas Container Inside an Enclosed Pickup Truck Bed. The pressure relief valve discharge line runs from the container to the roof of the topper.

outside and never be allowed to discharge inside the passenger compartment. Passenger cars must have this discharge directed upward within 45 degrees of the vertical and all other vehicles within 15 degrees of the vertical. Newer passenger cars have low profiles and lack top deck areas for directing them up as for other vehicles. Requirements for this pipeaway are specified to assure it has proper capacity, and that it can withstand fire conditions and be able to break away cleanly from the relief valve in the event a force is applied. Also, to assure that water does not get into the system and freeze, loose fitting raincaps are to be installed.

(1)   The relief valve discharge on fuel containers on vehicles other than passenger cars shall be directed upward within 15 degrees of vertical so that any gas released will not impinge upon containers or part of the vehicle, or on adjacent persons or vehicles or discharge inside of the passenger compartment. On passenger cars, the relief valve discharge on fuel containers shall be directed upward within 45 degrees of vertical so that gas may not be discharged inside of the passenger or luggage compartment and so that any gas released will not impinge upon a container, part of the vehicle or on an adjacent vehicle.

(2)   Pressure relief valve discharge lines shall be metallic and have a melting point over 1500°F (816°C). Discharge lines and adaptors shall be sized, located and secured so as to minimize the possibility of physical

damage and to permit required pressure relief valve discharge capacity. When the relief valve discharge must be piped away from the container, the relief valve shall be fitted with an approved break-away type adaptor or designed such that in the event of excessive stress the piping will break away without impairing the function of the relief valve. Flexible metal hose or tubing used shall be able to withstand the pressure from the relief vapor discharge when the relief valve is in full open position. A means shall be provided (such as loose fitting caps) to minimize the possibility of the entrance of water or dirt into either the relief valve or its discharge piping. The protecting means shall remain in place except when the relief valve operates. In this event, it shall permit the relief valve to operate at required capacity.

(3)   Relief valve adaptors installed directly in the relief valve to deflect the flow upward shall be metallic and have a melting point over 700°F (371°C).

3-6.2.7 Containers Mounted in the Interior of Vehicles.

(a) Containers mounted in the interior of vehicles shall be installed so that any LP-Gas released from container appurtenances due to operation, leakage or connection of the appurtenances will not be in an area communicating directly with the driver or passenger compartment or with any space containing radio transmitters or other spark producing equipment. This may be accomplished by 3-6.2.7(a)(1) or (2).

(1)   Locating the container, including its appurtenances, in an enclosure which is securely mounted to the vehicle, is gastight with respect to driver or passenger compartments and to any space containing radio transmitters or other spark producing equipment, and which is vented outside the vehicle.

a. the luggage compartment (trunk) of a vehicle may constitute such an enclosure provided it meets all these requirements.

(2)   Enclosing the container appurtenances and their connections in a structure which is securely mounted on the container, is gastight with respect to the driver or passenger compartments or with any space carrying radio transmitters or other spark producing equipment, and which is vented to outside the vehicle.

(b) Fuel containers shall be installed and fitted so that no gas from fueling and gauging operations can be released inside of the passenger or luggage compartments, by permanently installing the remote filling connections (double backflow check valve), see 3-6.2.3(a)(2), and fixed liquid level gauging device to the outside of the vehicle.

(c) Container pressure relief valve installation shall comply with 3-6.2.6(i).

Figure 3.20(a), (b) and (c) Containers Mounted in the Interior of Vehicles.

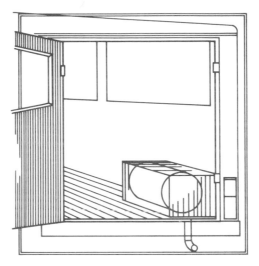

Figure 3.20(a) Gastight Box Built around the Container to Provide a Vapor Barrier in Van-Type Vehicles.

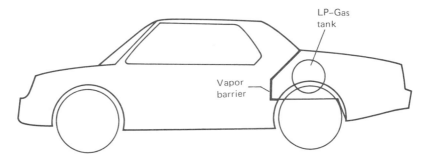

Figure 3.20(b) For Trunk Mounted Fuel Containers, a Vapor Barrier Must Be Created Between the Passenger and Luggage Compartments. The barrier must be 100 percent gastight so no propane vapor can enter the car.

(d) Enclosures, structures, seals and conduits used to vent enclosures shall be fabricated of durable materials and be designed to resist damage, blockage or dislodgement through movement of articles carried in the vehicle or by the closing of luggage compartment enclosures or vehicle doors, and shall require the use of tools for removal.

An alternate to mounting the fuel containers on the outside is the installation of them in the interior of the vehicle (trunk compartment, etc.). Certain conditions are specified to assure that no LP-Gas is released into the passenger compartment. Four options are given for installation of containers mounted in the interior as far as sealing the container appurtenances and

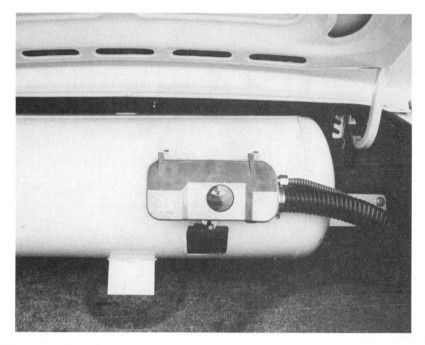

Figure 3.20(c) A Gastight Box Installed over the Valve Fittings on the Tank Itself Can also Provide a Vapor Barrier.

their connections from the passenger space is concerned. Essentially, these are: locating the container and its appurtenances in the luggage compartment (trunk) and sealing the trunk from the passenger carrying space; putting the entire container and its appurtenances in a compartment that is sealed; using a gastight box totally enclosing the container appurtenances; or mounting the container so appurtenances and their connections are outside of the passenger space. Spark producing equipment (i.e., radio transmitters, truck lid motors) are to be isolated from container appurtenances and their connections.

Remote filling is to be accomplished by permanently installing a double backflow check valve on the exterior of the vehicle in addition to one installed in the container. Remote fixed liquid level gauges are also to be permanently mounted and orificed to a No. 54 drill size in addition to the orificed connection on the container. This clearly points out that coiled filling hoses in trunks are not permitted.

3-6.2.8 Pipe and Hose Installation.

(a) The piping system shall be designed, installed, supported, and

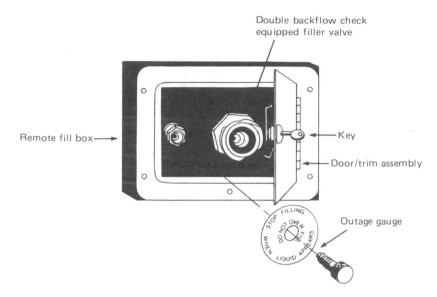

Double backflow check
equipped filler valve

Remote fill box→

Key

Door/trim assembly

Outage gauge

Figure 3.21 Remote Filling Connection with Fixed Liquid Level Gauge.

secured in such a manner as to minimize the possibility of damage due to expansion, contraction, vibration, strains or wear, and to preclude any working loose while in transit.

(b) Piping (including hose) shall be installed in a protected location. If outside, piping shall be under the vehicle and below any insulation or false bottom. Fastening or other protection shall be installed to prevent damage due to vibration or abrasion. At each point where piping passes through sheet metal or a structural member, a rubber grommet or equivalent protection shall be installed to prevent chafing.

(c) Fuel line piping which must pass through the floor of a vehicle shall be installed to enter the vehicle through the floor directly beneath, or adjacent to, the container. If a branch line is required, the tee connection shall be in the main fuel line under the floor and outside the vehicle.

(d) When liquid service lines of two or more individual containers are connected together, a spring loaded backflow check valve or equivalent shall be installed in each of the liquid lines prior to the point where the liquid lines tee together to prevent the transfer of LP-Gas from one container to another.

This requirement, added in the 1989 Edition, is intended to prevent transfer of LP-Gas between containers when more than one are used, including possible overfilling by transfer.

(e) Exposed parts of the piping system shall either be of corrosion-resistant material or adequately protected against exterior corrosion.

(f) Piping systems, including hose, shall be tested and proven free of leaks at not less than normal operating pressure.

(g) There shall be no fuel connection between a tractor and trailer or other vehicle units.

(h) A hydrostatic relief valve shall be installed in each section of piping (including hose) in which liquid LP-Gas can be isolated between shutoff valves so as to relieve to a safe atmosphere the pressure which could develop from the trapped liquid. This hydrostatic relief valve shall have a pressure setting not less than 400 psig (2.8 MPa gauge) or more than 500 psig (3.5 MPa gauge).

3-6.2.9 Equipment Installation.

(a) Installation shall be made in accordance with the manufacturer's recommendations and, in the case of listed or approved equipment, it shall be installed in accordance with the listing or approval.

(b) Equipment installed on vehicles shall be considered a part of the LP-Gas system on the vehicle and shall be protected against vehicular damage in accordance with 3-6.2.6(a).

(c) The gas regulator and the approved automatic shutoff valve shall be installed as follows:

(1)   Approved automatic pressure reducing equipment, properly secured, shall be installed between the fuel supply container and the carburetor to regulate the pressure of the fuel delivered to the carburetor.

(2)   An approved automatic shutoff valve shall be provided in the fuel system in compliance with 3-6.2.4(d).

(d) Vaporizers shall be securely fastened in position.

3-6.2.10 Marking.
Each over-the-road general purpose vehicle powered by LP-Gas shall be identified with a weather-resistant diamond shaped label located on an exterior vertical or near vertical surface on the lower right rear of the vehicle (on the trunk lid of a vehicle so equipped, but not on the bumper of any vehicle) inboard from any other markings. The label shall be approximately 4-¾ in. (120 mm) long by 3-¼ in. (83 mm) high. The marking shall consist of a border and the letters "PROPANE" [1 in. (25 mm) minimum height centered in the diamond] of silver or white reflective luminous material on a black background. (*See Figure 3-6.2.10.*)

This provision (new in the 1986 Edition) comprised TIA 58-83-2 (issued October 26, 1983). It is primarily to aid fire fighters and other emergency personnel in recognizing that the vehicle is not powered by more common gasoline or diesel fuel.

Figure 3-6.2.10 Example of Vehicle Identification Marking.

3-6.3 Industrial (and Forklift) Trucks Powered by LP-Gas.

NFPA 505, *Firesafety Standard for Powered Industrial Trucks Including Type Designations, Areas of Use, Maintenance, and Operation,* is a companion standard to NFPA 58 with respect to LP-Gas fueled industrial trucks. While NFPA 58 sets forth installation provisions and certain conditions for their use, NFPA 505 sets forth type designations, where they are to be used, dual fuel trucks, truck maintenance and, to a certain extent, fuel handling and storage.

The basic provisions for engine fuel systems on general purpose vehicles are also applicable to those for industrial truck engines except for some differences peculiar to this type of service.

3-6.3.1 This subsection applies to LP-Gas installation on industrial trucks (including forklift trucks) both to propel them and to provide the energy for their materials handling attachments. LP-Gas fueled industrial trucks shall comply with NFPA 505, *Firesafety Standard for Powered Industrial Trucks.*

3-6.3.2 ASME and DOT fuel containers shall comply with 3-6.2.2 and 3-6.2.3(a)(1) through (7).

Fuel containers for industrial trucks need not comply with 3-6.2.3(a)(8) (that is, be equipped with an automatic means for over-fill prevention) as required for general purpose vehicles.

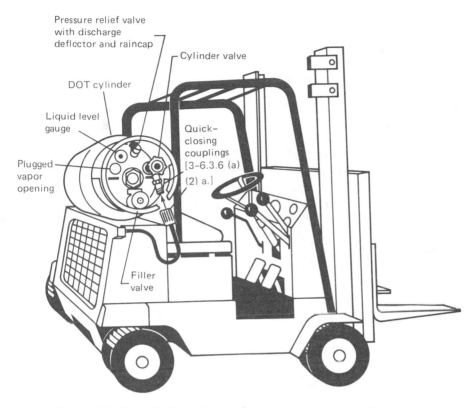

Figure 3.22 Typical LP-Gas Powered Industrial Truck (Forklift Truck).

These containers are refilled under more closely controlled conditions and experience has shown that over-filling has not been as much of a problem.

(a) Portable containers may be designed, constructed, and fitted for filling in either the vertical or horizontal position, or if of the portable universal type [*see 2-3.4.2(c)(2)*], in either position. The container shall be in the appropriate position when filled or, if of the portable universal type, may be loaded in either position, provided:

In addition to portable industrial truck containers designed for refilling and used specifically in a horizontal or a vertical position, a universal type is available where the same container can be filled and used in either position. With this type of container, the fixed liquid level gauge must indicate the maximum filling level in either the correct horizontal or vertical position (for horizontal position, the relief valve must be at 12 o'clock) and, also, the pressure relief valve must communicate with the vapor space in either position.

(1)   The fixed level gauge indicates correctly the maximum permitted filling level in either position.

(2)   The pressure relief valves are located in, or connected to, the vapor space in either position.

3-6.3.3   The container relief valve shall be vented upward within 45 degrees of vertical and otherwise comply with 3-6.2.6(i).

The pressure relief valve must be vented upward within 45 degrees of the vertical as required for passenger cars. Generally, industrial trucks do not require pressure relief valve pipeaways and adapters can be used for deflecting the discharge in view of the configurations of these vehicles.

3-6.3.4   Gas regulating and vaporizing equipment shall comply with 3-6.2.4(b)(1) through (5) and 3-6.2.4(c), (d), and (e).

3-6.3.5   Piping and hose shall comply with 3-6.2.5(a) through (d) except that hose 60 in. (1.5 m) in length or less need not be of stainless steel wire braid construction.

3-6.3.6   Industrial trucks (including forklift trucks) powered by LP-Gas engine fuel systems shall comply as to operation with NFPA 505, *Firesafety Standard for Powered Industrial Trucks*, and with the following:

(a) Refueling of such trucks shall be accomplished as follows:

(1)   Trucks with permanently mounted containers shall be refueled out-of-doors.

(2)   Exchange of removable fuel containers preferably should be done out-of-doors, but may be done indoors. If done indoors, means shall be provided in the fuel piping system to minimize the release of fuel when containers are exchanged, using one of the following methods:

a. Use of an approved quick-closing coupling (a type closing in both directions when uncoupled) in the fuel line, or

b. Closing the shutoff valve at the fuel container, and allowing the engine to run until the fuel in the line is exhausted.

(b) LP-Gas fueled industrial trucks may be used in buildings or structures as follows:

(1)   The number of fuel containers on such a truck shall not exceed two.

(2) With the approval of the authority having jurisdiction, industrial trucks may be used in buildings frequented by the public, including the times when such buildings are occupied by the public. The total water capacity of the fuel containers on an individual truck shall not exceed 105 lb (48 kg) [nominal 45 lb (20 kg) LP-Gas capacity].

(3) Trucks shall not be parked and left unattended in areas occupied by or frequented by the public except with the approval of the authority having jurisdiction. If so left, the fuel system shall be checked to be sure there are no leaks and that the container shutoff valve is closed.

(4) In no case shall industrial trucks be parked and left unattended in areas of excessive heat or near sources of ignition.

3-6.4 General Provisions for Vehicles Having Engines Mounted on Them (including floor maintenance machines).

This section was revised in the 1989 Edition with the addition of floor maintenance machines. This was because a new product — a propane powered floor buffer — was introduced into the market and resulted in confusion among users and enforcers because this new device was not specifically covered in the standard.

3-6.4.1 This subsection includes provisions for the installation of equipment on vehicles to supply LP-Gas as a fuel for engines mounted on these vehicles. The term "vehicles" includes floor maintenance and any other readily portable mobile unit, whether the engine is used to propel it or is mounted on it for other purposes.

3-6.4.2 Gas vaporizing, regulating and carburetion equipment to provide LP-Gas as a fuel for engines shall be installed in accordance with 3-6.2.8 and 3-6.2.9.

(a) In the case of industrial trucks (including forklift trucks) and other engines on vehicles operating in buildings other than those used exclusively to house engines, an approved automatic shutoff valve shall be provided in the fuel system in compliance with 3-6.2.4(d).

(b) The source of air for combustion shall be completely isolated from the driver and passenger compartment, ventilating system or air conditioning system on the vehicle.

While approved automatic shutoff valves are required, this recognizes the use of atmospheric type regulators (zero governors) for this purpose with portable engines of 12 horsepower or less with magneto ignition used exclusively outdoors. An atmospheric type regulator, as its name implies, depends upon the atmospheric pressure for its control. It is not an approved vacuum lockoff which is considered an approved automatic shutoff valve. It does operate on the principle that when there is no vacuum in the carburetor venturi the regulator shuts off the flow of fuel. Atmospheric type regulators have been used for many years on outdoor engine applications.

3-6.4.3   Piping and hose shall comply with 3-6.3.5.

3-6.4.4   Non-self-propelled floor maintenance machinery (floor polishers, scrubbers, buffers) and other similar portable equipment shall be listed and comply with 3-6.4.4(a) and (b).

(a)   The provisions of 3-6.3.2 through 3-6.3.5 and 3-6.3.6(a) and (b) shall apply.

(b)   The storage of LP-Gas containers mounted or used on such machinery or equipment shall comply with Chapter 5.

(1)   A label shall be affixed to the machinery or equipment, with the label facing the operator, denoting that the container or portion of the machinery or equipment containing the LP-Gas container, must be stored in accordance with Chapter 5.

There is great concern that a propane powered floor buffer with its 20 lb (9 kg) LP-Gas container will be stored inside buildings in concealed areas, such as closets. The Technical Committee, consistent with its long term prohibition (with certain exceptions) of LP-Gas containers larger than 1 lb (0.5 kg) in buildings wished to insure that containers are stored properly in accordance with Chapter 5.

The Technical Committee was especially concerned that the introduction of engine fuel containers would lead to improper storage and required that the container be labeled, with the label facing the operator, with proper storage instructions.

There is a similarity between propane powered floor buffers and forklift trucks in that they are both categorized as "industrial trucks." However, it is highly improbable that a forklift truck will be left in a closet. If a floor buffer and its 20 lb (9 kg) LP-Gas container were stored in a closet and a fire occurred, the propane in the cylinder could be a threat to fire fighters due to the potential for a BLEVE and the fuel that can accelerate a fire.

Figure 3.23 Floor Maintenance Machine.

3-6.5 Engine Installation Other than on Vehicle.

3-6.5.1 Stationary engines and gas turbines installed in buildings, including portable engines used in lieu of, or to supplement, stationary engines, shall comply with NFPA 37, *Standard for the Installation and Use of Stationary Combustion Engines and Gas Turbines*, and the applicable provisions of Chapters 1 and 2 and Section 3-2 of this standard.

3-6.5.2 Portable engines, except as provided in 3-6.5.1, may be used in buildings only for emergencies and the following shall apply:

(a) The capacity of the LP-Gas containers used with such engines and the equipment used to provide fuel to them shall comply with the applicable provisions of Section 3-4.

(b) An approved automatic shutoff valve shall be provided in the fuel system in compliance with 3-6.2.4(d). Atmospheric type regulators (zero governors) used for portable engines of 12 horsepower or less with magneto ignition and used exclusively outdoors shall be considered as in compliance with 3-6.2.4(d).

(c) Provision shall be made to supply sufficient air for combustion and cooling. Exhaust gases shall be discharged to a point outside the building, or to an area in which they will not constitute a hazard.

> For additional information involving the provisions of Section 3-6, refer to the following publications of the National Propane Gas Association (formerly National LP-Gas Association):
> NPGA 602, *Safe Use of LP-Gas in Industrial Trucks* (Also Spanish version); NPGA 803, *Safety Considerations in Converting Passenger Carrying Vehicles from Gasoline to LP-Gas*; NPGA 804, *Safety Bulletin for Drivers of LP-Gas Powered Vehicles*; NPGA 5909, *Safety Considerations for Motor Fuel Systems*; NPGA 5915, *Safety Factors in Converting School Buses to LP-Gas*; NPGA 5825, *Preventive Maintenance and Filling Procedures*; NPGA 4006, *Training Guidebook: LP-Gas Carburetion.*

3-6.5.3 Piping and hose shall comply with 3-6.2.5(a) through (d).

3-6.5.4 Gas regulating, vaporizing, and carburetion equipment shall comply with 3-6.2.4(b)(1) through (5), 3-6.2.4(c) and 3-6.2.4(e).

3-6.5.5 Installation of piping, carburetion, vaporizing, and regulating equipment for the engine fuel system shall comply with 3-6.2.8 and 3-6.2.9.

3-6.5.6 Engines installed or operated exclusively outdoors shall comply with 3-6.5.3, 3-6.5.4 and 3-6.5.5.

(a) Atmospheric type regulators (zero governor) shall be considered as automatic shutoff valves only in the case of completely outdoor operations, such as farm tractors, construction equipment or similar outdoor engine applications.

3-6.5.7 Engines used to drive portable compressors shall be equipped with exhaust system spark arrestors and shielded ignition systems.

3-6.6 Garaging of Vehicles.

3-6.6.1 Vehicles with LP-Gas engine fuel systems mounted on them and general purpose vehicles propelled by LP-Gas engines may be stored or serviced inside garages, provided:

(a) The fuel system is leak free and the container(s) is not filled beyond the limits specified in Chapter 4.

(b) The container shutoff valve is closed when vehicles or engines are under repair except when engine is operated.

(c) The vehicle is not parked near sources of heat, open flames, or similar sources of ignition, or near inadequately ventilated pits.

3-7 Vaporizer Installation.

3-7.1 Application.

3-7.1.1 This section applies to the installaton of vaporizing devices covered in 2-5.4. It does not apply to engine fuel vaporizers, or to integral vaporizing-burners such as those used for weed burners or tar kettles.

From an operational and maintenance standpoint, it should be recognized that some vaporizers (those used for standby systems) are used only in the colder time of year. It is imperative that they be given a thorough check and provisional operation before the season in which they will be needed so that they will be in the best condition to handle the vaporizing load. Periodic testing during idle periods is recommended as well as a program to alert the proper individuals to the need for this essential maintenance program. In the summer season, spiders, mud daubers and so forth can get into the burner area, pilot area, and regulator vents creating operating problems. It

is important to verify that the rain cap is always kept on the pressure relief valve outlet and also that such outlet is piped to a proper safe point for discharge.

3-7.2 Installation of Indirect-Fired Vaporizers.

3-7.2.1   Indirect-fired vaporizers shall comply with 2-5.4.2, and shall be installed as provided in 3-7.2.2 through 3-7.2.9.

3-7.2.2   Indirect vaporizers may be installed out-of-doors, in buildings used exclusively for gas manufacturing or distribution, or in separate structures constructed in accordance with Section 7-2. Any such buildings shall be well ventilated near the floor line and roof.

By definition, indirect vaporizers derive heat for operation from a remote source. The term "remote" is not defined. The strictest definition of "remote" would be any device not part of the unit itself. This could be a water or steam boiler mounted immediately adjacent to, or on the same skid or package with, the indirect vaporizer. In that event, 3-7.2.5(b) specifies that such a combination be sited in the same manner as direct-fired vaporizers since a source of ignition is present. This "source of ignition" can have positive or negative effects. For example, a source of ignition close to an indirect vaporizer installed outside has the effect of preventing a large buildup of gas from a leak because a leak would be ignited by the adjacent heat source before it became a more severe problem. With no source of ignition in the immediate area, on a calm day a large amount of gas could escape before finally reaching an ignition source. This could result in an unconfined vapor cloud explosion. On the other hand, if the indirect vaporizer were to be installed inside an enclosure, an adjacent source of ignition would cause a confined explosion which is considerably more devastating. Accordingly, indirect vaporizers are generally used where the source of heat is a plant facility, such as steam or hot water from a plant heating or processing system.

3-7.2.3   Indirect vaporizers may also be installed in structures attached to, or rooms within, buildings not used for gas manufacturing or distribution, provided such attached structures or rooms comply with Section 7-3, and that there are no openings of any sort from the vaporizer room into the building or structure of which it is a part.

To prevent passage of vapor from a vaporizer room into an adjacent room the connecting partition should be caulked to be gastight and contain no openings. When it is necessary to have

Figure 3.24 Indirect-Fired Steam Vaporizer. This vaporizer utilizes steam provided from a central steam supply and the steam flow and pressure is regulated at the vaporizer. Dual condensate traps are provided on the unit to prevent condensate buildup. A pressure relief valve is installed in the vapor space of the unit and the outlet of the relief valve is vented outside the building. A liquid carry-over control is provided to shut off the supply of liquid if the liquid level exceeds the height of the steam tubes. This unit is ASME Code stamped.

a window in such a partition it must be non-openable and should be of shatterproof plastic or wired glass material. Service piping should leave the vaporizer room and go to the outside before re-entering the adjacent room or structure. Electric conduit, which may be necessary for controls, connecting between a vaporizer room and adjacent control room may be imbedded in a concrete floor common to the two rooms provided that a conduit fitting for sealing is installed and the installation is otherwise in accordance with Section 501-5 of NFPA 70, *National Electrical Code®*. Any conduit passing through the vaportight wall should be near the ceiling of the room, fastened to the wall in such a manner that expansion and contraction will not break any caulking or sealing material, or the conduit must leave the area to the outside prior to going into the adjacent room.

3-7.2.4 The housing for the vaporizer covered by 3-7.2.2 or 3-7.2.3 shall not have any unprotected drains to sewers or sump pits. Pressure relief valves on vaporizers within buildings in industrial or gas manufacturing plants shall be piped to a point outside the building and shall discharge vertically upward.

Because LP-Gases are heavier than air and will seek low places, it is important that there be no open drains, sewers, or sump pits in the building enclosure. Any drain that is piped away from a vaporizer room or location is suspect in that it might connect with a general drainage system and convey flammable gases to a source of ignition in another building or open area. Where it is necessary to provide a drain, it should be protected with a trap that will not permit vapors to pass through, or should be discrete to the vaporizer room and terminate outside, in open air, well away from other drains or sources of ignition.

The relief valve piping (which is required to discharge vertically upward) should include a rain cap, weep hole, or other protection to prevent the piping from filling with water and freezing. (Refer to Chapter 2 for further information on relief valves.)

3-7.2.5 The device supplying the heat necessary for producing steam, hot water, or other heating medium may be installed out-of-doors, in a separate building, or in a structure attached to, or room within, another gas manufacturing or distributing building (but not buildings used for other purposes), provided:

(a) The housing provided shall comply with either Section 7-2 or 7-3, and shall be well ventilated near the floor line and roof.

(b) The heat supplying device, if out-of-doors, or the housing in which it is installed, shall be located with respect to other LP-Gas facilities and operations as required by Section 3-8. If the heat supplying device is gas-fired and is packaged with the vaporizer, or installed within 15 ft (5 m) of the vaporizer, it shall be subject to the provisions of 3-7.3 covering installation of direct gas-fired vaporizers.

Exception: The requirements of 3-7.2.5 are not applicable to domestic water heaters supplying heat for domestic system vaporizers.

3-7.2.6 The heating medium piping into and from the vaporizer shall be provided with a suitable means for preventing the flow of gas into a heating system which is supplying heat to areas other than the LP-Gas facility in the event of a tube rupture in the vaporizer. If the device supplying the heat to the vaporizer is for that purpose only, the device, or the piping to and from the device, shall contain a relief valve, vented to the outside, to relieve excessive pressure in the event of a tube rupture in the vaporizer.

Normally, the pressure in the vaporizer will be higher than the pressure in the heating medium piping. If there is leakage between the two, LP-Gas can flow into the heating medium piping and back to the heating appliance.

3-7.2.7 Gas-fired heating systems supplying heat for vaporization purposes shall be equipped with automatic safety devices to shut off gas to the main burners if the pilot light should fail.

To prevent raw, unburned gas from escaping into the building or the atmosphere.

3-7.2.8 Vaporizers may be an integral part of a fuel storage container, directly connected to either the liquid or vapor space, or to both. A limit control shall be provided to prevent the heater from raising the product pressure above the design pressure of the vaporizer equipment, or the pressure within the storage container above the pressure shown in the first column of Table 2-2.2.2 corresponding with the design pressure of the container (or its 1980 Code equivalent — *see Note 1 of Table 2-2.2.2*).

This applies to direct gas-fired tank heaters (*see definition*) and allows vaporizers to be a part of an integral unit, including storage container and mixer. This can be built as a mobile unit that can be moved into place where there is emergency need or other special needs for it to serve a particular plant or section of a pipeline.

3-7.2.9 Atmospheric vaporizers employing heat from the ground or surrounding air shall be installed as follows:

As pointed out in 2-5.4, some vaporizers use the atmospheric temperature for vaporizing, sometimes through tubing installed underground in order to maintain a relatively warm temperature to provide vaporizing heat. In other cases, the tubing or piping is installed in basements or in crawl spaces to gain some heat for vaporization. It is basically not desirable to have a vaporizer inside the building, but if it is restricted to 1 quart (0.9 L) maximum size, the hazard is felt to be reasonable. It is important however, to maintain such equipment properly so as to preclude the possibility of leakage. Such installations should be made only with the approval of the authority having jurisdiction because there are several safety factors that have to be taken into consideration to ensure a reasonably safe installation.

(a) Buried underground, or

(b) Located inside a building close to the point of entry of the supply pipe, provided the capacity of the unit does not exceed one quart (0.9 L).

(c) Vaporizers of less than one quart (0.9 L) capacity, not equipped with pressure relief valves [see 2-5.4.2(e)], may be installed provided one of the authorities listed in 1-3.1.1 certifies that it is safe without such a valve.

(d) Vaporizers designed primarily for domestic service shall be protected against tampering and physical damage.

3-7.3 Installation of Direct Gas-Fired Vaporizers.

3-7.3.1   Direct gas-fired vaporizers shall comply with 2-5.4.3, and shall be installed as provided in 3-7.3.2 through 3-7.3.6.

3-7.3.2   Direct gas-fired vaporizers may be installed out-of-doors or in separate structures constructed in accordance with Section 7-2. Any such buildings shall be well ventilated near the floor line and roof.

It is important to have the direct gas-fired vaporizer building or room well ventilated so there is not an accumulation of gas within the structure. The reason the ventilation is needed at both the floor line and the roof is to get proper circulation of air which will carry out the gases from the building. Normally, natural ventilation will take place, but in certain instances it may be advisable to provide for forced ventilation.

3-7.3.3   Direct gas-fired vaporizers may also be installed in structures attached to, or in rooms within, a gas manufacturing or distributing structure (but not buildings used for other purposes), provided:

(a) The housing provided shall comply with Section 7-3, and shall be well ventilated near the floor line and roof.

(b) The wall separating it from all other compartments or rooms containing LP-Gas vaporizers, pumps, and central gas mixing devices shall have no openings.

3-7.3.4 The housing for the vaporizer covered in 3-7.3.2 and 3-7.3.3 shall not have unprotected drains or sump pits. Pressure relief valves on vaporizers within buildings in industrial or gas manufacturing plants shall be piped to a point outside the building and shall discharge vertically upward.

See commentary on 3-7.2.4.

3-7.3.5 Direct gas-fired vaporizers may be connected to the liquid space or to both the liquid and the vapor space of the container, but in any case there shall be a manually operated shutoff valve in each connection at the container, to permit completely shutting off all flow of vapor or liquid.

3-7.3.6 Direct gas-fired vaporizers of any capacity shall be located in accordance with Table 3-7.3.6.

Table 3-7.3.6

Exposure	Minimum Distance Required
Container	10 ft (3 m)
Container shutoff valves	15 ft (5 m)
Point of transfer	15 ft (5 m)
Nearest important building or group of buildings or line of adjoining property which may be built upon (except buildings in which vaporizer is installed; see 3-7.3.2 and 3-7.3.3).	25 ft (7.6 m)
Ch. 7 building or room housing gas-air mixer	10 ft (3 m)
Cabinet housing gas-air mixer outdoors	0 ft (0 m)

The distance requirements in Table 3-7.3.6 have been developed based upon experience. It should be noted that they vary depending upon the exposure. For instance, a direct gas-fired vaporizer can be closer to a container than to the shutoff valves on the container because the possibility of leakage from the container shell is practically zero. Leakage potential is much greater from a connection to the container or at a valve. Also, with the making and breaking of connections at a transfer point, the hazard is greater than when the vaporizer is near the bare container without connections. The distance is greater from important buildings, a group of buildings, and the line of adjoining property which may be built upon to preclude mutual exposure in the event of a fire.

Figure 3.25 Installation of Multiple Direct-Fired Vaporizers. The vaporizers were installed 75 ft (23 m) from the storage tank. Each vaporizer includes inlet and outlet shutoff valves for servicing. A first stage pressure regulator station is installed directly on the outlet of the vaporizer to reduce the gas pressure below 20 psig (138 kPa gauge).

3-7.4 Installation of Direct Gas-Fired Tank Heaters.

3-7.4.1 Gas-fired tank heaters shall comply with 2-5.4.6, and shall be installed as follows:

(a) The container heated by a direct gas-fired tank heater shall be located in accordance with Table 3-7.4.1 with respect to the nearest important building, group of buildings or line of adjoining property which may be built upon.

(b) Direct gas-fired tank heaters shall be attached to aboveground containers only.

(c) If a point of transfer is located within 15 ft (5 m) of a direct gas-fired tank heater, the heater burner and pilot shall be shut off during the product transfer and a caution notice shall be displayed immediately adjacent to the filling connections, stating the following:

Table 3-7.4.1

Container Water Capacity	Minimum Distance Required
500 gal or less	10 ft (3 m)
501 to 2000 gal	25 ft (7.6 m)
2001 to 30,000 gal	50 ft (15 m)
30,001 to 70,000 gal	75 ft (23 m)
70,001 to 90,000 gal	100 ft (30 m)
90,001 to 120,000 gal	125 ft (38 m)

For SI Units: 1 gal = 3.785 L

"A gas-fired device which contains a source of ignition is connected to this container. Burner and pilot must be shut off before filling tank."

Note that the distance requirements in Table 3-7.4.1 are the same as the requirements for aboveground containers shown in Table 3-2.2.2, and that there are no references to the size of the tank heater but only to the size of the associated container. Table 3-7.4.1 assures that spacing criteria are properly applied to direct-fired tank heaters. Of utmost importance is the point of transfer (filling connection, etc.) in relation to the pilot and burner on the tank heater. Most tank heaters on the market are used on containers of 1000 gal (3.8 m³) and smaller, which have the point of transfer located at the top center of the container. It is essential that the gas supply to the pilot and burner be shut off during container filling.

3-7.5 Installation of Vaporizing-Burners.

3-7.5.1 Vaporizing-burners shall comply with 2-5.4.7 and shall be installed as follows:

Vaporizing-burners are burners that are fed liquid LP-Gas and vaporize it internally prior to burning. It is important that vaporizing-burners not be used inside buildings because they are generally large capacity units and both liquid and vapor are present. The distance requirements in Table 3-7.5.1 reflect the potential for ignition of any gas leakage at the container and the hazard to the container if there is impinging flame from the vaporizing-burner.

(a) Vaporizing-burners shall be installed outside of buildings. The minimum distance between any container and a vaporizing-burner shall be in accordance with Table 3-7.5.1.

(b) Manually operated positive shutoff valves shall be located at the containers to shut off all flow to the vaporizing-burners.

Table 3-7.5.1

Container Water Capacity	Minimum Distance Required
500 gal or less	10 ft (3 m)
501 to 2000 gal	25 ft (7.6 m)
Over 2000 gal	50 ft (15 m)

For SI Units: 1 gal = 3.785 L

3-7.6 Installation of Waterbath Vaporizers.

Figure 3.26 Waterbath Vaporizer with Venturi Gas-Air Mixer. Installation of a waterbath vaporizer with gas-fired burner. This vaporizer was installed in conjunction with a venturi gas-air mixer as standby for natural gas. The vaporizer burner control includes Factory Mutual and Industrial Risk Insurers approved flame safeguard controls. The vaporizer also includes a shutoff to prevent liquid carry-over in the event of burner failure or over-capacity. A water and antifreeze mixture with a rust inhibitor was installed in the waterbath of the vaporizer. The vaporizing coil within the vaporizer waterbath is ASME Code stamped.

3-7.6.1 Waterbath vaporizers shall comply with 2-5.4.4 and shall be installed as follows:

(a) If a waterbath vaporizer is electrically heated and all electrical equipment is suitable for Class 1, Group D locations, the unit shall be treated as indirect-fired and installed in accordance with 3-7.2.

(b) All others shall be treated as direct-fired vaporizers and installed in accordance with 3-7.3.

3-7.7 Installation of Electric Vaporizers.

3-7.7.1 Electric vaporizers whether direct immersion or indirect immersion shall be treated as indirect-fired and installed in accordance with 3-7.2.

3-7.8 Installation of Gas-Air Mixers.

3-7.8.1 Gas-air mixing equipment shall comply with 2-5.4.8(a) through (e) and shall be installed as follows:

(a) When used without vaporizer(s), mixer(s) may be installed out-of-doors or in buildings complying with Chapter 7.

(b) When used with indirect heated vaporizer(s), mixer(s) may be installed out-of-doors, or in the same compartment or room with the vaporizer(s), in building(s) complying with Chapter 7, or may be installed remotely from the vaporizer(s) and shall be located in accordance with 3-7.2.

(c) When used with direct-fired vaporizer(s), mixer(s) shall be installed as follows:

(1) Listed or approved in a common cabinet with the vaporizer(s) out-of-doors in accordance with 3-7.3.6.

(2) Out-of-doors on a common skid with the vaporizer(s) in accordance with 3-7.3.

(3) Installed adjacent to the vaporizer(s) to which it is connected in accordance with 3-7.3.

(4) In a building complying with Chapter 7 with no direct-fired vaporizer in the same room.

3-7.8.2 Listed vaporizer-mixers in a common cabinet having a direct-fired type vaporizer shall be installed outdoors in accordance with the distance provisions in 3-7.3. Listed vaporizer-mixers not in a common cabinet having an indirect-fired type vaporizer may be installed in a building or structure complying with Chapter 7 provided there is no source of ignition in such building or structure.

3-8 Ignition Source Control.

3-8.1 Application.

3-8.1.1 This section includes provisions to minimize the possibility of ignition of flammable LP-Gas-air mixtures resulting from the normal or

Figure 3.27 Vaporizer/Gas-Air Mixers Installed on Common Pad.

accidental release of nominal quantities of liquid or vapor from LP-Gas systems installed and operated in accordance with this standard.

3-8.1.2 Liquefied petroleum gas storage containers do not require lightning protection (*see NFPA 78, Lightning Protection Code*).

If a container or its piping is associated with a building or other structure which is equipped with lightning protection, the LP-Gas system may have to be integrated into the lightning protection system. Reference should be made to NFPA 78, *Lightning Protection Code*—especially Chapter 3—in such instances.

3-8.1.3 Since liquefied petroleum gas is contained in a closed system of piping and equipment, the system need not be electrically conductive or electrically bonded for protection against static electricity (*see NFPA 77, Recommended Practice on Static Electricity*).

This does not imply that a flammable LP-Gas/air mixture cannot be ignited by static sparks. There have been a number of fires and explosions in which a static spark has been the source of ignition. In these cases, however, liquid LP-Gas was released at high velocity and a mixture of liquid drops, vapor, air and water drops (due to condensation of water vapor in the air from the refrigerating effect of vaporizing liquid) was created. Such a mixed phase discharge will generate static electricity which may cause ignition. See NFPA 77, *Recommended Practice on Static Electricity*, for a discussion of this phenomenon.

Figure 3.28 Test Flare. Test flare normally installed with a gas-air system which is used as standby for natural gas. This flare permits the system equipment to be checked during the non-operational months and may also be used to train new people in the operation of the gas-air mixing system.

3-8.2 Electrical Equipment.

3-8.2.1 Electrical equipment and wiring shall be of a type specified by and shall be installed in accordance with NFPA 70, *National Electrical Code®*, for ordinary locations except that fixed electrical equipment in classified areas shall comply with 3-8.2.2.

3-8.2.2 Fixed electrical equipment and wiring installed within classified areas specified in Table 3-8.2.2 shall comply with Table 3-8.2.2 and shall be installed in accordance with NFPA 70, *National Electrical Code*. This provision does not apply to fixed electrical equipment at residential or commercial installations of LP-Gas systems or to systems covered by Section 3-9.

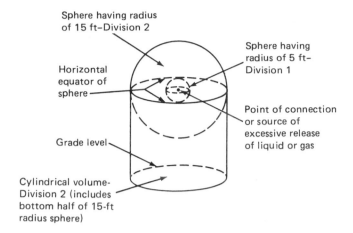

Figure 3-8.2.2 (*See Table 3-8.2.2.*)

The classified areas in Table 3-8.2.2 are based upon experience with the types of LP-Gas system installations covered by NFPA 58. These may not be necessarily consistent with the concepts expressed in Article 500 of NFPA 70, *National Electrical Code.* The latter are general precepts which must be applicable to a very broad range of hazardous materials and their applications. This has led to disputes in the field.

It should be noted that NFPA policy is to rely upon the document most specifically applicable to the material or applications involved. This is reflected in a note in Section 500-2 in NFPA 70 which references a number of NFPA standards covering flammable liquids and gases, including NFPA 58.

When using Table 3-8.2.2 it is important to understand the definitions of Division 1 and 2 areas, which are drawn from the National Electrical Code. A Division 1 area is one where combustible gases are normally present during operation, and a Division 2 area is one where combustible gases are present only under abnormal conditions. Therefore a point of transfer where a hose is connected for filling is a Division 1 area because escape of some liquid is normal when the hose is disconnected.

The exception for residential and commercial installations recognizes that the smaller containers usually involved at such locations can be very close to structures and that these structures contain ordinary electrical equipment and wiring. Consideration of both the fire experience and the expense of the electrical installations specified in NFPA 70, Article 501, justify this exception.

It is noted that, while residential occupancies are rather homogeneous in character, the nature of commercial occupancies is not. For example, an LP-Gas service station could be considered a commercial occupancy, but is addressed in Part F of Table 3-8.2.2. Therefore, commercial occupancies should be considered with judgment.

With regard to other provisions of Table 3-8.2.2, there has been some confusion with regard to the requirements of parts D and E, specifically with regard to relief devices on vaporizers other than fired units. Part E provides that indirect or electric vaporizers (or any other type other than direct-fired or indirect-fired with an attached or adjacent gas fired heat source) installed outdoors in open air at or above grade be classified as Division 2 within 15 ft (5 m) of the point of connection or source of excessive release of liquid or gas. Part D, on the other hand, specifies that space within 5 ft (2 m) on the point of discharge of the relief valve be classified as Division 1. Many have asked if this is an inconsistency in the standard. It is not, but recognizes that the small containers covered in part D can be installed close to buildings, hence the stricter area classification. In addition, all vaporizers [except atmospheric units holding less than 1 quart (0.9 L)] include pressure relief devices whose discharge is of concern when considering location of electrical equipment.

To clear up misconceptions regarding part E, the terms "pumps," "vapor compressors," "gas-air mixers," and "vaporizers" refer to those items handling LP-Gas. A water pump or air compressor in a separate room or building from the gas equipment is not covered by this requirement. Also, gas-air mixers mixing noncombustible gases, or noncombustible gas with air are not covered by this provision, nor are vaporizers handling noncombustible refrigerants and medical and industrial gases.

The exception for Section 3-9 concerns electrical installations on vehicles.

In addition to similarities with residential occupancies noted earlier, an internal combustion engine powered vehicle has inherent non-electrical ignition sources—e.g., an exhaust system—and obviating its electrical system as an ignition source would be of little real value.

3-8.2.3 Electrical equipment installed on LP-Gas cargo vehicles shall comply with 6-1.1.4.

3-8.3 Other Sources of Ignition.

3-8.3.1 Open flames or other sources of ignition shall not be permitted in pump houses, container filling rooms or other similar locations. Direct-fired vaporizers or indirect-fired vaporizers attached or installed adjacent to gas-fired heat sources shall not be permitted in pump houses or container filling rooms.

Table 3-8.2.2

Part	Location	Extent of Classified Area[1]	Equipment Shall Be Suitable for National Electrical Code, Class 1, Group D[2]
A	Storage Containers Other Than DOT Cylinders and ASME Vertical Containers of Less Than 1000 lb Water Capacity.	Within 15 feet in all directions from connections, except connections otherwise covered in Table 3-8.2.2.	Division 2
B	Tank Vehicle and Tank Car Loading and Unloading.[3]	Within 5 feet in all directions from connections regularly made or disconnected for product transfer.	Division 1
		Beyond 5 feet but within 15 feet in all directions from a point where connections are regularly made or disconnected and within the cylindrical volume between the horizontal equator of the sphere and grade. (See Figure 3-8.2.2)	Division 2
C	Gage Vent Openings Other Than Those on DOT Cylinders and ASME Vertical Containers of Less Than 1000 lb Water Capacity.	Within 5 feet in all directions from point of discharge.	Division 1
		Beyond 5 feet but within 15 feet in all directions from point of discharge.	Division 2
D	Relief Device Discharge Other Than Those on DOT Cylinders and ASME Vertical Containers of Less Than 1000 lb Water Capacity.	Within direct path of discharge.	Division 1 *Note:* Fixed electrical equipment should preferably not be installed.

[1]The classified area shall not extend beyond an unpierced wall, roof, or solid vaportight partition.

[2]See Article 500 "Hazardous (Classified) Locations" in NFPA 70 (ANSI) for definitions of Classes, Groups, and Divisions.

[3]When classifying extent of hazardous area, consideration shall be given to possible variations in the spotting of tank cars and tank vehicles at the unloading points and the effect these variations of actual spotting point may have on the point of connection.

[4]Where specified for the prevention of fire or explosion during normal operation, ventilation is considered adequate where provided in accordance with the provisions of this standard.

SI Conversions for Table 3-8.2.2 18 in. = 256 mm   4 ft = 1.2 m   5 ft = 1.5 m   15 ft = 5 m   20 ft = 6 m

Table 3-8.2.2 (cont.)

Part	Location	Extent of Classified Area[1]	Equipment Shall Be Suitable for National Electrical Code, Class 1, Group D[2]
		Within 5 feet in all directions from point of discharge.	Division 1
		Beyond 5 feet but within 15 feet in all directions from point of discharge except within the direct path of discharge.	Division 2
E	Pumps, Vapor Compressors, Gas-Air Mixers and Vaporizers (other than direct-fired or indirect-fired with an attached or adjacent gas-fired heat source).		
	Indoors without ventilation.	Entire room and any adjacent room not separated by a gastight partition.	Division 1
		Within 15 feet of the exterior side of any exterior wall or roof that is not vaportight or within 15 feet of any exterior opening.	Division 2
	Indoors with adequate ventilation.[4]	Entire room and any adjacent room not separated by a gastight partition.	Division 2
	Outdoors in open air at or abovegrade.	Within 15 feet in all directions from this equipment and within the cylindrical volume between the horizontal equator of the sphere and grade. (See Figure 3-8.2.2.)	Division 2

[1]The classified area shall not extend beyond an unpierced wall, roof, or solid vaportight partition.

[2]See Article 500 "Hazardous (Classified) Locations" in NFPA 70 (ANSI) for definitions of Classes, Groups, and Divisions.

[3]When classifying extent of hazardous area, consideration shall be given to possible variations in the spotting of tank cars and tank vehicles at the unloading points and the effect these variations of actual spotting point may have on the point of connection.

[4]Where specified for the prevention of fire or explosion during normal operation, ventilation is considered adequate where provided in accordance with the provisions of this standard.

SI Conversions for Table 3-8.2.2 18 in. = 256 mm 4 ft = 1.2 m 5 ft = 1.5 m 15 ft = 5 m 20 ft = 6 m

Table 3-8.2.2 (cont.)

Part	Location	Extent of Classified Area[1]	Equipment Shall Be Suitable for National Electrical Code, Class 1, Group D[2]
F	Service Station Dispensing Units.	Entire space within dispenser enclosure, and 18 inches horizontally from enclosure exterior up to an elevation 4 feet above dispenser base. Entire pit or open space beneath dispenser.	Division 1
		Up to 18 inches above-grade within 20 feet horizontally from any edge of enclosure. *Note:* For pits within this area, see Part G of this table.	Division 2
G	Pits or Trenches Containing or Located Beneath LP-Gas Valves, Pumps, Vapor Compressors, Regulators, and Similar Equipment.		
	Without mechanical ventilation.	Entire pit or trench.	Division 1
		Entire room and any adjacent room not separated by a gastight partition.	Division 2
		Within 15 feet in all directions from pit or trench when located outdoors.	Division 2
	With adequate mechanical ventilation.	Entire pit or trench.	Division 2
		Entire room and any adjacent room not separated by a gastight partition.	Division 2

[1]The classified area shall not extend beyond an unpierced wall, roof, or solid vaportight partition.

[2]See Article 500 "Hazardous (Classified) Locations" in NFPA 70 (ANSI) for definitions of Classes, Groups, and Divisions.

[3]When classifying extent of hazardous area, consideration shall be given to possible variations in the spotting of tank cars and tank vehicles at the unloading points and the effect these variations of actual spotting point may have on the point of connection.

[4]Where specified for the prevention of fire or explosion during normal operation, ventilation is considered adequate where provided in accordance with the provisions of this standard.

SI Conversions for Table 3-8.2.2 18 in. = 256 mm 4 ft = 1.2 m 5 ft = 1.5 m 15 ft = 5 m 20 ft = 6 m

Table 3-8.2.2 (cont.)

Part	Location	Extent of Classified Area[1]	Equipment Shall Be Suitable for National Electrical Code, Class 1, Group D[2]
		Within 15 feet in all directions from pit or trench when located outdoors.	Division 2
I	Special Buildings or Rooms for Storage of Portable Containers.	Entire room.	Division 2
	Pipelines and Connections Containing Operational Bleeds, Drips, Vents or Drains.	Within 5 feet in all directions from point of discharge.	Division 1
		Beyond 5 feet from point of discharge, same as Part E of this table.	
	Container Filling:		
	Indoors with adequate ventilation[4]	Within 5 feet in all directions from connections regularly made or disconnected for product transfer.	Division 1
		Beyond 5 feet and entire room.	Division 2
	Outdoors in open air	Within 5 feet in all directions from connections regularly made or disconnected for product transfer.	Division 1
		Beyond 5 feet but within 15 feet in all directions from a point where connections are regularly made or disconnected and within the cylindrical volume between the horizontal equator of the sphere and grade. (See Figure 3-8.2.2.)	Division 2

[1]The classified area shall not extend beyond an unpierced wall, roof, or solid vaportight partition.

[2]See Article 500 "Hazardous (Classified) Locations" in NFPA 70 (ANSI) for definitions of Classes, Groups, and Divisions.

[3]When classifying extent of hazardous area, consideration shall be given to possible variations in the spotting of tank cars and tank vehicles at the unloading points and the effect these variations of actual spotting point may have on the point of connection.

[4]Where specified for the prevention of fire or explosion during normal operation, ventilation is considered adequate where provided in accordance with the provisions of this standard.

SI Conversions for Table 3-8.2.2 18 in. = 256 mm   4 ft = 1.2 m   5 ft = 1.5 m 15 ft = 5 m 20 ft = 6 m

3-8.3.2   Open flames (except as provided for in Section 3-7), cutting or welding, portable electric tools and extension lights capable of igniting LP-Gas shall not be permitted within classified areas specified in Table 3-8.2.2 unless the LP-Gas facilities have been freed of all liquid and vapor, or special precautions observed under carefully controlled conditions.

> NFPA 51B, *Standard for Fire Prevention in Use of Cutting and Welding Processes,* is useful in the context of "special precautions observed under carefully controlled conditions." The API, AGA, and American Welding Society also have publications pertinent to this provision.

3-8.4 Control of Ignition Sources during Transfer.

3-8.4.1   Sources of ignition shall be carefully controlled during transfer operations, while connections or disconnections are made, or while LP-Gas is being vented to the atmosphere. In addition to the other provisions of Section 3-8, the following shall apply:

(a) Internal combustion engines within 15 ft (5 m) of a point of transfer shall be shut down while such transfer operations are in progress, except as follows:

(1)   Engines of LP-Gas cargo vehicles constructed and operated in compliance with Chapter 6 while such engines are driving transfer pumps or compressors on these vehicles to load containers as provided in 4-3.2.1.

(2)   Engines installed in buildings as provided in 3-6.3.

(b) Smoking, open flames, metal cutting or welding, portable electrical tools, and extension lights capable of igniting LP-Gas shall not be permitted within 15 ft (5 m) of a point of transfer while filling operations are in progress. Care shall be taken to assure that materials which have been heated have cooled before the transfer is started.

Formal Interpretation
Reference   3-2.10.6(b)(3), 4-3.4.1, 3-8.4.1(a), 3-8.4.2(b)

Question: A 1600-gallon LP-Gas storage container and associated liquid transfer equipment is centrally installed in a parking area serving a suburban shopping center for the purpose of filling containers mounted on vehicles and portable containers. The vehicles and portable containers are in the immediate vicinity of the storage container while they are being filled. The lane for vehicles being serviced is also a driveway for parking of vehicles and for movement within the parking area. Is such an installation and operation specifically covered by the provisions of NFPA 58?

Answer: Yes. Specifically, such an installation is a "Distributing Point" as defined in Section 1-7 of NFPA 58 and is subject to all provisions applicable to Distributing Points.

Applicable provisions include Paragraphs 3-2.10.6(b)(3), 4-3.4.1, 3-8.4(a) and 3-8.4(b) and Parts 6 and 7 of Table 4-3.3.2. In the context of an installation in a parking area open to the public, the Committee is of the opinion that portions of such an area accessible to the public for activities normally associated with the movement and parking of vehicles should be considered as public ways, streets, sidewalks and thoroughfares as these terms are used in 3-2.10.6(b)(3), 4-3.4.1 and Parts 6 and 7 of Table 4-3.3.2.

The Committee noted that while it would be possible to install such a distributing point "centrally" in such a parking area, provisions for restricting movement of the public and vehicles when not associated with the functioning of the installation [especially to assure compliance with 3-8.4(b)] could make a less central location more practical.

Issue Edition: 1974
Reference: 3195(b)(3), 4012, 4060(a), 4060(b)
Date: October-November 1976

   (c) Sources of ignition, such as pilot lights, burners, electrical appliances, and engines, located on the vehicle being refueled shall be turned off during the filling of any LP-Gas container on the vehicle.

3-8.4.2 Transfers to containers serving agricultural or industrial equipment requiring refueling in the field shall comply with the following:

   (a) Air moving equipment, such as large blowers on crop driers or on space heaters, shall be shut down while containers are being refilled, unless the point of transfer is at least 50 ft (15 m) from the air intake of the blower.

   (b) Equipment employing open flames, or equipment with integral containers such as flame cultivators, weed burners, tractors, large blower type space heaters, or tar kettles shall be shut down while refueling.

3-9 LP-Gas Systems on Vehicles (Other than Engine Fuel Systems).

3-9.1 Application.

3-9.1.1 This section applies to non-engine fuel systems on commercial, industrial, construction, and public service vehicles such as trucks, semi-trailers, trailers, portable tar kettles, road surface heating equipment, mobile laboratories, clinics, and mobile cooking units (such as catering and canteen vehicles). LP-Gas systems on such vehicles may be either vapor-withdrawal or liquid-withdrawal type. Included are provisions for installations served by exchangeable (removable) container systems and by permanently mounted containers.

Road surface heating equipment was added to the list of vehicles covered under this section in the 1989 Edition. (*See commentary following 3-9.2.1(f) and the requirements of 3-9.2.4.*)

3-9.1.2   This section does not apply to:

(a) Systems installed on mobile homes.

(b) Systems installed on recreational vehicles [*see 3-1.1.4(d)*].

(c) Tank trucks, truck transports (trailers and semitrailers), and similar units used to transport LP-Gas as cargo, which are covered by Chapter 6.

3-9.1.3   LP-Gas engine fuel systems on the vehicles covered by Section 3-9 and those cited in 3-9.1.2 are covered by Section 3-6.

This section is the counterpart of Section 3-6 for vehicle propulsion engine systems, as it covers all other applications of LP-Gas using systems mounted on vehicles. There are three exceptions listed in 3-9.1.2, namely, mobile homes, recreational vehicles, and cargo tank vehicles. Standards for installations on mobile vehicles for cooking and heating first appeared in the 1950 Edition of NFPA 58. These were later separated into two different chapters: (1) systems for mobile homes and travel trailers and (2) systems for commercial vehicle uses. These provisions for mobile homes formed the basis for NFPA 501B, *Standard for Mobile Homes*. NFPA 501B was discontinued when the federal agency of Housing and Urban Development (HUD) used NFPA 501B and issued their "Mobile Home Construction and Safety Standards," Part 280 CFR 24, in 1976. NFPA 501C, *Standard on Firesafety Criteria for Recreational Vehicles*, utilized the former provisions in NFPA 58 extensively for standards on LP-Gas systems. Cargo tank systems are covered in Chapter 6 of NFPA 58. Thus, vehicular systems covered by the section are now limited to the type of applications listed in 3-9.1.

3-9.2 Construction, Location, Mounting and Protection of Containers and Systems.

3-9.2.1   Containers shall comply with Section 2-2 and appurtenances used to equip them for service shall comply with Section 2-3. In addition, 3-9.2.1(a) through (g) shall apply:

Provisions for the construction of containers parallel those in Section 3-6, where ASME container design pressure is 312.5

psig (2.2 MPa gauge) for those installed in enclosed spaces [otherwise 250 psig (1.7 MPa gauge) design pressure may be used], maximum sizes of containers are specified, etc. Paragraph (g) is particularly significant with respect to the need for protection of the container appurtenances and their connections. A standard stationary type container without the protection specified in 2-2.4 should not be used as these appurtenances would be vulnerable in the case of a vehicular accident.

(a) ASME containers shall be constructed for a minimum 250 psig (1.7 MPa gauge) design pressure.

(b) Containers installed in enclosed spaces on vehicles (including recesses or cabinets covered in 3-9.2.2) shall be constructed as follows:

(1) DOT cylinder specification containers shall be designed and constructed for at least a 240 psig (1.6 MPa gauge) service pressure.

(2) ASME containers shall be constructed for at least a 312.5 psig (2.2 MPa gauge) design pressure.

(c) Portable (removable) containers shall comply with 2-2.4.

(d) Containers to be permanently mounted shall be constructed so that, after mounting the protection of all container appurtenances and the connections to these appurtenances, they comply with 3-9.2.3(c).

(e) LP-Gas fuel containers used on passenger-carrying vehicles shall not exceed 200 gal (0.8 m³) aggregate water capacity.

(f) Individual LP-Gas containers used on other than passenger-carrying vehicles normally operating on the highway shall not exceed 300 gal (1 m³) water capacity. This shall not be construed as applying to the use of LP-Gas from the cargo tanks of vehicles covered by Chapter 6.

Exception: Containers on road surface heating equipment shall not exceed 1000 gal (3.0 m³) water capacity.

This exception was added in the 1989 Edition when it became evident that the requirement was overly restrictive. When the requirement was written in an early edition of the standard a 300 gal (1.1 m³) container was the largest the Technical Committee could envision being mounted on a vehicle. In recent years road surfacing equipment has been developed which melts road surfacing after removal from roadways to recover the asphalt content for immediate reuse. Such equipment has a very high propane consumption rate and the 300 gal (1.1 m³) limit on container size, along with the constraints of truck mounted equipment limiting multiple containers, forced frequent shutdown for refilling (and the required refilling vehicle to stand by during operation). As experience has shown that the chances of LP-Gas release (especially liquid) are greatest during liquid transfer operations, reducing the number of such operations by permitting a larger container to be used was believed to reduce the overall hazard.

The 1000 gal (3.8 m³) limit on container size is based on units in safe operation.

(g) Containers designed for stationary service only, and not in compliance with 2-2.4, shall not be used.

3-9.2.2 Containers utilized for the purposes covered by this section shall not be installed, transported, or stored (even temporarily) inside any vehicle covered by Section 3-9, except as provided in 3-9.2.3(d), Chapter 6, or as provided by applicable DOT regulations. The LP-Gas supply system, including the containers, may be installed on the outside of the vehicle, or in a recess or cabinet vaportight to the inside of the vehicle but accessible from and vented to the outside, with the vents located near the top and bottom of the enclosure, and 3 ft (1 m) horizontally away from any opening into the vehicle below the level of the vents.

Basically, containers are to be located on the outside of the vehicle or in a recess or compartment that is vaportight to the inside of the vehicle. The exceptions are those systems installed in the interior of a vehicle similar to engine fuel containers which are covered in 3-6.2.7; and the transportation of spare portable containers as outlined in Chapter 6 or by applicable DOT regulations. For example, portable cylinders used in connection with food warmers on delivery vehicles should be located outside or in a compartment and piped to the appliance inside. Requirements for supply systems on vehicles subject to DOT regulations are contained in Section 393.77 of the Bureau of Motor Carrier Safety Regulations, Part 393 CFR 49 and Section 177.834(d) of the Hazardous Materials Regulations, Part 177 CFR 49.

3-9.2.3 Containers shall be securely mounted on the vehicle, or within the enclosing recess or cabinet, and located and installed so as to minimize the possibility of damage to containers, their appurtenances or contents as follows:

The provisions of this subsection are also to assure the container mounting arrangement is strong enough to stay intact and to protect the container, its appurtenances and connections from damage due to collisions, road debris, and weather. They are identical to those for engine fuel supply systems with one exception: the direction of pressure relief valve discharge as set out for engine fuel containers is required only on those containers mounted inside passenger carrying vehicles. However, 3-9.2.5(a)(2)b stipulates that the basic premise of assuring the release of gas from relief valves is that the gas does not impinge upon a container, vehicle parts, or a vehicle in an adjacent line of traffic.

As with vehicle propulsion engine fuel containers, protection of exterior containers and appurtenances against material thrown up from the road is set forth. However, with non-engine fuel systems a pressure regulator may be mounted at the container for vapor withdrawal systems. The vent on this regulator must not be blocked with slush, etc., thrown up from the road. This may be prevented with a splash guard or by locating the regulator in a compartment. A blocked vent may lead to abnormally high pressures in the utilization system. Appurtenances on a container, particularly those mounted below the vehicle, should be installed so that they are readily accessible.

(a) Containers shall be installed with road clearance in accordance with 3-6.2.6(e).

(b) Fuel containers shall be securely mounted to prevent jarring loose and slipping or rotating, and the fastenings shall be designed and constructed to withstand without permanent visible deformation static loading in any direction equal to four times the weight of the container filled with fuel. When containers are mounted within a vehicle housing, the securing of the housing to the vehicle shall comply with this provision. Any hoods, domes, or removable portions of the housing or cabinet shall be provided with means to keep them firmly in place in transit. Field welding shall comply with 3-2.3.1(e).

(c) All container valves, appurtenances and connections shall be adequately protected to prevent damage due to accidental contacts with stationary objects, from loose objects, stones, mud or ice, thrown up from the ground or floor, and from damage due to overturned or similar vehicular accident. In the case of permanently mounted containers, this provision may be met by the location on the vehicle, with parts of the vehicle furnishing the protection. On portable (removable) containers the protection for container valves and connections shall be permanently attached to the container. (*See 2-2.4.1 and 2-2.4.2.*) Such weather protection as may be necessary to ensure safe operation shall be provided for containers and systems mounted on the outside of the vehicle.

(d) Containers mounted on the interior of passenger-carrying vehicles shall be installed in compliance with 3-6.2.7. Pressure relief valve installations for such containers shall comply with 3-6.2.6(i).

3-9.2.4 Containers installed on portable tar kettles alongside the kettle, or on the vehicle frame, or on road surface heating equipment, shall be protected against radiant or convected heat from open flame or other burners by the use of a heat shield or by the location of the container(s) on the vehicle so as to prevent the temperature of the fuel in the container from becoming abnormally high. In addition, the following shall apply:

> **Although somewhat in conflict with the basic format of NFPA 58 of not setting out provisions for specific applications, tar kettles are one of the more extensive vehicular applications of LP-Gas. This paragraph provides direction as to where to find the pertinent information in the standard.**
>
> **Tar kettles are generally standardized equipment today. However, attention should be given to these requirements to make certain they are applied. Mounting of cylinders should be such that no wear takes place (chains can cut grooves in cylinders due to vehicular motion). Most tar kettles use liquid burners so the right type of cylinder must be used. The word "liquid" is stamped on the valve handle or on the cylinder next to the valve. Valve protection for portable cylinders covered by Section 3-9 must be permanently attached, such as by the use of a collar welded to the container. However, with regard to portable containers used with tar kettles, these need not be permanently attached. The type of collar used is threaded to the cylinder valve boss.**

(a) Container location, mounting and protection shall comply with 3-9.2.3(a), (b) and (c) except that the protection for DOT container valves need not be permanently attached to the container; however, the protection shall comply with 2-2.4.1(a) and (b);

(b) Piping shall comply with 3-9.2.7(a), (b), (d), (e), (g), (h), and (i);

(c) Flexible connections shall comply with 2-4.6.1, 2-4.6.2 and 2-4.6.3;

(d) Container valves shall be closed when burner is not in use;

(e) Containers shall not be refilled while burners are in use as provided in 3-8.4.2(b).

3-9.2.5 Container appurtenances shall be installed in accordance with 3-9.2.5(a) through (f).

Three types of fuel container installations impact upon how the pressure relief valve is installed and located. These are: (1) portable containers in cabinet or recess, (2) containers mounted on the exterior, and (3) containers installed in the interior of passenger-carrying vehicles. Paragraph 3-9.2.2 specifies what to do with cabinets and recesses (excluding cabinets in the interior of passenger vehicles), 3-9.2.5(a)(2) describes what has to be done with exterior mounted containers, and 3-6.2.6(i) indicates what must be done with interior mounted containers on passenger vehicles. Subparagraph (4) draws attention to protecting the pressure relief valve outlet (valve or discharge piping) from being plugged with water (which can freeze), or dirt, asphalt, etc.

(a) Container pressure relief devices shall be located and installed as follows:

(1) Except as provided in 3-9.2.3(d), pressure relief devices on portable containers installed inside cabinets or recesses complying with 3-9.2.2 may discharge within the enclosure.

(2) Relief device discharge outlets on containers installed on the outside of the vehicle shall be located:

a. Outside of enclosed spaces, at least 3 ft (1 m) horizontally away from any opening into the vehicle below the level of such discharge, and as far as practicable from sources of ignition.

b. In such a manner as to minimize the possibility of impingement of escaping gas upon a container, vehicle parts or on other vehicles in adjacent lines of traffic.

(3) Pressure relief device discharge lines shall be metallic and have a melting point over 1500°F (816°C). Relief valve adaptors installed directly in the relief valve to deflect the flow upward shall be metallic and have a melting point over 700°F (371°C). Discharge lines and adaptors shall be sized, located and secured so as to permit sufficient pressure relief device relieving capacity. Flexible metal hose or tubing used shall be able to withstand the pressure from the relief device vapor discharge when the relief device is in full open position.

(4) On vehicles used outdoors or in industrial locations, means shall be provided (such as loose-fitting caps) to minimize the possibility of the entrance of water or dirt into either the relief device or its discharge piping. The protecting means shall remain in place except when the relief device operates. In this event, it shall permit the relief device to operate at sufficient capacity.

(b) The filling, withdrawal and equalizing connections of containers shall be equipped in compliance with 2-3.3.1 through 2-3.3.3 (see "Used as Fuel on Vehicles," Column 5 of Table 2-3.3.2).

(c) Main shutoff valves on container for liquid and vapor shall be readily accessible.

(d) Containers to be filled volumetrically shall be equipped with liquid level gauging devices as provided in 2-3.4. Portable containers may be designed, constructed, and fitted for filling in either the vertical or horizontal position or, if of the portable universal type [see 2-3.4.2(c)(2)], in either position. The container shall be in the appropriate position when filled or, if of the portable universal type, may be loaded in either position, provided:

(1)   The fixed level gauge indicates correctly the maximum permitted filling level in either position.

(2)   The pressure relief devices are located in, or connected to, the vapor space in either position.

(e) All container inlets and outlets, except pressure relief devices and gauging devices, shall be labeled to designate whether they communicate with the vapor or liquid space. Labels may be on valves.

(f) Containers from which only vapor is to be withdrawn shall be installed and equipped with suitable connections to minimize the possibility of the accidental withdrawal of liquid.

3-9.2.6 Regulators shall comply with 2-5.1 and 2-5.8 and shall be installed in accordance with 3-2.5. If in an enclosed space, the regulator relief device and the space above the regulator and relief device diaphragms shall be vented to the outside air. Such venting is not required if the regulator is located in a recess or cabinet as provided for in 3-9.2.2.

Provisions for venting regulators installed in cabinets to the outside are available through the louvers in the cabinet door, whereas an enclosed space does not have access to the outside. Therefore, provisions for venting this space must be provided.

(a) If a regulator(s) is (are) installed in conjunction with the vapor withdrawal portion of an LP-Gas system installed upon a vehicle, (a) approved two stage regulator(s) shall be used. Such regulator(s) shall have a capacity not less than the total input of all LP-Gas appliances installed in the vehicle. The regulator(s) shall be mounted with the vent position downward within 45 degrees of vertical with the diaphragm area being drained. Regulators not installed in compartments shall be equipped with a durable cover designed to protect the regulator vent opening from sleet, snow, freezing rain, ice, mud, and wheel spray.

NOTE: Durable means that the cover will not become brittle at temperatures as low as -40°F (-40°C).

(b) If a vehicle mounted regulator(s) is installed at or below the floor level, it shall be installed in a compartment which provides protection against the weather and wheel spray. The compartment shall be of sufficient size to permit tool operation for connection to and replacement of the regulator(s), shall be vaportight to the interior of the vehicle, shall have a 1 sq in. (6.5 cm^2) minimum vent opening to the exterior located within 1 in. (25 mm) of the bottom of the compartment, and shall not contain flame-or spark-producing equipment. A regulator vent outlet shall be at least 2 in. (51 mm) above the compartment vent opening.

Paragraphs (a) and (b) were added in the 1989 Edition to require two stage regulation and to use the same requirements as NFPA 501C, *Standard on Firesafety Criteria for Recreational Vehicles,* **where propane is also used as a fuel in vapor form.**

3-9.2.7 Piping shall comply with Section 2-4 as to material and design and shall be installed in accordance with 3-2.7, except that steel tubing shall have a minimum wall thickness of 0.049 in. Paragraphs 3-9.2.7(a) through (j) shall also apply to piping systems on vehicles covered by Section 3-9.

Steel tubing in vehicular installations has been specified as having a minimum 0.049 in. (1.2 mm) wall thickness to provide strength against vibration and also an additional tolerance for corrosion. This came about due to experience with steel tubing during World War II when copper tubing was not available. Subparagraphs (e) and (f) are the basic approach to piping of all vehicular type installations (i.e., keeping the main lines and branch connections outside). Even though flow control protection is required as per Table 2-3.3.2, no fuel lines are to be connected between two vehicular units in order to avoid compounding problems involved in collisions, overturns, disconnection of vehicles, etc. The risk here is greater than that of a system completely on one vehicle.

(a) A flexible connector or a tubing loop shall be installed between the regulator outlet and the piping system to protect against expansion, contraction, jarring, and vibration strains.

(b) In the case of removable containers, flexibility shall be provided in the piping between the container and the gas piping system or regulator.

(c) Flexible connectors shall comply with 2-4.6 and be installed in accordance with 3-2.7.8(a). Flexible connectors of more than 36 in. (914.4 mm) overall length, or fuel lines of essentially all hose, shall be used only with the approval of the authority having jursidiction.

(d) The piping system shall be designed, installed, supported and secured in such a manner as to minimize the possibility of damage due to vibration, strains or wear, and to preclude any working loose while in transit.

(e) Piping (including hose) shall be installed in a protected location. If outside, piping shall be under the vehicle and below any insulation or false bottom. Fastening or other protection shall be installed to prevent damage due to vibration or abrasion. At each point where piping passes through sheet metal or a structural member, a rubber grommet or equivalent protection shall be installed to prevent chafing.

(f) Gas piping shall be installed to enter the vehicle through the floor directly beneath, or adjacent to, the appliance served. If a branch line is required, the tee connection shall be in the main gas line under the floor and outside the vehicle.

(g) Exposed parts of the piping system shall either be of corrosion-resistant material or adequately protected against exterior corrosion.

(h) Hydrostatic relief valves, complying with 2-4.7.1, shall be installed in isolated sections of liquid piping as provided in 3-2.8.

(i) Piping systems, including hose, shall be tested and proven free of leaks in accordance with 3-2.9.

(j) There shall be no fuel connection between a tractor and trailer or other vehicle units.

3-9.3 Equipment Installation.

3-9.3.1   Equipment for installation on vehicles shall comply with Section 2-5 as to design and construction, and shall be installed in accordance with 3-2.10, and with the following:

(a) Installation shall be made in accordance with the manufacturer's recommendations and, in the case of listed or approved equipment, as provided in the listing or approval.

(b) Equipment installed on vehicles shall be considered as part of the LP-Gas system on the vehicle and shall be protected against vehicular damage as provided for container appurtenances and connections in 3-9.2.3(c).

3-9.4 Appliance Installation.

In addition to provisions for cargo heaters set out in this subsection, reference should be made to the U.S. Department of Transportation (DOT) requirements for those vehicles subject to their jurisdiction. Safety shutoffs are required on all heating appliances except tar kettle burners, hand torches, melting pots, and small heaters using a torch type cylinder, as these are attended appliances. The basic concept of isolating the combustion system of appliances — except ranges in vehicle interiors where passengers might be — is consistent with all other vehicle standards on this subject (i.e, those for recreational vehicles). No portable or conventional room heaters should be used inside. Isolation may be accomplished through the use of direct vent type heaters and water heaters, or separation by installation in a compartment with provisions for outside air. The range is an attended appliance and need not be isolated, but it should never be used for comfort heating.

3-9.4.1 The term "appliances" as used in this subsection shall include any commercial or industrial gas consuming device except engines.

3-9.4.2 All gas consuming devices (appliances), other than engines, installed on vehicles shall be approved as provided in 2-6.2, shall comply with 2-6.3 and shall be installed as follows:

(a) Whenever the device or appliance is of a type designed to be in operation while the vehicle is in transit, such as a cargo heater or cooler, suitable means to stop the flow of gas in the event of a line break, such as an excess-flow valve, shall be installed. Excess-flow valves shall comply with 2-4.5.3 and 2-3.3.3(b).

(b) All gas-fired heating appliances shall be equipped with safety shutoffs in accordance with 2-6.3.5(a) except those covered in 3-4.2.8(b).

(c) For installations on vehicles intended for human occupancy, all gas-fired heating appliances, except ranges and illuminating appliances, shall be designed or installed to provide for a complete separation of the combustion system from the atmosphere inside the vehicle. Combustion air inlets and flue gas outlets shall be listed or certified as components of the appliance.

(d) For installations on vehicles not intended for human occupancy, unvented-type gas-fired heating appliances may be used to protect the cargo. Provision shall be made to provide air for combustion [*see 3-9.4.2(f)*] and to dispose of the products of combustion to the outside.

(e) Appliances installed within vehicles shall comply with the following:

(1) If in the cargo space, they shall be located so as to be readily accessible whether the vehicle is loaded or empty.

(2) Appliances shall be so constructed or otherwise protected as to minimize possible damage or impaired operation due to cargo shifting or handling.

(3)   Appliances shall be located so that a fire at any appliance will not block egress of persons from the vehicle.

(f) Provision shall be made in all appliance installations to ensure an adequate supply of outside air for combustion.

(g) A permanent caution plate shall be provided, affixed either to the appliance, or to the vehicle outside of any enclosure and adjacent to the container(s), including the following items:

CAUTION

(1)   Be sure all appliance valves are closed before opening container valve.

(2)   Connections at the appliances, regulators and containers shall be checked periodically for leaks with soapy water or its equivalent.

(3)   Never use a match or flame to check for leaks.

(4)   Container valves shall be closed when equipment is not in use.

3-9.5 General Precautions.

3-9.5.1   Containers on vehicles shall be filled or refilled as provided by 4-2.2.2. See 2-2.1.3 for requalification requirements for continued use or reinstallation.

3-9.5.2   Mobile units containing hotplates and other cooking equipment, including mobile kitchens and catering vehicles, shall be provided with at least one approved portable fire extinguisher rated in accordance with NFPA 10, *Standard for Portable Fire Extinguishers*, at not less than 10-B:C.

3-9.6 Parking, Servicing, and Repair.

This section was added in the 1989 Edition of the standard to provide specific coverage for the parking, servicing, and repair of vehicles with non-engine fuel LP-Gas systems on them. Note that this section does not cover recreational vehicles (which are covered by NFPA 501C), mobile homes (which are covered under federal regulations), or vehicles used to transport LP-Gas as cargo (which are covered under Chapter 6).

3-9.6.1   Vehicles with LP-gas fuel systems mounted on them for purposes other than propulsion may be parked, serviced, or repaired inside buildings, in accordance with 3-9.6.1(a) through (d).

(a) The fuel system shall be leak free and the container(s) shall not be filled beyond the limits specified in Chapter 4.

(b) The container shutoff valve shall be closed except when fuel is required for test or repair.

(c) The vehicle shall not be parked near sources of heat, open flames, or similar sources of ignition, or near unventilated pits.

(d) Vehicles having containers with water capacities larger than 300 gallons shall comply with the requirements of Section 6-6.

Note that these requirements are similar to those for vehicles using LP-Gas as an engine fuel (3-6.6), and that vehicles having a total LP-Gas storage of over 300 lb (136 kg) are required to follow the more detailed requirements of Chapter 6.

3-10 Fire Protection.

3-10.1 Application.

3-10.1.1 This section includes provisions for fire protection to augment the leak control and ignition source control provisions in this standard.

With the exception of Section 3-10, all other provisions in NFPA 58 are leak control and ignition source control provisions reflecting equipment design, fabrication, installation, maintenance, and facility personnel performance. If these provisions were complied with initially and maintained subsequently, there would be little need for Section 3-10. However, the accident record rather overwhelmingly reveals failures to comply with one or more provisions of the standard. Also, equipment doesn't always function as intended and people make mistakes. For these reasons, fire protection is required to augment the other provisions.

As noted in the commentary on 3-2.2.2, consideration of fire protection can often be a significant factor in determination of the actual location of a facility.

3-10.2 General.

3-10.2.1 The wide range in size, arrangement and location of LP-Gas installations covered by this standard precludes the inclusion of detailed fire protection provisions completely applicable to all installations. Provisions in this section are subject to verification or modification through analysis of local conditions.

When it is considered that an installation covered by NFPA 58 can range from one or two 100 lb (45 kg) LP-Gas capacity cylinders at a single-family residence in a rural area that are

filled once a month (or maybe only exchanged and not filled at the installation), to several 30,000 gal (114 m³) LP-Gas containers at a bulk plant or industrial plant (and larger storage facilities at marine and pipeline terminals) in a heavily populated or congested area with liquid transfer and associated transportation operations being conducted more or less continually, it is apparent that the need for and character of fire protection provisions can vary widely. In point of fact, these can be determined only by studying each installation.

3-10.2.2* The planning for effective measures for control of inadvertent LP-Gas release or fire shall be coordinated with local emergency handling agencies, such as fire and police departments. Such measures require specialized knowledge and training not commonly present in the training programs of emergency handling agencies. Planning shall consider the safety of emergency personnel.

The supplement *Guidelines for Conducting a Firesafety Analysis* contains a thorough discussion of the factors needed to plan for an LP-Gas emergency.

Training facilities are expensive to install and operate, therefore their numbers in the U.S. are limited. Three of the largest are conducted by the Texas A & M Fire College, the Massachusetts Firefighting Academy, and Nassau County on Long Island, N.Y., on an ongoing basis each year. The need for this hands-on training is being recognized increasingly, however, and new facilities can be anticipated.

The NFPA film, *LP-Gas: Emergency Planning and Response,* is a useful tactical training instrument. The National Propane Gas Association has produced an audio-visual slide production, *Handling LP-Gas Leaks and Fires.*

A-3-10.2.2 The National Fire Protection Association, American Petroleum Institute and National Propane Gas Association publish material, including visual aids, useful in such planning.

3-10.2.3 Except as provided in 3-10.2.4 or 3-10.2.5, fire protection shall be provided for installations having storage containers with an aggregate water capacity of more than 4000 gal (15 m³) subject to exposure from a single fire. The mode of such protection shall be arrived at through competent fire safety analysis of local conditions of hazard within the container site, exposure to or from other properties, water supply, the probable effectiveness of plant fire brigades, and the time of response and probable effectiveness of fire departments. (*See 3-2.2.8.*)

Unless the circumstances described in 3-10.2.4 (*see commentary*) exist, it must be recognized that a degree of exposure exists to the facility, its employees, and the persons and property of neighbors, and that fire protection is needed beyond that specifically required by NFPA 58. This fire protection could be provided by a plant fire brigade and/or fire department.

Such a facility will undoubtedly be within an area served by one or more emergency services—e.g., public fire, police departments, and ambulance services. Facility management has a right to expect assistance from these sources. Management has a duty to obtain this assistance without requiring the emergency personnel to accept undue risks. The resolution of this risk factor is a key element in the firesafety analysis process.

It is evident that a fire department is suited to accomplish Objectives 2 and 3 cited in the commentary on 3-10.2.2. Water in the proper form can disperse gas clouds and cool containers and equipment. Aside from its rescue function, a fire department is essentially a person-machine system designed to apply water. These are the reasons that 3-10.2.3(a) states that the first consideration in the analysis should be directed at the fire department's capabilities.

The citation of 4,000 gal (15 m³) total water capacity reflects fire experience in that the most severe accidents revealing fire protection problems have involved facilities larger than this. The exception for smaller facilities is intended to reduce the burden on regulatory officials, installers, and the facility operators. However, it is not intended that smaller facilities be totally excluded and there may well be circumstances where smaller facilities should be considered.

A discussion of the many aspects of the six fire analysis factors given in 3-10.2.3 is covered in the supplement.

(a) The first consideration in such an analysis shall consist of the use of water applied by hose streams by the fire brigade or fire department for the effective control of hazardous leakage or fire exposing storage tanks, cargo vehicles or railroad tank cars which may be present.

NOTE: Experience has indicated that hose stream application of water in adequate quantities as soon as possible after the initiation of flame contact is an effective way to prevent container failure from fire exposure. The majority of large containers exposed to sufficient fire to result in container failure have failed in from 10 to 30 minutes after the start of the fire when water was not applied. Water in the form of a spray can also be used to control unignited gas leakage.

3-10.2.4 If the analysis specified in 3-10.2.3 indicates a serious hazard does not exist, the fire protection provisions of 3-10.2.3 need not apply.

There may be situations in which no fire protection is needed beyond what is provided in the facility in compliance with specific provisions in NFPA 58. An example of such a situation would be a well-isolated facility.

If an adequate water supply is not available, agreement should be reached with public safety authorities to limit their emergency forces to control of onlookers.

3-10.2.5 If the analysis specified in 3-10.2.3 indicates that a serious hazard exists and the provisions of 3-10.2.3 cannot be met, special protection (*see definition*) shall be provided in accordance with 3-10.3.

Special protection addresses only the BLEVE hazard — because this hazard usually represents the greatest exposure to the facility, its neighbors, and emergency personnel. It must be recognized that special protection only greatly simplifies, and thus enhances the effectiveness of, fire department activities such as rescue and controlling unignited leaks. Therefore, the reference in 3-10.2.3 to 3-10.2.5 as an exception to 3-10.2.3 is misleading in that the provision of special protection does not obviate the need to consider other fire protection measures.

3-10.2.6 Suitable roadways or other means of access for emergency equipment, such as fire department apparatus, shall be provided.

3-10.2.7 Each industrial plant, distributing plant and distributing point shall be provided with at least one approved portable fire extinguisher having a minimum capacity of 20 lb of dry chemical with a B:C rating.

See commentary on 5-5.1.1.

3-10.2.8 LP-Gas fires shall not normally be extinguished until the source of the burning gas has been shut off or can be shut off.

3-10.2.9 Emergency controls shall be conspicuously marked and the controls shall be located so as to be readily accessible in emergencies.

3-10.3 Special Protection.

This provision describes five modes of special protection to minimize the chances of a BLEVE from fire exposure. Special protection is not limited to these five modes, as noted in the definition in Section 1-7 which also permits "any means listed for this purpose." There has been some interest in designs for water application systems other than those given in 3-10.3.4 and 3-10.3.5. However, the contributors are unaware of any such systems that have been listed for this purpose.

Although not specifically cited in this provision, any combination of these five modes (or another listed for the purpose) is also acceptable. For example, a common combination consists of earth mounding for the portion of a container not fitted with appurtenances or connections (which is customarily at least 75 percent of the container surface area) and water spray or monitor protection for the remainder. Because of the rather high water demand for water spray and many monitor nozzle systems, such a combination can preclude expensive water supply improvements.

3-10.3.1* If insulation is used, it shall be capable of limiting the container temperature to not over 800°F (427°C) for a minimum of 50 minutes as determined by test with insulation applied to a steel plate and subjected to a

Figure 3.29 Storage Containers with Special Protection. Installation of two insulated 90,000 gal (341 m³) containers with steel saddles. In addition, monitor nozzles, automatically activated by heat sensors, located at the tank openings were installed.

test flame substantially over the area of the test plate. The insulation system shall be inherently resistant to weathering and the action of hose streams. (*See Appendix H.*)

Several types of cementitious, ablative, or intumescent coatings and steel-jacketed insulation systems have been used or proposed for LP-Gas container special protection. However, it is important that they meet the LP-Gas exposure fire, weathering, and hose stream resistance criteria specified (the latter because it is unrealistic to expect hose streams not to be applied even though an insulated container does not require it). A new Appendix was added in the 1989 Edition, containing a test method for evaluating insulation based on the above criteria. *(See Appendix H.)*

The most extensive testing and experience has been for insulation of railroad tank cars. The U.S. Department of Transportation (DOT) approves a number of both coatings and jacketed systems on the basis of a large scale liquid LP-Gas torch fire exposure test. This test does not involve hose stream application because it is unlikely that hose streams will be used in the environment associated with typical multicar derailment pileups. (It is recognized that hose stream resistance is not as significant with jacketed systems.) Such insulated tank cars have performed well in actual fires.

The Factory Mutual Engineering Corporation lists coating systems for special protection on the basis of a smaller scale modified furnace test, and their test includes hose stream application. The contributors are unaware of any actual fires in insulated stationary LP-Gas containers.

A-3-10.3.1   For LP-Gas fixed storage facilities of 60,000 gal (227 m³) water capacity or less, a competent firesafety analysis (*see 3-10.2.3 and 3-10.2.5*) could indicate that applied insulating coatings are quite often the most practical solution for special protection.

It is recommended that insulation systems be evaluated on the basis of experience or listings by an approved testing laboratory.

The information in the Appendix to 3-10.3.1 was included in the 1976 Edition based upon cost analyses at that time. There is incomplete information to verify the current validity of this.

3-10.3.2   If mounding is utilized, the provisions of 3-2.3.7 shall constitute adequate protection.

This states that a container mounded in accordance with the basic installation criteria automatically complies with the requirements for special protection.

3-10.3.3   If burial is utilized, the provisions of 3-2.3.8 shall constitute adequate protection.

This states that a container buried in accordance with the basic installation criteria automatically complies with the requirements for special protection.

3-10.3.4 If water spray fixed systems are used, they shall comply with NFPA 15, *Standard for Water Spray Fixed Systems for Fire Protection*. Such systems shall be automatically actuated by fire responsive devices and also have a capability for manual actuation.

NFPA 15, *Standard for Water Spray Fixed Systems for Fire Protection*, provides criteria for water spray fixed systems for a number of purposes. The purpose pertinent to special protection is known as vessel exposure protection and is addressed in 4-4.3.2 in NFPA 15.

It should be noted that Appendix A-4-4.3.2(b) of NFPA 15 points out that where the temperature of a vessel or its contents should be limited, higher densities than 0.25 gpm/sq ft [10.2 (L/min)/m²] may be required. Several years ago, a survey of the Technical Committee on Water Spray Fixed Systems was conducted to ascertain the BLEVE experience with systems designed on the basis of 0.25 gpm/sq ft [10.2 (L/min)/m²]. There proved to be very little data available but indications were that this density could not be relied upon to prevent a crack from developing but would probably prevent a crack from growing to the point of container dismemberment. In one actual fire, the crack stopped after 2 ft (0.61 m).

Automatic actuation is required because the nozzles and much of the piping are close to the container where they can be exposed to fire. Water must be introduced quickly to protect the system as well as the container.

Such systems must be carefully maintained by qualified personnel.

3-10.3.5 If monitor nozzles are used, they shall be located and arranged so that all container surfaces likely to be exposed to fire will be wetted. Such systems shall otherwise comply with NFPA 15, *Standard for Water Spray Fixed Systems for Fire Protection*, and shall be automatically actuated by fire responsive devices and also have a capability for manual actuation.

NFPA standards do not provide water application design criteria for monitor nozzles for any application, including for special protection. Because of the hydraulics of fire streams, however, the pressure required to throw an effective stream from the nozzle to the container will probably result in a density on the container well in excess of 0.25 gpm/sq ft [10.2 (L/min)/m²]. Because of the requirement that all container surfaces likely to be exposed to fire must be wetted, considerable ingenuity may be needed to avoid the need for an impractically large water supply.

These systems must be carefully maintained by qualified personnel.

As is the case with fixed water spray systems, the use of other modes—e.g., mounding or insulation—can reduce the water demands greatly.

REFERENCES CITED IN COMMENTARY

The following publications are available from the National Fire Protection Association, Batterymarch Park, Quincy, MA 02269.

NFPA 15, *Standard for Water Spray Fixed Systems for Fire Protection.*
NFPA 30, *Flammable and Combustible Liquids Code.*
NFPA 37, *Standard for the Installation and Use of Stationary Combustion Engines and Gas Turbines.*
NFPA 51B, *Standard for Fire Prevention in Use of Cutting and Welding Processes.*
NFPA 54, *National Fuel Gas Code.*
NFPA 59, *Standard for the Storage and Handling of Liquefied Petroleum Gases at Utility Gas Plants.*
NFPA 70, *National Electrical Code.*
NFPA 77, *Recommended Practice on Static Electricity.*
NFPA 78, *Lightning Protection Code.*
NFPA 501C, *Standard on Firesafety Criteria for Recreational Vehicles.*
NFPA 505, *Firesafety Standard for Powered Industrial Trucks Including Type Designations, Areas of Use, Maintenance, and Operation.*
NFPA Film, *LP-Gas: Emergency Planning and Response.*

The following publications are available from the U.S. Government Printing Office, Washington, DC.

Mobile Home Construction and Safety Standards, Part 280 CFR 24.
Section 393.77 of the Bureau of Motor Carrier Safety Regulations, Part 393 CFR 49.
Section 177.834(d) of the Hazardous Materials Regulations, Part 177 CFR 49.

The following publications are available from the National Propane Gas Association (Formerly National LP-Gas Association), 1301 West 22nd St., Oak Brook, IL 60521.

NPGA 602, *Safe Use of LP-Gas in Industrial Trucks (also Spanish version).*

NPGA 603, *How to Use LP-Gas Safely at Construction Sites.*

NPGA 604, *Safe Use of LP-Gas for Heating Tar.*

NPGA 605, *Safe Use of LP-Gas in Temporary Space Heating with Portable Containers.*

NPGA 606, *Safe Use of LP-Gas with Portable Cylinders for Cutting, Brazing.*

NPGA 800, *Recommended Procedures for the Temporary Use of LP-Gas in Places of Public Assembly Indoors.*

NPGA 803, *Safety Considerations in Converting Passenger Carrying Vehicles from Gasoline to LP-Gas.*

NPGA 804, *Safety Bulletin for Drivers of LP-Gas Powered Vehicles.*

NPGA 4006, *Training Guidebook: LP-Gas Carburetion.*

NPGA 5909, *Safety Considerations for Motor Fuel Systems.*

NPGA 5915, *Safety Factors in Converting School Buses to LP-Gas.*

NPGA 5825, *Preventive Maintenance and Filling Procedures.*

NPGA Film, *Handling LP-Gas Leaks and Fires.*

The following publication is available from the American National Standards Institute, 1430 Broadway, New York, NY 10018.

ANSI B31, *Code for Pressure Piping.*

4

LP-GAS LIQUID TRANSFER

4-1 Scope.

4-1.1 Application.

4-1.1.1 This chapter covers transfers of liquid LP-Gas from one container to another whenever this transfer involves connections and disconnections in the transfer system, or the venting of LP-Gas to the atmosphere. Included are provisions covering operational safety, location of transfer operations, and methods for determining the quantity of LP-Gas permitted in containers.

Experience has shown that the chances of LP-Gas release (especially liquid) are greatest during liquid transfer operations (release is an operational necessity to a degree), and that accurate container filling is necessary to avoid release through pressure relief devices after the container has been filled. The standard, therefore, has always thoroughly addressed these hazards. However, prior to the 1972 Edition, the provisions were scattered throughout the standard, leading to inconsistent treatment of the same basic hazard. By placing them in a single chapter, the inconsistencies were reduced and a focus on this major hazard was provided.

4-1.1.2 Provisions for ignition source control at transfer locations are covered in Section 3-8. Fire protection shall be in accordance with Section 3-10.

4-2 Operational Safety.

4-2.1 Transfer Personnel.

4-2.1.1 Transfer operations shall be conducted by qualified personnel meeting the provisions of 1-6.1.1. At least one qualified person shall remain in attendance at or near the transfer operation from the time connections are made until the transfer is completed, shutoff valves are closed, and lines are disconnected.

The individual doing the transfer should not only be fully qualified in such work, but should also be an individual who

will not panic under emergency conditions. The person must be able to analyze quickly what to do and then do it when an emergency arises. The individual should also be alert to the transfer operation and not just assume that everything is going to move along smoothly without incident. Knowledge of what is happening is most important in the liquid transfer.

The words "in attendance" were added in the 1989 Edition to clarify the Technical Committee's frequently misinterpreted intent which is that transfer operations be continuously attended.

4-2.1.2 Transfer personnel shall exercise precaution to assure that the LP-Gases transferred are those for which the transfer system and the containers to be filled are designed.

Accidents have occurred because propane or a mixture of propane and butane was placed into containers suitable only for butane. The lower setting for the pressure relief device has resulted in its operation, leading to injury and property damage.

4-2.2 Containers to Be Filled.

4-2.2.1 Containers shall be filled only by the owner or upon the owner's authorization.

It is important that containers be filled only by the owner or upon the owner's authorization because if indiscriminate filling is done, the individuals having the container filled might not be properly qualified in the storage, handling, and use of LP-Gas. Unqualified persons getting containers filled may use them inside of buildings, may not have the proper connections to prevent leakage, and may be using the product in improper appliances.

4-2.2.2 Valve outlets on containers of 108 lb (49 kg) water capacity [nominal 45 lb (20 kg) propane capacity] or less shall be equipped with an effective seal such as a plug, cap or an approved quick closing coupling. This seal shall be in place whenever the container is not connected for use. Single trip nonrefillable, disposable, and new unused containers are excluded from this requirement.

The increasing public consumer ownership of portable LP-Gas containers [especially the nominal 20 lb (9 kg) propane cylinder] has resulted in a number of fires due to leakage of gas from cylinder valve outlets. Several have been due to the valve partially (and unintentionally) opening—often during transportation in passenger vehicles. This provision was added in the 1986 Edition and is intended to minimize such accidents.

This provision was placed in Chapter 4 because its execution is best assured in conjunction with the filling operation. However, it is also referenced in Chapters 5 ("Storage of Portable Containers Awaiting Use or Resale") and 6 ("Vehicular Transportation of LP-Gas").

4-2.2.3 Containers shall be filled only after determination that they comply with the design, fabrication, inspection, marking, and requalification provisions of this standard. (*See 2-2.1.2, 2-2.1.3, and 2-2.1.4.*)

As a practical matter, the individual filling a container is usually more qualified than the owner to make this determination. In many instances, this person is the only one in contact with the container who can be expected to possess this ability.

This provision places a great deal of responsibility upon the filler, which may not always be adequately recognized. The filler must be familiar with the marking and inspection requirements and the latter, especially, require considerable training. With the proliferation of ownership of 20 lb (9 kg) DOT specification cylinders by the general public, the necessary requalification 12 years after the date of manufacture and every 5 to 12 years thereafter (*see Appendix C*) is often overlooked. When filling DOT cylinders, the filler should be careful to verify that the cylinder requalification period has not been exceeded.

4-2.2.4 DOT specification cylinders authorized as "single trip," "nonrefillable," or "disposable" containers shall not be refilled with LP-Gas.

These containers are not designed for extended service. Their small size increases the chances for them to be overfilled using commonly available filling equipment and procedures. They are equipped with filling connections that common filling equipment will not fit in an effort to control these hazards.

4-2.2.5 Containers into which LP-Gas is to be transferred shall comply with the following as to service or design pressure in relation to the vapor pressure of the LP-Gas:

(a) For DOT specification cylinders, the service pressure marked on the container shall not be less than 80 percent of the vapor pressure of the LP-Gas at 130°F (54.4°C). For example, if the vapor pressure of a commercial propane is 300 psig (2.0 MPa gauge) at 130°F (54.4°C), the service pressure must be at least 80 percent of 300, or 240 psig (1.6 MPa gauge).

(b) For ASME containers, the minimum design pressure shall comply with Table 2-2.2.2 in relation to the vapor pressure of the LP-Gas.

This is an elaboration upon 4-2.1.2.

4-2.2.6   The provisions of 5-2.2.1 and 6-2.2.4(a) shall apply.

4-2.3 Arrangement and Operation of Transfer Systems.

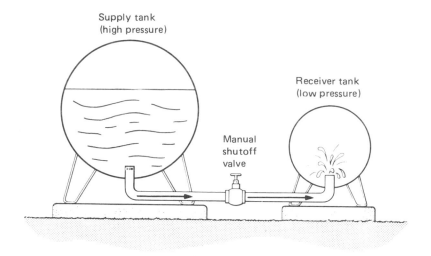

Figure 4.1 Liquid Transfer by Pressure Differential or Gravity.

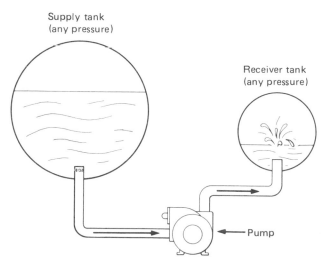

Figure 4.2 Liquid Transfer by Pump.

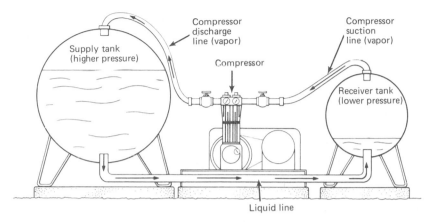

Figure 4.3 Liquid Transfer by Compressor.

Figure 4.4 Cylinder Filling Room with Automatic Cylinder Filling Scales. When the cylinder (not shown) is filled to its correct filling weight, the automatic shutoff valve on the hose inlet will shut off and automatically stop the flow.

4-2.3.1 Liquid transfer may be accomplished by pressure differential, by gravity or by the use of pumps or compressors complying with Section 2-5.

The pressure differential method is used mostly in unloading tank cars and, occasionally, transport trucks. Vapor is taken by a compressor from the container that will be receiving the product and placed into the container from which the liquid is being unloaded. This creates a pressure in the container being unloaded which is higher than the pressure in the receiving container; therefore, liquid will flow into the receiving container. When the liquid has all been transferred, the residual vapors in the railroad or transport vehicle being unloaded can be recovered and placed into the receiving vessel by reversing the flow through the compressor.

It is best to have vapor enter at the bottom of the receiving container so that it will bubble up through the liquid and be cooled and more easily condensed. If the vapor is put into the top of the receiving container, heat of compression will create a higher pressure in the receiving container.

Care also has to be taken not to open valves too rapidly because the closing ("slugging") of any excess-flow check valves present can result.

If unloading by gravity, the receiving container must be at a lower level than the container from which the liquid is being removed. Here again, care has to be taken to prevent closing of an excess-flow check valve.

Pumping of the liquid is the usual method of liquid transfer in operations such as delivery to a customer's container, the transfer from bulk storage into delivery truck containers, and the unloading of transport vehicles. In both methods, it is important to have proper correlation of the design size of the piping and an excess-flow check valve so that there will not be premature closing. This is also important from a safety standpoint because if there is a break or rupture in the line, the size of the line must permit the excess-flow valve to operate.

4-2.3.2 Compressors used for liquid transfer normally shall take suction from the vapor space of the container being filled and dicharge into the vapor space of the container from which the withdrawal is being made.

4-2.3.3 Transfer systems using positive displacement pumps shall comply with 2-5.2.2.

4-2.3.4 Transfer hose equipped with a shutoff valve at the free end, so that the hose normally contains liquid (called "wet hose" by the industry), shall be protected against excessive hydrostatic pressure by the use of hydrostatic relief valves. (*See 3-2.8.*)

NOTE: When hose is to be used for liquid transfer, this arrangement is recommended.

"Wet hose" is recommended so that it is not necessary to evacuate the liquid from the hose when the delivery has been completed thereby creating a hazard in the atmosphere in that area.

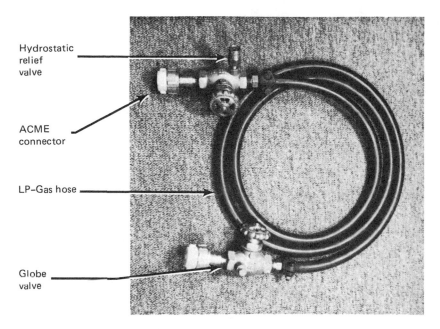

Hydrostatic
relief
valve

ACME
connector

LP-Gas hose

Globe
valve

Figure 4.5 "Wet" Liquid Transfer Hose.

4-2.3.5 The provisions of 3-2.7.9 shall apply.

4-2.3.6 When a hose or swivel type piping is used for loading or unloading railroad tank cars, an emergency shutoff valve complying with 2-4.5.4 shall be used at the tank car end of the hose or swivel type piping.

See 3-2.7.9. Note that 4-2.3.6 requires the use of an emergency shutoff valve regardless of any of the qualifications in 3-2.7.9.

4-2.3.7 Transfer hoses larger than ½ in. (12 mm) internal diameter shall not be used for making connections to individual containers being filled indoors.

This is a leakage rate limiting provision.

4-2.3.8 During the time the tank cars are on sidings for loading or unloading, the following shall apply:

(a) A caution sign, such as "STOP. TANK CAR CONNECTED," shall be placed at the active end(s) of the siding while car is connected as required by DOT Regulations.

(b) A wheel at each end of car shall be blocked on the rails.

Figure 4.6 Truck Unloading Station. This station includes a complete bulkhead to prevent valve breakage in the event of a truck pulling away with the hoses connected. In addition the installation includes a shutoff valve, strainer, sight flow and backflow check valve in the liquid fill line. The vapor equalizing line includes a shutoff valve, emergency shutoff valve with remote cable release and excess-flow valve. Also installed are two crash posts for vehicular protection.

This provision reiterates DOT regulations. There have been cases where tank cars have been moved when the unloading hose or connections were in place and accidents have occurred. There have been cases also where a tank car has been partially unloaded and the hoses disconnected, but the railroad personnel moved the car thinking that it was empty. Therefore, the warning sign should always be in place until the car has been completely unloaded and the railroad personnel advised to move the car. The value of chock blocks has been illustrated by incidents where trespassers have released the brake on the car and it has moved from its unloading point. In one such instance, a string of such cars ran down a grade and collided with a locomotive, killing two railroad workers.

4-3 Location of Transfer Operations.

4-3.1 General.

4-3.1.1   Liquid shall be transferred into containers, including containers mounted on vehicles, only outdoors or in structures especially designed for this purpose.

Figure 4.7 Liquid Transfer Chocking and Warning Sign for Tank Car Loading and Unloading.

Figure 4.8 Rail Car Unloading Tower. This system utilizes a vapor compressor, located in cabinet at base of the tower, to withdraw vapor from the stationary storage tank and discharge the higher pressure vapor into the tank car. This will provide a higher vapor pressure in the tank car and thereby liquid will be discharged from the tank car to the stationary storage tank via a separate liquid fill line. These railcar unloading facilities require check valves and emergency shutoff valves as described in 3-2.7.9. In addition, the track must be grounded, bonded and insulated to meet current American Railway Engineering Association requirements. A derailer, "Stop. Tank Car Connected" sign, and wheel blocks are also required.

(a) Structures housing transfer operations or converted for such use after December 31, 1972, shall comply with Chapter 7.

(b) The transfer of liquid into containers on the roofs of structures is prohibited.

The location provisions reflect not only the hazards where the transfer takes place, but also the fact that the use of hose is common. For example, 4-3.1.1(b) considers that a delivery hose may be taken up through a building. Although not stated, a hose should not be taken through a building to reach a container.

The key to the location of transfer operations is the use of the "point of transfer" concept. This is defined (*see Section 1-7*) as "the location where connections and disconnections are made or where LP-Gas is vented to the atmosphere in the course of transfer operations." From a frequency point of view, a flammable mixture is most often present at and near the point of transfer. This concept was introduced in the 1972 Edition.

4-3.2 Containers in Stationary Installations.

4-3.2.1 Containers located outdoors in stationary installations (*see definition*) in accordance with Section 3-2, and equipped with appurtenances for filling at, or adjacent to, the container may be filled at that location, provided that a cargo vehicle is used for the delivery which complies with Chapter 6 as to construction and method of operation.

This paragraph recognizes that containers fitted with filling connections that are actually container appurtenances—or that are very close to the container—are smaller containers and lower transfer rates are involved. Under these circumstances, the siting criteria for the container are considered adequate for the transfer operation.

4-3.2.2 If the point of transfer (*see definition*) is not located at the container, it shall be located in accordance with 4-3.3.

4-3.3 Containers in Nonstationary Installations.

4-3.3.1 This subsection includes provisions for filling of portable containers not a part of a stationary installation (*see definition*), including containers mounted on vehicles (including recreational vehicles) and industrial and agricultural equipment.

4-3.3.2 The point of transfer (*see definition*) or the nearest part of a structure housing transfer operations, whichever is closer, shall be located in accordance with Table 4-3.3.2 with respect to various types of exposures.

It must be recognized that the containers being filled, covered by 4-3.3, are not fixed in place. Therefore, neither are their filling connections and, as these connections are a component of the point of transfer, neither is the point of transfer. The siting provisions in 4-3.3.2 do fix these points of transfer and, although indirectly, therefore establish the location of these containers and any vehicles they may be installed upon during transfer.

Figure 4.9 Portable Container Filling at Distributing Plant or Industrial Plant.

Figure 4.10 Portable Container Filling at Distributing Point.

(a) If the point of transfer is a component of a system covered by Sections 3-6 or 3-9 or part of a system installed in accordance with standards referenced in 3-1.1.4, Parts 1, 2 and 3 of Table 4-3.3.2 do not apply to the structure containing the point of transfer.

Table 4-3.3.2[1]
Distance Between Point of Transfer and Exposures

Part	Exposure	Min. Horizontal Distance, Feet (Meters)
1.	Buildings[2] with fire resistive walls[3]	10 (3)
2.	Buildings[2] with other than fire resistive walls	25 (7.6)
3.	Building wall openings or pits at or below the level of the point of transfer	25 (7.6)
4.	Line of adjoining property which can be built upon	25 (7.6)
5.	Outdoor places of public assembly, including school yards, athletic fields and playgrounds	50 (15)
6.	Public ways, including public streets, highways, thoroughfares and sidewalks (a) From points of transfer in distributing points	10 (3)
	(b) From points of transfer in all other locations	25 (7.6)
7.	Driveways	5 (1.5)
8.	Mainline railroad track centerlines	25 (7.6)
9.	Containers[4] other than those being filled	10 (3)

[1]Table 4-3.3.2 is not applicable to the transfer operations covered in 4-3.2.1.

[2]"Buildings," for the purpose of this table, include structures such as mobile homes, recreational vehicles, modular homes, tents and box trailers at construction sites.

[3]Walls constructed of noncombustible materials having, as erected, a fire resistance of at least one hour as determined by NFPA 251, *Standard Methods of Fire Tests of Building Construction and Materials*.

[4]Not applicable to filling connections at the storage container or to dispensing units of 2000 gal (7.6 m³) water capacity or less when used for filling containers not mounted upon vehicles.

Parts 1, 2, and 3 of Table 4-3.3.2 pertain to buildings containing general occupancies for which transfer operations can be considered as exposures. If the point of transfer is in a building, it is apparent that the building can not be separated from it and 4-3.3.2(a) simply states the obvious.

(b) If LP-Gas is vented to the atmosphere under the conditions stipulated in 4-4.1.1(d), the distances in Table 4-3.3.2 shall be doubled.

(c) If the point of transfer is housed in a structure complying with Chapter 7, the distances in Table 4-3.3.2 may be reduced provided either the exposing wall(s) or the exposed wall(s) complies with 7-3.1.1(a).

Subparagraph 7-3.1.1(a) describes a pressure resistant wall having no openings. Such a wall reduces the liquid transfer hazards substantially and lesser distances are acceptable.

In Part 6 of Table 4-3.3.2, the lesser distance for a distributing point considers that these facilities are used for filling portable containers and containers on vehicles. These containers are small and are filled at low rates.

The containers cited in Part 9 of Table 4-3.3.2 are containers from which liquid is being withdrawn. These containers must be filled themselves; Note 4 excludes the point of transfer represented by their filling connections (if these are on or at the container). This is simply a restatement of 4-3.2.1.

The other point of transfer present in Part 9 is the filling connection on the container being filled. The remainder of Note 4 is intended to permit portable containers (or any containers not mounted upon vehicles) to be closer than 10 ft (3 m) to the container from which the liquid is being withdrawn [provided the latter is 2,000 gal (7.6 m³) water capacity or less]. In most such installations, the containers being filled are nominal 20 lb (9 kg) LP-Gas capacity filled at low rates and the hazard is minimal. It is suggested, however, that the 10 ft (3 m) distance be maintained whenever possible because, in spite of the operator qualification provisions of 4-2.1.1, the number of such installations is proliferating, especially as adjuncts to establishments whose principal business is other than LP-Gas liquid transfer.

4-3.4 Cargo Vehicles.

4-3.4.1 Cargo vehicles (*see Section 6-3*) unloading into storage containers shall be at least 10 ft (3 m) from the container and so positioned that the shutoff valves on both the truck and the container are readily accessible. In the case of distributing points, such as LP-Gas service stations, the truck or transport shall not be parked on a public way.

Formal Interpretation
Reference   3-2.10.6(b)(3), 4-3.4.1, 3-8.4.1(a), 3-8.4.1(b)

Question: A 1600-gallon LP-Gas storage container and associated liquid transfer equipment are centrally installed in a parking area serving a suburban shopping center for the purpose of filling containers mounted on vehicles and portable containers. The vehicles and portable containers are in the immediate vicinity of the storage container while they are being filled. The lane for vehicles being serviced is also a driveway for parking of vehicles and for movement within the parking area. Is such an installation and operation specifically covered by the provisions of NFPA 58?

Answer: Yes. Specifically, such an installation is a "Distributing Point" as defined in Section 1-7 of NFPA 58 and is subject to all provisions applicable to Distributing Points.

Applicable provisions include 3-2.10.6(b)(3), 4-3.4.1, 3-8.4.1(a) and 3-8.4.1(b), and Parts 6 and 7 of Table 4-3.3.2. In the context of an installation in a parking area open to the public, the Committee is of the opinion that portions of such an area accessible to the public for activities normally associated with the movement and parking of vehicles should be considered as public ways, streets, sidewalks, and thoroughfares as these terms are used in 3-2.10.6(b)(3), 4-3.4.1, and Parts 6 and 7 of Table 4-3.3.2.

The Committee noted that while it would be possible to install such a distributing point "centrally" in such a parking area, provisions for restricting movement of the public and vehicles when not associated with the functioning of the installation [especially to assure compliance with 3-8.4.1(b)] could make a less central location more practical.

Issue Edition   1974
Reference   3195(b)(3), 4012, 4060(a), 4060(b)
Date   October-November 1976

The unloading cargo vehicle should be a distance away from the container receiving the product so that if something happens at either point the other will not be involved to the extent that it would be if it were in close proximity. Also, it is important to have the cargo vehicle so located that it is easy to get to the valves on both the truck and the container so that they can quickly be shut off if there is an emergency need to do so.

4-3.5   Marine Shipping and Receiving.

This section was added in the 1989 Edition in conjunction with the inclusion of marine terminals to the scope of the standard. Much of the material was drawn from NFPA 59A, *Standard for the Storage and Handling of Liquefied Natural Gas (LNG)*, and the existing practices at marine terminals as required by the U.S. Coast Guard and other authorities having jurisdiction. This section is an outgrowth of the requirements of NFPA 30, *Flammable and Combustible Liquids Code*, and NFPA 59A which reflect the experience and practices of the marine industry.

4-3.5.1 Design, construction, and operation of piers, docks, and wharves shall comply with relevant regulations and the requirements of the authorities having jurisdiction.

NOTE: NFPA 30, *Flammable and Combustible Liquids Code,* may be used for guidance as appropriate.

NFPA 30 provides requirements for locating wharves handling flammable liquids.

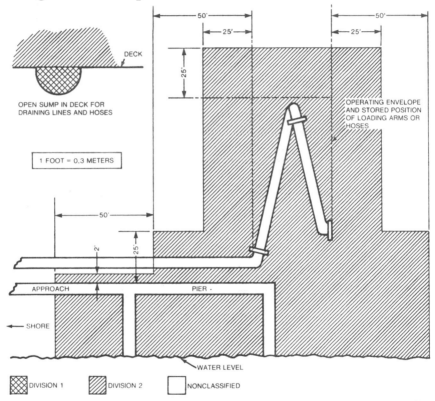

NOTES: (1) The "source of vapor" shall be the operating envelope and stored position of the outboard flange connection of the loading arm (or hose).

(2) The berth area adjacent to tanker and barge cargo tanks is to be Division 2 to the following extent:
a. 25 ft (7.6m) horizontally in all directions on the pier side from that portion of the hull containing cargo tanks.
b. From the water level to 25 ft (7.6m) above the cargo tanks at their highest position.

(3) Additional locations may have to be classified as required by the presence of other sources of flammable liquids (or flammable gases) on the berth, or by Coast Guard or other regulations.

Figure 4.11 Marine Terminal Handling Flammable Liquids.

4-3.5.2 General cargo, other than ships' stores for the LPG tank vessel, shall not be handled over a pier or dock within 100 ft (30 m) of the point of transfer connection while LPG or flammable fluids are being transferred through piping systems. Ship bunkering may be done provided that bunkering is from a pipeline, rather than a barge.

Bunkering is the process of refueling the ship. To eliminate the generation and release of any flammable vapors, this refuel-

ing is usually conducted during cargo transfer if the bunkers are handled by a pipeline and not transferred from a barge.

4-3.5.3 Vehicle traffic shall be prohibited on the pier or dock within 100 ft (30 m) of the loading and unloading manifold while transfer operations are in progress. Suitable warning signs or barricades shall be used to indicate when transfer operations are in progress.

4-3.5.4 Pipelines shall be located on the dock or pier so that they are not exposed to damage from vehicular traffic or other possible cause of physical damage. Underwater pipelines shall be located or protected so that they are not exposed to damage from marine traffic and their locations shall be posted or identified in accordance with federal regulations.

NOTE: Refer to *Code of Federal Regulations*, Title 49, Part 195.

The *Code of Federal Regulations*, Title 49, Part 195 provides safety standards for pipeline facilities used in the transportation of hazardous liquids.

The Coast Guard recognizes the special problems presented by operations involving loading, unloading, handling, and storage of bulk cargoes of certain hazardous materials, especially if the operation occurs at a general cargo marine terminal. For this reason the Coast Guard has defined "cargoes of particular hazard" and "facilities of particular hazard" in its regulations (33 CFR 126.10 and 126.05, respectively). Included as cargoes of particular hazard are flammable liquids, flammable compressed gases, and liquefied natural gas. These cargoes may only be handled at a designated facility.

4-3.5.5 Isolation valving and bleed connections shall be provided at the loading or unloading manifold for both liquid and vapor return lines so that hoses and arms can be blocked off, drained or pumped out, and depressurized before disconnecting. Liquid isolation valves, regardless of size, and vapor valves 8 in. (20 mm) and larger in size shall be equipped with powered operators in addition to means for manual operation. Power operated valves shall be capable of being closed from a remote control station located at least 50 ft (15 m) from the manifold area, as well as locally. Unless the valve will automatically fail closed on loss of power, the valve actuator and its power supply within 50 ft (15 m) of the valve shall be protected against operational failure due to fire exposure of at least 10 minutes duration. Valves shall be located at the point of hose or arm connection to the manifold. Bleeds or vents shall discharge to a safe area.

4-3.5.6 In addition to the isolation valves at the manifold, each vapor return and liquid transfer line shall be provided with a readily accessible isolation valve located on shore near the approach to the pier or dock. Where more than one line is involved, the valves shall be grouped in one location. Valves shall be identifed as to their service. Valves 8 in. (20 mm) and larger in size shall be equipped with power operators. Means for manual operation shall be provided.

4-3.5.7   Pipelines used for liquid unloading only shall be provided with a check valve located at the manifold adjacent to the manifold isolation valve.

4-3.5.8   Marine terminals used for loading ships or barges shall be equipped with a vapor return line designed to connect to the vessel's vapor return connections.

4-3.5.9   Prior to transfer, the officer in charge of vessel cargo transfer and the person in charge of the shore terminal shall inspect their respective facilities to ensure that transfer equipment is in proper operation condition. Following this inspection, they shall meet and determine the transfer procedure, verify that adequate ship-to-shore communications exist, and review emergency procedures.

> The *Code of Federal Regulations*, Title 33, Part 126.15 provides guidelines for the control of liquid cargo transfer systems, including actions required prior to the transfer of cargo.

4-4 Venting LP-Gas to the Atmosphere.

4-4.1 General.

4-4.1.1   Except as provided in 4-4.2 and 4-4.3, LP-Gas, in either liquid or vapor form, shall not be vented to the atmosphere except under the following conditions:

(a) Venting for the operation of fixed liquid level, rotary or slip tube gauges, provided the maximum flow does not exceed that from a No. 54 drill orifice.

(b) Venting the LP-Gas between shutoff valves before disconnecting the liquid transfer line from the container. When necessary, suitable bleeder valves shall be used.

(c) LP-Gas may be vented for the purposes described in 4-4.1.1(a) and (b) within structures designed for container filling as provided in 4-3.1.1 and Chapter 7.

(d) Venting vapor from listed liquid transfer pumps using such vapor as a source of energy, provided the rate of discharge does not exceed that from a No. 31 drill size opening. (*See 4-3.3.2 as to location of such transfer operations.*)

> While it is necessary to vent both vapor and a small amount of liquid from a fixed liquid level, rotary, or slip tube gauge, it should be done with consideration for the surroundings. For instance, discharge should not be close to or under a window into a building. Serious accidents have occurred when this situation was not taken into consideration and the vapors entered the opening. Accidents have occurred when the liquid

trapped between shutoff valves was not bled off before disconnecting the liquid transfer hose. The bleeding off must be done slowly and carefully. If this is not done, and the disconnection is made, a potential hazard is created because the rapid discharge could result in liquid propane being squirted into a person's eyes as well as increasing the area of vapor ignition hazard.

There are a few applications where propane is used as the source of energy in pumps and the vapors are vented to the air. Even though the amount of vented vapor is restricted by a No. 31 drill size opening, the extent of the vapor ignition hazard is sizeable as indicated by the distances cited in 4-3.3.2(b).

4-4.2 Purging.

4-4.2.1 Venting of gas from containers for purging or for other purposes shall be accomplished as follows:

(a) If indoors, containers may be vented only in structures designed and constructed for container filling in accordance with 4-3.1.1 and Chapter 7 and with the following provisions:

(1) Piping shall be provided to carry the vented product outside and to a point at least 3 ft (1 m) above the highest point of any building within 25 ft (7.6 m).

(2) Only vapors shall be exhausted to the atmosphere.

(3) If a vent manifold is used to allow for the venting of more than one container at a time, each connection to the vent manifold shall be equipped with a backflow check valve.

(b) When out-of-doors, container venting shall be done under conditions that will result in rapid dispersion of the product being released. Consideration shall be given to such factors as distance to buildings, terrain, wind direction and velocity, and use of a vent stock so that a flammable mixture will not reach a point of ignition.

(c) If conditions are such that venting into the atmosphere cannot be accomplished safely, LP-Gas may be burned off providing such burning is done under controlled conditions remote from combustibles or a hazardous atmosphere.

The term "purging" is generally applied to the process of completely emptying a container of both liquid and vapor or of replacing the air in a container with liquid and vapor. The most common other purpose referred to in 4-4.2.1 is the removal of LP-Gas from an overfilled container. Whatever the purpose, the release should be only vapor whenever it is feasible to do so.

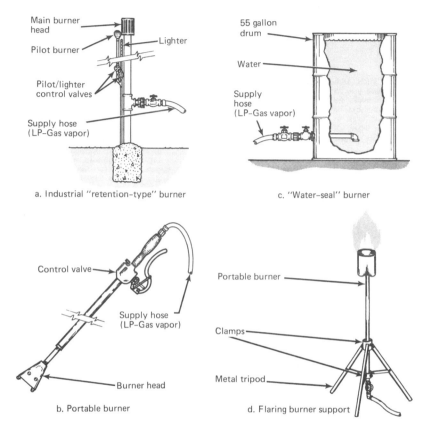

Figure 4.12 Some Burner Arrangements for Flaring LP-Gas.

This not only considers the liquid-to-vapor expansion ratio but recognizes that liquid discharge can present a static ignition hazard (*see 3-8.1.3*).

If a container has been grossly over-filled, it is impossible to prevent liquid discharge from the container outlet initially when correcting an over-fill condition. In these circumstances, compliance with 4-4.2.1(a)(1) is especially vital because the piping provides a safe place for vaporization.

The backflow check valve required in each connection by 4-4.2.1(a)(3) is needed to minimize the possibility of a flammable mixture being created in any one of the connections. When a container is almost completely empty of vapor and the pressure is practically at atmospheric, there is a possibility of a flammable mixture being created. While the container could withstand the combustion explosion pressure developed, the piping could be subjected to detonation pressures it could not withstand.

When venting off the vapor, it must be kept in mind that the evaporation of the liquid vapor results in refrigeration, and the temperature of the liquid may get down to approximately the normal boiling point. Because there will be no pressure, the operator may erroneously assume that the container is empty. When the liquid warms up there will be a pressure condition and, if the valves have been left open, there will be a flow of vapor into the atmosphere.

When venting is being done and the vapors are being discharged, it is important to check on conditions such as direction of wind, lack of wind, speed of wind, and the general layout of the area and structures. A given location might be perfectly satisfactory for venting of the vapors under certain wind conditions, but be unsafe under other atmospheric conditions.

In some situations it is safer to burn off the vapors that are being vented rather than disperse them into the atmosphere. However, when this is done a properly engineered flare stack should be used.

4-4.3 Emergency Venting.

4-4.3.1 The procedure to be followed for the disposal of LP-Gas in an emergency will be dictated by the conditions present, requiring individual judgment in each case and using, where practical, the provisions of this standard.

The need for the emergency venting cited here is most common in cargo vehicle and railroad tank car accidents. These often require the services of experts. In recent years, the industry has increased the availability of such experts and there are teams available that have been organized on a state or regional basis.

4-5 Quantity of LP-Gas in Containers.

Filling containers to a proper capacity is important because LP-Gas liquids expand when heated and contract when cooled. Also, the pressure increases as the temperature of the liquid rises. However, the degree of expansion varies for the LP-Gases. Table 4-5.2.1 was developed to stipulate the filling density for different liquids, as shown by the different specific gravities in the left-hand column. Propane generally falls within the 0.504-0.510 specific gravity. (*See Appendix B.*)

Determination of the maximum quantity that can be placed in a container is expressed as a percent of the water weight capacity of the container. As shown in the second column of Table 4-5.2.1, propane may be 42 percent of the water weight capacity for containers up to 1,200 gal (4.5 m³). For containers over 1,200 U.S. Gallons (4.5 m³), the amount will be 45 percent of the water weight capacity. For example, a "100 lb" (45 kg) propane cylinder which has a marked water capacity of 239 lb (108 kg), may contain 239 × 0.42, or 100.38 lb (45 kg) of propane.

A higher filling density is allowed for the containers over 1,200 U.S. Gallons (4.5 m³) capacity because the larger quantity of LP-Gas is less subject to temperature rise. Buried containers are insulated from atmospheric and solar heating and a higher percent of the water capacity is allowed for such containers. This is shown in the right-hand column of Table 4-5.2.1. Appendix F, "Liquid Volume Tables, Computations, and Graphs," can be of material assistance in determining the proper filling of containers and should be included in the training of all personnel filling containers.

Proper filling is especially important for containers that are going to be used inside a building, e.g., industrial truck containers. The cylinder usually will be filled outdoors and the temperature might be extremely cold, even below zero. When it is taken inside the building, the temperatures may be in the 60°F (15.5°C) range. When these extreme differential temperatures are involved, the container must be very carefully filled to avoid having it become liquid-full and create a hydrostatic pressure situation within the building which would cause the pressure relief device to open. [*See 2-3.4.4(a)*.] The qualifications of personnel handling such containers should include the importance of the liquid expansion factor.

The most accurate method for filling small portable containers is by weight, provided the scales are accurate. A second method is by volume, using a fixed liquid level dip tube gauge. If properly designed and installed, this gauge is accurate. Other volume gauges, such as a rotary gauge, should not be used because they are not accurate enough. The rotary gauge is used to check on gas availability and not for specific filling accuracy gauging.

When using the volumetric method of filling, it is necessary to correct the gauge reading by reference to the liquid volume temperature correction table in Appendix F to determine the true amount of liquid at 60°F (15.5°C). If the volumetric method of filling is used, utilizing the fixed liquid level gauge, or a variable liquid level gauge without liquid volume temperature correction, the liquid level indicated by such gauges must be computed on the basis of the maximum permitted filling density when the liquid is at 40°F (4.4°C) for aboveground containers and at 50°F (10°C) for underground containers. [*See 2-3.4.4(a)*.]

It is also suggested that when a variable type gauging device such as the rotary gauge is used, its accuracy be checked periodically with the fixed liquid level gauge.

4-5.1 Application.

4-5.1.1 This section includes provisions covering the maximum permissible LP-Gas content of containers and the methods of verifying this quantity.

4-5.2 LP-Gas Capacity of Containers (see Appendix F).

4-5.2.1 The maximum LP-Gas content of any container shall be that quantity which equals the maximum permitted filling density given in Table 4-5.2.1.

4-5.2.2 Filling density is defined as the ratio of the weight of LP-Gas in a container to the weight of water at 60°F (15.6°C) that the container will hold. The maximum permitted filling density shown in Table 4-5.2.1 is in percent of the water weight capacity (WWC) for the specific gravity of the particular LP-Gas, the size of the container and the container location.

4-5.2.3 Determination that the LP-Gas content of a container complies with Table 4-5.2.1 may be either by weight or by volume in accordance with 4-5.3. If by volume, the volume having a weight equal to the maximum permitted filling density shall be calculated by the formula in 4-5.2.3(b). These equivalent volumes are shown in Tables 4-5.2.3(a), (b), and (c).

Table 4-5.2.1
Maximum Permitted Filling Density

Specific Gravity at 60°F (15.6°C)	Aboveground Containers		Underground Containers all Capacities
	0 to 1200 U.S. Gal (1000 Imp. gal, 4.5 m³) Total Water Cap.	Over 1200 U.S. Gal (1000 Imp. gal, 4.5 m³) Total Water Cap.	
.496-.503	41%	44%	45%
.504-.510	42	45	46
.511-.519	43	46	47
.520-.527	44	47	48
.528-.536	45	48	49
.537-.544	46	49	50
.545-.552	47	50	51
.553-.560	48	51	52
.561-.568	49	52	53
.569-.576	50	53	54
.577-.584	51	54	55
.585-.592	52	55	56
.593-.600	53	56	57

(a) The maximum liquid LP-Gas content of any container depends upon the size of the container, whether it is installed aboveground or underground, the maximum permitted filling density and the temperature of the liquid [see Tables 4-5.2.3(a), (b), and (c)].

(b) The maximum volume "V_t" (in percent of container capacity) of an LP-Gas at temperature "t," having a specific gravity "G" and a filling density of "L," shall be computed by use of the formula (see Appendix F-4.1.2 for example):

$$V_t = \frac{L}{G} \div F, \text{ or } V_t = \frac{L}{G \times F} \text{ where:}$$

V_t = percent of container capacity which may be filled with liquid.

L = filling density.
G = specific gravity of particular LP-Gas.
F = correction factor to correct volume at temperature "t" to 60°F (15.6°C).

4-5.3 Compliance with Maximum Permitted Filling Density Provisions.

4-5.3.1   The maximum permitted filling density for any container, where practical, may be determined by weight.

4-5.3.2   The volumetric method may be used for the following containers if designed and equipped for filling by volume:

(a) DOT specification cylinders of less than 200 lb (91 kg) water capacity which are not subject to DOT jurisdiction (such as, but not limited to, motor fuel containers on vehicles not in interstate commerce or cylinders filled at the installation).

(b) DOT specification cylinders of 200 lb (91 kg) water capacity or more. (See DOT regulations requiring spot weight checks.)

(c) Cargo tanks or portable tank containers complying with DOT Specifications MC-330, MC-331 or DOT 51.

(d) ASME and API-ASME containers complying with 2-2.1.3 or 2-2.2.2.

4-5.3.3   When the volumetric method is used, it shall be in accordance with 4-5.3.3(a) through (c).

(a) If a maximum fixed liquid level gauge, or a variable liquid level gauge without liquid volume temperature correction is used, the liquid level indicated by these gauges must be computed on the basis of the maximum permitted filling density when the liquid is at 40°F (4.4°C) for aboveground containers or at 50°F (10°C) for underground containers.

Table 4-5.2.3(a)
Maximum Permitted Liquid Volume
(Percent of Total Water Capacity)

Aboveground Containers
0 to 1200 Gallons (0 to 4.5 m³)

Liquid Temperature °F (°C)	Specific Gravity												
	.496 to .503	.504 to .510	.511 to .519	.520 to .527	.528 to .536	.537 to .544	.545 to .552	.553 to .560	.561 to .568	.569 to .576	.577 to .584	.585 to .592	.593 to .600
−50 (−45.6)	70	71	72	73	74	75	75	76	77	78	79	79	80
−45 (−42.8)	71	72	73	73	74	75	76	77	77	78	79	80	80
−40 (−40)	71	72	73	74	75	75	76	77	78	79	79	80	81
−35 (−37.2)	71	72	73	74	75	76	77	77	78	79	80	80	81
−30 (−34.4)	72	73	74	75	76	76	77	78	78	79	80	81	81
−25 (−31.5)	72	73	74	75	76	77	77	78	79	80	80	81	82
−20 (−28.9)	73	74	75	76	76	77	78	79	79	80	81	81	82
−15 (−26.1)	73	74	75	76	77	77	78	79	80	80	81	82	83
−10 (−23.3)	74	75	76	76	77	78	79	79	80	81	81	82	83
− 5 (−20.6)	74	75	76	77	78	78	79	80	80	81	82	82	83
0 (−17.8)	75	76	76	77	78	79	79	80	81	81	82	83	84
5 (−15)	75	76	77	78	78	79	80	81	81	82	83	83	84
10 (−12.2)	76	77	77	78	79	80	80	81	82	82	83	84	84
15 (−9.4)	76	77	78	79	80	80	81	81	82	83	83	84	85
20 (−6.7)	77	78	78	79	80	80	81	82	83	84	84	84	85
25 (3.9)	77	78	79	80	80	81	82	82	83	84	84	85	85
30 (−1.1)	78	79	79	80	81	81	82	83	83	84	85	85	86
35 (1.7)	78	79	80	81	81	82	83	83	84	85	85	86	86
*40 (4.4)	79	80	81	81	82	82	83	84	84	85	86	86	87
45 (7.8)	80	80	81	82	82	83	84	84	85	85	86	87	87
50 (10)	80	81	82	82	83	83	84	85	85	86	86	87	88
55 (12.8)	81	82	82	83	84	84	85	85	86	86	87	87	88
60 (15.6)	82	82	83	84	84	85	85	86	86	87	87	88	88
65 (18.3)	82	83	84	84	85	85	86	86	87	87	88	88	89
70 (21.1)	83	84	84	85	85	86	86	87	87	88	88	89	89
75 (23.9)	84	85	85	85	86	86	87	87	88	88	89	89	90
80 (26.7)	85	85	86	86	87	87	87	88	88	89	89	90	90
85 (29.4)	85	86	87	87	88	88	88	89	89	89	90	90	91
90 (32.2)	86	87	87	88	88	88	89	89	90	90	90	91	91
95 (35)	87	88	88	88	89	89	89	90	90	91	91	91	92
100 (37.8)	88	89	89	89	89	90	90	90	91	91	92	92	92
105 (40.4)	89	89	90	90	90	90	91	91	91	92	92	92	93
110 (43)	90	90	91	91	91	91	92	92	92	92	93	93	93
115 (46)	91	91	92	92	92	92	92	92	93	93	93	94	94
120 (49)	92	92	93	93	93	93	93	93	93	94	94	94	94
125 (51.5)	93	94	94	94	94	94	94	94	94	94	94	95	95
130 (54)	94	95	95	95	95	95	95	95	95	95	95	95	95

*See 4-5.3.3(a).

Table 4-5.2.3(b)
Maximum Permitted Liquid Volume
(Percent of Total Water Capacity)

Liquid Temperature °F (°C)	Aboveground Containers Over 1200 Gallons (4.5 m³)												
	Specific Gravity												
	.496 to .503	.504 to .510	.511 to .519	.520 to .527	.528 to .536	.537 to .544	.545 to .552	.553 to .560	.561 to .568	.569 to .576	.577 to .584	.585 to .592	.593 to .600
−50 (−45.6)	75	76	77	78	79	80	80	81	82	83	83	84	85
−45 (−42.8)	76	77	78	78	79	80	81	81	82	83	84	84	85
−40 (−40)	76	77	78	79	80	80	81	82	83	83	84	85	85
−35 (−37.2)	77	78	78	79	80	81	82	82	83	84	84	85	86
−30 (−34.4)	77	78	79	80	80	81	82	83	83	84	85	85	86
−25 (−31.5)	78	79	79	80	81	82	82	83	84	84	85	86	86
−20 (−28.9)	78	79	80	81	81	82	83	83	84	85	85	86	87
−15 (−26.1)	79	79	80	81	82	82	83	84	85	85	86	87	87
−10 (−23.3)	79	80	81	82	82	83	84	84	85	86	86	87	87
− 5 (−20.6)	80	81	81	82	83	83	84	85	85	86	87	87	88
0 (−17.8)	80	81	82	82	83	84	84	85	86	86	87	88	88
5 (−15)	81	82	82	83	84	84	85	86	86	87	87	88	89
10 (−12.2)	81	82	83	83	84	85	85	86	87	87	88	88	89
15 (−9.4)	82	83	83	84	85	85	86	87	87	88	88	89	90
20 (−6.7)	82	83	84	85	85	86	86	87	88	88	89	89	90
25 (−3.9)	83	84	84	85	86	86	87	88	88	89	89	90	90
30 (−1.1)	83	84	85	86	86	87	87	88	89	89	90	90	91
35 (1.7)	84	85	86	86	87	87	88	89	89	90	90	91	91
*40 (4.4)	85	86	86	87	87	88	88	89	90	90	91	91	92
45 (7.8)	85	86	87	87	88	88	89	89	90	91	91	92	92
50 (10)	86	87	87	88	88	89	90	90	91	91	92	92	92
55 (12.8)	87	88	88	89	89	90	90	91	91	92	92	92	93
60 (15.6)	88	88	89	89	90	90	91	91	92	92	93	93	93
65 (18.3)	88	89	90	90	91	91	91	92	92	93	93	93	94
70 (21.1)	89	90	90	91	91	91	92	92	93	93	94	94	94
75 (23.9)	90	91	91	91	92	92	92	93	93	94	94	94	95
80 (26.7)	91	91	92	92	92	93	93	93	94	94	95	95	95
85 (29.4)	92	92	93	93	93	93	94	94	95	95	95	96	96
90 (32.2)	93	93	93	94	94	94	95	95	95	95	96	96	96
95 (35)	94	94	94	95	95	95	95	96	96	96	96	97	97
100 (37.8)	94	95	95	95	95	96	96	96	96	97	97	97	98
105 (40.4)	96	96	96	96	96	97	97	97	97	97	98	98	98
110 (43)	97	97	97	97	97	97	97	98	98	98	98	98	99
115 (46)	98	98	98	98	98	98	98	98	98	99	99	99	99

*See 4-5.3.3(a)

Table 4-5.2.3(c)
Maximum Permitted Liquid Volume
(Percent of Total Water Capacity)

All Underground Containers

Liquid Temperature °F (°C)	Specific Gravity												
	.496 to .503	.504 to .510	.511 to .519	.520 to .527	.528 to .536	.537 to .544	.545 to .552	.553 to .560	.561 to .568	.569 to .576	.577 to .584	.585 to .592	.593 to .600
−50 (−45.6)	77	78	79	80	80	81	82	83	83	84	85	85	86
−45 (−42.8)	77	78	79	80	81	82	82	83	84	84	85	86	87
−40 (−40)	78	79	80	81	81	82	83	83	84	85	86	86	87
−35 (−37.2)	78	79	80	81	82	82	83	84	85	85	86	87	87
−30 (−34.4)	79	80	81	81	82	83	84	84	85	86	86	87	88
−25 (−31.5)	79	80	81	82	83	83	84	85	85	86	87	87	88
−20 (−28.9)	80	81	82	82	83	84	84	85	86	86	87	88	88
−15 (−26.1)	80	81	82	83	84	84	85	86	86	87	87	88	89
−10 (−23.3)	81	82	83	83	84	85	85	86	87	87	88	88	89
− 5 (−20.6)	81	82	83	84	84	85	86	86	87	88	88	89	89
0 (−17.8)	82	83	84	84	85	85	86	87	87	88	89	89	90
5 (−15)	82	83	84	85	85	86	87	87	88	88	89	90	90
10 (−12.2)	83	84	85	85	86	86	87	88	88	89	90	90	91
15 (−9.4)	84	84	85	86	86	87	88	88	89	89	90	91	91
20 (−6.7)	84	85	86	86	87	88	88	89	89	90	90	91	91
25 (−3.9)	85	86	86	87	87	88	89	89	90	90	91	91	92
30 (−1.1)	85	86	87	87	88	89	89	90	90	91	91	92	92
35 (1.7)	86	87	87	88	88	89	90	90	91	91	92	92	93
40 (4.4)	87	87	88	88	89	90	90	91	91	92	92	93	93
45 (7.8)	87	88	89	89	90	90	91	91	92	92	93	93	94
*50 (10)	88	89	89	90	90	91	91	92	92	93	93	94	94
55 (12.8)	89	89	90	91	91	91	92	92	93	93	94	94	95
60 (15.6)	90	90	91	91	92	92	92	93	93	94	94	95	95
65 (18.3)	90	91	91	92	92	93	93	94	94	94	95	95	96
70 (21.1)	91	91	92	93	93	93	94	94	94	95	95	96	96
75 (23.9)	92	93	93	93	94	94	94	95	95	95	96	96	97
80 (26.7)	93	93	94	94	94	95	95	95	96	96	96	97	97
85 (29.4)	94	94	95	95	95	95	96	96	96	97	97	97	98
90 (32.2)	95	95	95	95	96	96	96	97	97	97	98	98	98
95 (35)	96	96	96	96	97	97	97	97	98	98	98	98	99
100 (37.8)	97	97	97	97	97	98	98	98	98	99	99	99	99
105 (40.4)	98	98	98	98	98	98	99	99	99	99	99	99	99

*See 4-5.3.3(a).

(b) When a variable liquid level gauge is used and the liquid volume is corrected for temperature, the maximum permitted liquid level shall be in accordance with Tables 4-5.2.3(a), (b), and (c).

(c) In the case of containers fabricated after December 31, 1965, with a water capacity of 2,000 gal (7.6 m³) or less and which are filled at consumer sites, gauging shall comply with the following:

(1) The variable gauge shall have been checked for accuracy by comparison with the liquid level indicated by the fixed maximum liquid level gauge.

(2) If the container is to be filled beyond the level indicated by the fixed maximum liquid level gauge, the reading of the variable gauge, adjusted for the error indicated by the check with the fixed maximum liquid level gauge, shall be corrected for the LP-Gas liquid temperature.

4-5.3.4 When containers are to be filled volumetrically by a variable liquid level gauge in accordance with 4-5.3.3(b), provisions shall be made for determining the liquid temperature (*see F-3.1.2*).

REFERENCES CITED IN COMMENTARY

The following publications are available from the National Fire Protection Association, Batterymarch Park, Quincy, MA 02269.

NFPA *30, Flammable and Combustible Liquids Code..*
NFPA *59A, Standard for the Storage and Handling of Liquified Natural Gas (LNG).*

The following publication is available from the U.S. Government Printing Office, Washington, DC.
Code of Federal Regulations, Title 49, Part 195, and Title 33, Part 126.15.

5
STORAGE OF PORTABLE CONTAINERS AWAITING USE OR RESALE

5-1 Scope.

5-1.1 Application.

5-1.1.1 The provisions of this chapter are applicable to the storage of portable containers of 1,000 lb (454 kg) water capacity, or less, whether filled, partially filled or empty (if they have been in LP-Gas service) as follows:

(a) At consumer sites or distributing points, but not connected for use.

(b) In storage for resale by dealer or reseller.

5-1.1.2 The provisions of this chapter do not apply to:

(a) Containers stored at distributing plants.

Provisions for the storage of portable containers on the premises of users and at locations for resale first appeared in the standard in the early 1940s. By that time, many filled cylinders were being resold at hardware stores, etc. and it was felt that it was necessary to regulate the manner in which these containers should be stored. Similarly, many users, particularly industrial users, stored cylinders on their premises awaiting use, and standards for their storage were also needed. The current provisions of NFPA 58 do not differentiate between storage at a reseller or user location, but rather, stipulate what is to be done if portable containers are stored in conventional buildings (whether frequented by the public or not), special buildings or rooms, and outside of buildings. The provisions of this chapter apply only to the storage of portable containers on the premises of consumers, at distributing points, and at locations for resale by the cylinder dealer or reseller. They do not apply to storage of portable containers at distributing plants.

5-2 General Provisions.

5-2.1 General Location of Containers.

5-2.1.1 Containers in storage shall be so located as to minimize exposure to excessive temperature rise, physical damage or tampering.

Because they are infrequently occupied by persons, the temperature of storage locations is often not rigorously controlled. Because of the smaller size of the containers covered by Chapter 5, their contents' temperatures tend to fluctuate more directly with ambient air temperatures or solar radiation. Even though [as noted in Appendix F, F-2.1(a)] these containers will not relieve LP-Gas through their pressure relief devices until the temperature of their contents exceeds 130°F (54°C), such temperatures could be reached in some poorly located, constructed, or ventilated storage locations. Some storage locations have considerable vehicular traffic (e.g., forklift trucks) and experience has shown the need to consider physical damage protection.

5-2.1.2 Containers in storage having individual water capacity greater than 2 ½ lb (1 kg) [nominal 1 lb (0.45 kg) LP-Gas capacity] shall be positioned such that the pressure relief valve is in direct communication with the vapor space of the container.

This is to assure that discharge is largely vapor rather than liquid, which is important from the standpoint of both the quantity of potential flammable vapor/air mixture and the relieving capacity of the pressure relief devices.

5-2.1.3 Containers stored in buildings in accordance with Section 5-3 shall not be located near exits, stairways, or in areas normally used, or intended to be used, for the safe egress of people.

This provision is consistent with NFPA 101®, *Life Safety Code*®.

5-2.1.4 Empty containers which have been in LP-Gas service shall preferably be stored in the open. If stored inside, they shall be considered as full containers for the purposes of determining the maximum quantities of LP-Gas permitted in 5-3.1.1, 5-3.2.1, and 5-3.3.1.

Once filled, an LP-Gas container is seldom ever truly empty. At the least, it will usually be full of vapor and may contain some liquid or a residue that could contain the flammable odorant. If empty containers were not counted as full containers it would be difficult for an enforcing authority to determine if the storage limits were being exceeded without lifting or tilting the cylinders.

Figure 5.1 Storage Rack for Engine Fuel (Industrial Truck) Containers. Containers are so-called "universal" type in which a curved tube connects the relief valve inlet to the vapor space, thus permitting them to be stored on their sides. All enclosure surfaces, including the roof, are open mesh both for ventilation and so that cooling water can be applied in the event of leakage or fire.

5-2.1.5 Containers not connected for use shall not be stored on roofs.

Rooftops are largely "out-of-sight and out-of-mind," and are often places where combustible materials accumulate. Even an "empty" container can complicate fire control activities in a location that is difficult to handle.

5-2.2 Protection of Valves on Containers in Storage.

5-2.2.1 Container valves shall be protected as required by 2-2.4.1.

Screw-on type caps or collars shall be securely in place on all containers stored regardless of whether they are full, partially full or empty, and container outlet valves shall be closed and plugged or capped. The provisions of 4-2.2.2 shall apply.

See commentary on 4-2.2.2.

5-3 Storage within Buildings.

5-3.1 Storage within Buildings Frequented by the Public.

5-3.1.1 DOT specification cylinders with a maximum water capacity of 2 ½ lb (1 kg), used with completely self-contained hand torches and similar applications, may be stored or displayed in a building frequented by the public. The quantity of LP-Gas shall not exceed 200 lb (91 kg) except as provided in 5-3.3.

> **Earlier editions of the standard could be interpreted to mean that 20 lb (9 kg) LP-Gas cylinders for wallpaper removal (steamers) could be stored at paint and wallpaper stores. A formal interpretation was issued to indicate that the intent of these provisions was that the size of the cylinder be limited to 2½ lb (1 kg) water capacity or about 1 lb (0.5 kg) LP-Gas capacity torch-type cylinders. Storage of containers larger than this is covered in 5-3.2, 5-3.3, or Section 5-4. With respect to storage at hardware or paint stores, see 5-3.3 or Section 5-4.**
>
> **In earlier days, the storage of 2½ lb (1 kg) water capacity torch-type cylinders was appropriately covered by limiting such storage to 24 units of each brand of display with a maximum of 200 lb (91 kg) of LP-Gas. However, with the popularity of discount stores and mass merchandizing, this limitation was not practical. These types of retail establishments needed more inventory on hand. Consequently, the current standards continue to limit the amount on display, but provide for additional storage if located in accordance with 5-3.3, "Storage within Special Buildings or Rooms."**

5-3.2 Storage within Buildings Not Frequented by the Public (Such as Industrial Buildings).

5-3.2.1 The maximum quantity allowed in one storage location shall not exceed 735 lb (334 kg) water capacity [nominal 300 lb (136 kg) LP-Gas]. If additional storage locations are required on the same floor within the same building, they shall be separated by a minimum of 300 ft (91 m). Storage beyond these limitations shall comply with 5-3.3.

> **It became apparent that the previous provision of the standard limiting the storage in industrial type buildings to 300**

lb (136 kg) LP-Gas capacity was not realistic for large industrial complexes where buildings may be as large as 300,000 sq ft (28 000 m²). As a consequence, the standards were revised to permit more than one storage location with a maximum of 300 lb (136 kg) LP-Gas if such locations had a separation of 300 ft (91 m). If additional storage was needed, a facility described by 5-3.3 would have to be used.

5-3.2.2 Containers carried as part of the service equipment on highway mobile vehicles are not to be considered in the total storage capacity in 5-3.2.1 provided such vehicles are stored in private garages and carry no more than 3 LP-Gas containers with a total aggregate capacity per vehicle not exceeding 100 lb (45 kg) of LP-Gas. Container valves shall be closed when not in use.

The provision was developed originally to recognize the carrying of portable containers on telephone and electric utility service vehicles and the garaging of such vehicles in buildings owned and occupied by such firms. It was amended in the 1989 Edition to permit up to 3 containers on a vehicle while maintaining the total quantity at 100 lb (45 kg) LP-Gas because the Technical Committee was made aware that many utility vehicles had been carrying multiple containers to meet service requirements and the fire experience has been satisfactory.

5-3.3 Storage within Special Buildings or Rooms.

5-3.3.1 The maximum quantity of LP-Gas which may be stored in special buildings or rooms shall be 10,000 lb (4 540 kg).

5-3.3.2 Special buildings or rooms for storing LP-Gas containers shall not be located adjoining the line of property occupied by schools, churches, hospitals, athletic fields, or other points of public gathering.

5-3.3.3 The construction of all such special buildings, and rooms within, or attached to, other buildings, shall comply with Chapter 7 and the following:

(a) Adequate vents, to the outside only, shall be provided at both top and bottom, located at least 5 ft (1.5 m) away from any building opening.

(b) The entire area shall be classified for purposes of ignition source control in accordance with Section 3-8.

5-3.4 Storage Within Residential Buildings.

5-3.4.1 Storage of containers within a residential building including the basement or any storage area in a common basement storage area in multiple family buildings and attached garages shall be limited to 2 containers each with a maximum water capacity of 2 ½ lb (1.1 kg) and not exceed 5 lbs (2.3 kg) total water capacity for smaller containers per each living space unit. Each container shall meet DOT specifications.

> This provision was added in the 1989 Edition to clarify the Committee's intent on the restriction of storage of LP-Gas containers in buildings. It clearly limits storage to two 1 lb (0.5 kg) (net) containers. Previously, some users of the standard have interpreted that a 20 lb (9 kg) barbeque grill cylinder could be stored in a building if it were not connected for use. There have been many incidents resulting in injury, death, and property damage due to fire and explosion resulting from leakage from 20 lb (9 kg) barbeque grill and recreational vehicle cylinders being stored in buildings. It has been the Committee's long standing position that LP-Gas cylinder storage in buildings be limited to small containers, necessary attended uses, and industrial buildings.

5-4 Storage Outside of Buildings.

5-4.1 Location of Storage Outside of Buildings.

5-4.1.1 Storage outside of buildings, for containers awaiting use or resale, shall be located in accordance with Table 5-4.1.1 with respect to:

(a) Nearest important building or group of buildings.

(b) Line of adjoining property which may be built upon.

(c) Busy thoroughfares or sidewalks.

(d) Line of adjoining property occupied by schools, churches, hospitals, athletic fields, or other points of public gathering.

> Prior to the 1967 Edition of the standard, the provisions for outside storage could be interpreted to mean that any amount less than 10,000 lb (4 536 kg) of LP-Gas could be stored next to a building or located adjoining a line of property occupied by schools, churches, hospitals, public gatherings, or thoroughfares. Also, many years ago it was a custom for hardware stores in small communities to store cylinders on the sidewalk adjacent to the store. Consequently, the standards were revised to delineate clearly how much storage could be located with respect to certain types of exposures. Also, in the case of construction work it is sometimes impractical to limit the

Table 5-4.1.1

Quantity of LP-Gas Stored	Distance to:	
	(a) and (b)	(c) and (d)
500 lb (227 kg) or less	0	0
501 (227+ kg) to 2,500 lb (1134 kg)	0	10 ft (3 m)
2,501 (1134+ kg) to 6,000 lb (2721 kg)	10 ft (3 m)	10 ft (3 m)
6,001 (2721+ kg) to 10,000 lb (4540 kg)	20 ft (6 m)	20 ft (6 m)
Over 10,000 lb (4540 kg)	25 ft (7.6 m)	25 ft (7.6 m)

amount of storage necessary for operations according to NFPA 58 provisions. In that case, the authority having jurisdiction must determine the conditions for storage.

5-4.2 Protection of Containers.

5-4.2.1 Containers shall be stored within a suitable enclosure or otherwise protected against tampering.

5-4.3 Alternate Location and Protection of Storage.

5-4.3.1 When the provisions of 5-4.1.1 and 5-4.2.1 are impractical at construction sites, or at buildings or structures undergoing major renovation or repairs, the storage of containers shall be acceptable to the authority having jurisdiction.

5-5 Fire Protection.

5-5.1 Fire Extinguisher Requirements.

5-5.1.1 Storage locations, other than supply depots at separate locations apart from those of the dealer, reseller, or user's establishments, shall be provided with at least one approved portable fire extinguisher having a minimum capacity of 20 lb dry chemical with a B:C rating. (*Also see NFPA 10, Standard for Portable Fire Extinguishers.*)

Prior to the 1983 Edition, any type of extinguisher with a B:C rating was acceptable. They could have as little as 2 lb (0.9 kg) of agent and very short discharge times. Considering that it may not always be possible to stop the flow of escaping gas after extinguishing the fire, it was felt advisable for the safety of the operator to provide enough agent and a longer discharge time to enable him to recognize that he had a problem. It also recognizes that "B" ratings are not derived from tests on gas fires but are based upon flammable liquid pan fires.

REFERENCES CITED IN COMMENTARY

The following publication is available from the National Fire Protection Association, Batterymarch Park, Quincy, MA 02269.

NFPA *101, Life Safety Code.*

6

VEHICULAR TRANSPORTATION OF LP-GAS

Little or no information existed in NFPA standards on the transportation of LP-Gas during the early years of the LP-Gas industry, since bulk transportation was generally by rail and delivery to the consumer site was usually by cylinder delivery. Good practice requirements for the construction and operation of tank trucks for bulk transportation on highways were adopted in 1935 and published by the National Board of Fire Underwriters as Regulations for the Design, Construction, and Operation of Automobile Tank Trucks and Tank Trailers for the Transportation of Liquefied Petroleum Gases, NBFU 59. These provisions were incorporated into the 1940 Edition of NFPA 58 as Division III. Division III related solely to tank trucks and trailers until expanded in the 1961 Edition to cover all types of truck transportation of LP-Gas, including transportation in portable containers, bulk tanks to the consumer site, movable fuel storage tenders, or farm carts. Also included were the first provisions for parking and garaging of LP-Gas tank vehicles. The current Chapter 6 has further modified these earlier provisions to take into account the increased transportation of cylinders by the general public consumer from the filling facility to his location, and also to recognize that DOT and CTC regulations are being used increasingly at the local level, whether or not the transportation is in interstate commerce.

6-1 Scope.

6-1.1 Application.

6-1.1.1 This chapter includes provisions applying to containers, container appurtenances, piping, valves, equipment, and vehicles used in the transportation of LP-Gas, as follows:

(a) Transportation of portable containers.

The primary purpose of Chapter 6 is to set forth provisions for the transportation of LP-Gas for subsequent use. As noted in the exception to 6-1.1.1(a) and in 6-1.1.2, this chapter does not apply to those installations where the gas is being used on transportation equipment for engine fuel, tar kettles, catering vehicles, etc. These installations are covered in Chapter 3.

Exception: The provisions of this chapter are not applicable to LP-Gas containers and related equipment incident to their use on vehicles as covered in Sections 3-6 and 3-9.

(b) Transportation in cargo vehicles, whether fabricated by mounting cargo tanks on conventional truck or trailer chassis, or constructed as integral cargo units in which the container constitutes in whole, or in part, the stress member of the vehicle frame. Transfer equipment and piping, and the protection of such equipment and the container appurtenances against overturn, collision, or other vehicular accidents are also included.

(c) Most truck transportation of LP-Gas is subject to regulation by the U.S. Department of Transportation. Many of the provisions of this chapter are identical or similar to DOT regulations and are intended to extend these provisions to areas not subject to DOT regulation. Vehicles and procedures under the jurisdiction of DOT shall comply with DOT regulations.

Until the late 1960s, the provisions on "tank trucks" (now "cargo vehicles") were quite detailed and were essentially specifications. After World War II, the U.S. and Canadian governmental agencies concerned with interstate and interprovence transportation of hazardous materials expanded their regulations. Eventually, treatment in these regulations and in Chapter 6 of NFPA 58 became very similar. For a time, this dual coverage was justified on the basis that some transportation was not in interstate or interprovence commerce and that NFPA 58 was needed to cover such vehicles. However, in spite of efforts to keep NFPA 58 consonant with DOT/CTC regulations, some differences remained. These led to confusion and, because it makes little economic sense to build a cargo vehicle that cannot be used in a different state or which cannot cross a state line, the federal regulations became dominant.

Many of the earlier provisions on cargo vehicle construction have been deleted and references made to the federal regulations, since 80 percent of state laws are based on adoption of these regulations and 10 percent of state laws have similar requirements. However, certain provisions have been retained which are not included in the federal regulations, such as those for emergency shutoff valves, flexible connectors, chock blocks, etc.

Today, cargo vehicles in the U.S. are constructed to Specification MC-331 of the Code of Federal Regulations, Title 49, Part 178. Older units constructed to Specification MC-330 have undergone some retrofitting and are also in operation today. No further construction to the MC-330 specification is permitted. Additionally, the U.S. Department of Transportation (DOT) has issued a ruling to recognize those vehicles constructed to NFPA 58 provisions for continued operation as long as certain minimum requirements are met. In addition to regulations issued by the Office of Hazardous Materials of DOT, pertinent regulations promulgated by the Bureau of Motor Carrier Safety should be complied with.

6-1.1.2 The provisions of this chapter are not applicable to the transportation of LP-Gas on vehicles incident to its use on these vehicles as covered in 3-6.5, 3-6.6, 3-6.7, and Section 3-9.

6-1.1.3 If LP-Gas is used for engine fuel, the supply piping and regulating, vaporizing, gas-air mixing and carburetion equipment, shall be designed, constructed, and installed in accordance with Section 3-6. Fuel systems (including fuel containers) shall be constructed and installed in accordance with Section 3-9. Fuel may be used from the cargo tank of tank trucks, but not from cargo tanks on trailers or semitrailers.

Many retail bulk delivery trucks (referred to as "bobtails" by the industry) utilize LP-Gas as an engine fuel, although some newer vehicles use diesel fuel for fuel economy. Although provisions for the installation of the engine fuel system are covered in Chapter 3, it is important to note that fuel for the engine may be obtained from the cargo container on this type of truck with one chassis, but cannot be obtained from cargo containers on trailers or semitrailers. An LP-Gas fueled trailer or semitrailer requires a separate LP-Gas fuel container.

6-1.1.4 No artificial light other than electrical shall be used with the vehicles covered by this chapter. Wiring used shall have adequate mechanical strength and current-carrying capacity with suitable overcurrent protection (fuses or automatic circuit breakers) and shall be properly insulated and protected against physical damage.

6-2 Transportation in Portable Containers.

6-2.1 Application.

6-2.1.1 This section applies to the vehicular transportation of portable containers filled with LP-Gas delivered as "packages," including containers built to DOT cylinder specifications and of other portable containers (such as DOT portable tank containers and skid tanks). The design and construction of these containers is covered in Chapter 2.

6-2.2 Transportation of DOT Specification Cylinders or Portable ASME Containers.

6-2.2.1 Portable containers having an individual water capacity not exceeding 1,000 lb (454 kg) [nominal 420 lb (191 kg) LP-Gas capacity], when filled with LP-Gas, shall be transported in accordance with 6-2.2.2 through 6-2.2.9

The maximum size of an individual cylinder permitted under the DOT regulations is 1,000 lb (454 kg) water capacity [nominal 420 lb (191 kg) of LP-Gas]. Portable ASME containers which generally serve the same purpose as DOT cylinders are also limited to this size for the purpose of this section.

6-2.2.2 Containers shall be constructed as provided in Section 2-2 and equipped in accordance with Section 2-3 for transportation as portable containers.

6-2.2.3 The quantity of LP-Gas in containers shall be in accordance with Chapter 4.

6-2.2.4 Valves of containers shall be protected in accordance with 2-2.4.1. Screw-on type protecting caps or collars shall be secured in place.

(a) The provisions of 4-2.2.2 shall apply.

Paragraph 4-2.2.2 referenced here relates to the sealing of valve outlets with a plug, cap, or approved quick closing coupling on cylinders with sizes up to 45 lb (20 kg) of LP-Gas (except single trip nonrefillable and new, unused containers). The reference here is also to assure that this seal is in place while the cylinder is being transported — primarily by the general public.

Note that cylinders over 45 lb (20 kg) are not required to be plugged as they are almost exclusively handled by employees of LP-Gas distribution firms who are trained in proper practice. The requirement for plugs was added following a tragic accident with a consumer transported cylinder.

6-2.2.5 The cargo space of the vehicle shall be isolated from the driver's compartment, the engine and its exhaust system, except as provided in 6-2.2.5(a). Open-bodied vehicles shall be considered as in compliance with this provision. Closed-bodied vehicles having separate cargo, driver's, and engine compartments shall be considered as in compliance with this provision.

(a) Closed-bodied vehicles such as passenger cars, vans, and station wagons shall not be used for transporting more than 215 lb (98 kg) water capacity [nominal 90 lb (41 kg) LP-Gas capacity] but not more than 108 lb (49 kg) water capacity [nominal 45 lb (20 kg) LP-Gas capacity] per container (*see 6-2.2.6 and 6-2.2.7*), unless the driver's and engine compartments are separated from the cargo space by a vapor-tight partition which contains no means of access to the cargo space.

These container capacity provisions rule out the transportation of the commonly used 100 lb (45 kg) LP-Gas cylinder in this type of vehicle. This provision evolved as a result of a serious accident where fire fighters responded to a fire in a van carrying a large quantity of unseparated cylinders with no marking or other indication that it had such a cargo.

6-2.2.6 Containers and their appurtenances shall be determined to be leak-free before being loaded into vehicles. Containers shall be loaded into vehicles with substantially flat floors or equipped with suitable racks for holding containers. Containers shall be securely fastened in position to minimize the possibility of movement, tipping over, or physical damage.

Figure 6.1 Open Portable Container Cargo Vehicle (Delivery Truck).

Racks are commonly used for the delivery of industrial truck cylinders which hold the cylinder in a position tilted upward slightly from horizontal. The safety experience with this type of transportation has been good and is the basis for the treatment in 6-2.2.7 and 6-2.2.8, as far as cylinders with a maximum size of 45 lb (20 kg) of LP-Gas are concerned.

Generally, this is the maximum size for a portable cylinder for industrial truck use.

6-2.2.7   Containers having an individual water capacity exceeding 108 lb (49 kg) [nominal 45 lb (20 kg) LP-Gas capacity] transported in open vehicles shall be transported with the relief devices in direct communication with the vapor spaces. Containers having an individual water capacity exceeding 10 lb (4.5 kg) [nominal 4.2 lb (2 kg) LP-Gas capacity] transported in enclosed spaces of the vehicle shall be transported with the relief device in direct communication with the vapor spaces.

See commentary on 6-2.2.8.

6-2.2.8   Containers having an individual water capacity not exceeding 108 lb (49 kg) [nominal 45 lb (20 kg) LP-Gas capacity] transported in open vehicles may be transported in other than the upright position. Containers having an individual water capacity not exceeding 10 lb (4.5 kg) [nominal 4.2 lb (2 kg) LP-Gas capacity] transported in enclosed spaces of the vehicle may be transported in other than the upright position.

The primary purpose of 6-2.2.7 and 6-2.2.8 is to assure the containers are transported upright and with the relief valve in communication with the vapor space. This minimizes the quantity of LP-Gas released under abnormal conditions. The exception for lift truck cylinders [maximum 45 lb (20 kg) LP-Gas] in open vehicles has been referred to in 6-2.2.8. According to these provisions, 20 lb (9 kg) cylinders — e.g., for gas grills — cannot be transported inside a vehicle on their sides.

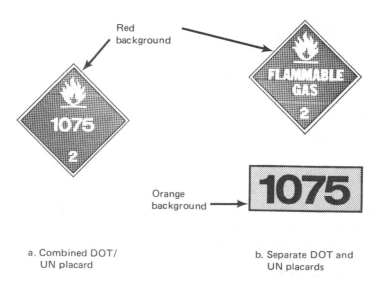

a. Combined DOT/
UN placard

b. Separate DOT and
UN placards

Figure 6.2 DOT Placards.

6-2.2.9 Vehicles transporting more than 1,000 lb (454 kg) of LP-Gas, including the weight of the containers, shall be placarded as required by DOT regulations and/or state law.

Note that the weight of the container must be considered. A load of six typical 100 lb (45 kg) propane capacity DOT cylinders would weigh over 1,000 lb (454 kg) (and require placarding) even though the LP-Gas weight is only 600 lb (272 kg).

6-2.3 Transportation of Portable Containers of More than 1,000 lb (454 kg) Water Capacity.

Portable containers of this size are used when a need exists to transport LP-Gas in bulk to a point of use where the container is not part of a vehicle such as in tank trucks (Section 6-3) and trailer types (Section 6-4). If these portable containers are permanently mounted on a vehicle and used for making deliveries, then they become cargo tanks and must comply with Section 6-3. They are definitely not storage containers used for a permanent installation and such a storage container should never be used for this purpose. The primary difference between the two is that portable containers of this size are designed and the fittings are of a type and so protected that the container may be transported filled to its maximum quantity of LP-Gas. Storage containers do not have such features.

DOT regulations provide for a portable container designed for transporting quantities of LP-Gas greater than 420 lb (191 kg). This container is a Specification 51 container and its construction and fitting arrangement are outlined in the *Code of Federal Regulations* (CFR), Title 49, Parts 100-199. A counterpart constructed in accordance with Section 2-2 and equipped for portable use according to Section 2-3 may be used. Vehicles transporting this size portable containers must be placarded as required by DOT regulations.

Portable containers of this size are used temporarily at construction sites, for emergency power generation, and delivery in bulk to remote areas. Certain applications would dictate the use of a Specification 51 container, particularly if transported by vessel under the jurisdiction of the U.S. Coast Guard, as this is the only type this DOT agency recognizes.

6-2.3.1 Portable containers having an individual water capacity exceeding 1000 lb (454 kg) [nominal 420 lb (191 kg) LP-Gas capacity] when filled with LP-Gas shall be transported in compliance with 6-2.3.2 through 6-2.3.9.

6-2.3.2 Containers shall be constructed in accordance with Section 2-2 and equipped in accordance with Section 2-3 for portable use, or shall comply with DOT portable tank container specifications for LP-Gas service.

6-2.3.3 The quantity of LP-Gas put into containers shall be in accordance with Chapter 4.

6-2.3.4 Valves and other container appurtenances shall be protected in accordance with 2-2.4.2.

6-2.3.5 Containers and their appurtenances shall be determined to be leak-free before being loaded into vehicles. Containers shall be loaded into vehicles with substantially flat floors or equipped with suitable racks for holding containers. Containers shall be securely fastened in position to minimize the possibility of movement, tipping over, or physical damage.

6-2.3.6 Containers and their appurtenances shall be determined to be leak-free before being loaded into vehicles. Containers shall be loaded onto a flat vehicle floor or platform, or onto a suitable vehicle frame. In either case, containers shall be securely blocked or held down to minimize movement, relative to each other or to the supporting structure, while in transit.

6-2.3.7 Containers shall be transported with relief devices in communication with the vapor space.

6-2.3.8 Vehicles carrying more than 1,000 lb (454 kg) of LP-Gas, including the weight of the containers, shall be placarded as required by DOT regulations and/or state law.

6-2.3.9 When portable containers complying with 6-2.3.1 through 6-2.3.8 are permanently or semipermanently mounted on vehicles to serve as cargo tanks, so that the assembled vehicular unit can be used for making liquid deliveries to other containers at points of use, the provisions of Section 6-3 shall apply.

6-2.4 Fire Extinguishers.

6-2.4.1 Each truck or trailer transporting portable containers as provided by 6-2.2 or 6-2.3 shall be equipped with at least one approved portable fire extinguisher having a minimum capacity of 20 lb dry chemical with a B:C rating. (*Also see NFPA 10, Standard for Portable Fire Extinguishers.*)

See commentary on 5-5.1.1.

6-3 Transportation in Cargo Vehicles.

Essentially, cargo vehicles are considered to comply with DOT regulations, CFR 49, Parts 100-199 and the Federal Motor

Carrier Safety Regulations. Added requirements are contained in NFPA 58 as noted in the commentary on 6-1.1.1(c).

6-3.1 Application.

6-3.1.1 This section includes provisions for cargo vehicles used for the transportation of LP-Gas as liquid cargo, normally loaded into the cargo container at the distributing or manufacturing point, and transferred into other containers at the point of delivery. Transfer may be made by a pump or compressor mounted on the vehicle or by a transfer means at the delivery point.

6-3.1.2 All LP-Gas cargo vehicles, whether used in interstate or intrastate service, shall comply with the applicable portion of the US Department of Transportation Hazardous Materials Regulations (Title 49 Code of Federal Regulations Parts 171-179) and Parts 393, 396, and 397 of the DOT Federal Motor Carrier Safety Regulations and shall also comply with the added requirements of this standard.

This paragraph was changed in the 1989 Edition by adding a reference to the relevant parts of the *Code of Federal Regulations*, parts 171-179. These cover:
CFR 49 Part 171/172 - Hazardous Materials
173 - Shipping Requirements
174 - Rail Transportation
175 - Air Transportation

Figure 6.3   Typical Retail Delivery Cargo Vehicle (Bobtail).

176 - Transportation by Vessel
177 - Highway Transportation
178 - Shipping Container Specifications
179 - Tank Car Specifications
393/397 - Truck Safety Requirements

6-3.2 Containers Mounted on, or a Part of, Cargo Vehicles.

6-3.2.1 Containers mounted on, or comprising in whole, or in part, the stress member used in lieu of a frame for cargo vehicles shall comply with DOT cargo tank specifications for LP-Gas service. Such containers shall also comply with Section 2-2, be equipped with appurtenances as provided in Section 2-3 for cargo service, and comply with 6-3.2.1(a):

(a) Liquid hose of 1 ½ in. (nominal size) and larger size and vapor hose of 1 ¼ in. (nominal size) and larger size shall be protected with an emergency shutoff valve complying with 2-4.5.4, except that:

(1) If an internal valve meets the functional provisions for an emergency shut-off valve in compliance with 2-4.5.4 and 3-2.7.9(a)(1), an emergency shutoff valve shall not be required in the cargo container piping.

(2) A backflow check valve may be used in the cargo container piping or container in lieu of an emergency shutoff valve if the flow is only into the cargo container.

Emergency shutoff valves (ESVs), as described in 2-4.5.4, must be installed at both ends of the transfer hose [1½ in. (38 mm) or larger for liquid hose and 1¼ in. (32 mm) or larger for vapor hose]. The provisions of 6-3.2.1(a) recognize the use of an internal valve on the truck for this purpose if it meets the provisions of 2-4.5.4 and 3-2.7.9(a)(1). DOT regulations require the installation of ESVs on tank trucks and these valves meet those requirements. If flow is only in one direction, a backflow check valve may be used as an alternative to the ESV.

6-3.3 Piping (Including Hose), Fittings, and Valves.

6-3.3.1 Pipe, tubing, pipe and tubing fittings, valves, hose and flexible connectors shall comply with Section 2-4, with the provisions of DOT cargo tank specifications for LP-Gas, and shall be suitable for the working pressure specified in 6-3.3.2. In addition, 6-3.3.1(a) through (e) shall apply:

(a) Pipe shall be wrought iron, steel, brass or copper in accordance with 2-4.2.1.

(b) Tubing shall be steel, brass or copper in accordance with 2-4.3.1(a), (b), or (c).

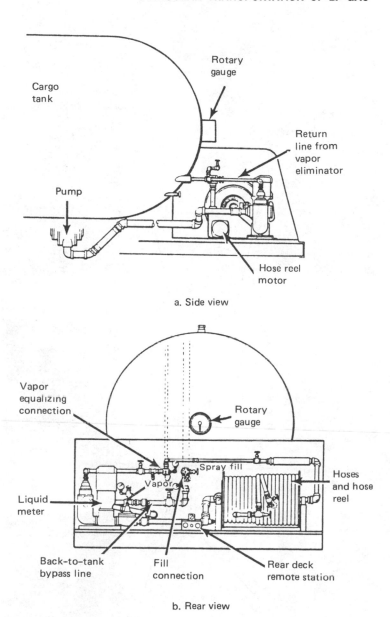

a. Side view

b. Rear view

Figure 6.4   Typical Retail Delivery Cargo Vehicle (Bobtail) Transfer System.

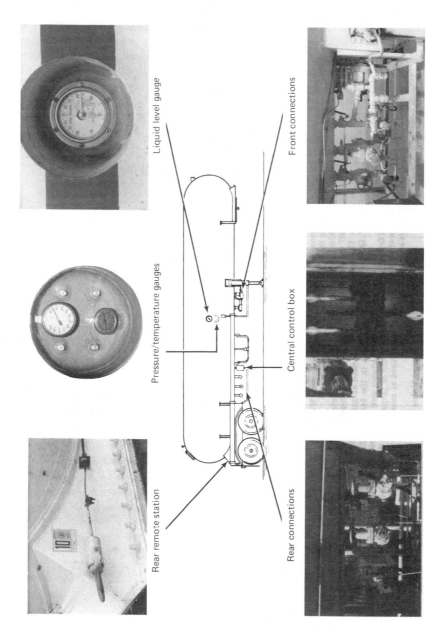

Figure 6.5   Typical Semitrailer Cargo Vehicle (Transport).

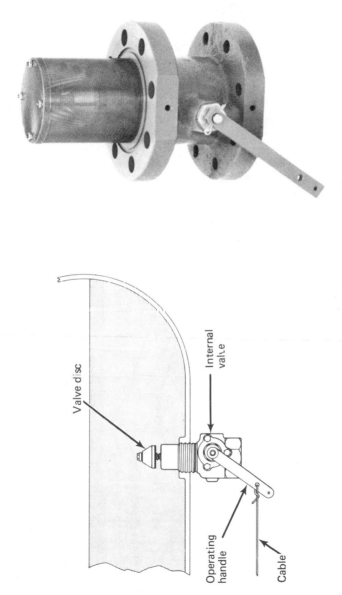

Valve disc

Internal valve

Operating handle

Cable

Figure 6.6 Internal Valve in Cargo Vehicle Container.

(c) Pipe and tubing fittings shall be steel, brass, copper, malleable iron or ductile (nodular) iron suitable for use with the pipe or tubing used as specified in 6-3.3.1(a) or (b).

(d) Pipe joints may be threaded, flanged, welded or brazed. Fittings when used shall comply with 6-3.3.1(c).

(1) When joints are threaded, or threaded and back welded, pipe and nipples shall be Schedule 80 or heavier. Copper or brass pipe and nipples shall be of equivalent strength.

(2) When joints are welded or brazed, the pipe and nipples shall be Schedule 40 or heavier. Fittings or flanges shall be suitable for the service. (*See 6-3.3.2.*)

(3) Brazed joints shall be made with a brazing material having a melting point exceeding 1,000°F (538°C).

(e) Tubing joints shall be brazed, using a brazing material having a melting point of at least 1,000°F (538°C).

These provisions for piping, fittings, and valves are more detailed than those in the DOT regulations. They also specify the working pressure of such materials, particularly the 350 psig (2.4 MPa gauge) minimum working pressure on the discharge of pumps.

6-3.3.2 Pipe, tubing, pipe and tubing fittings, valves, hose and flexible connectors, and complete cargo vehicle piping systems including connections to equipment (*see 6-3.4*), after assembly, shall comply with 2-5.1.2.

See commentary on 6-3.3.1(e).

6-3.3.3 Valves, including shutoff valves, excess-flow valves, backflow check valves and remotely controlled valves, used in piping shall comply with the applicable provisions of DOT cargo tank specifications for LP-Gas service, and with 2-4.5, provided, however, that their minimum design pressure shall comply with 6-3.3.2.

See commentary on 6-3.3.1(e).

6-3.3.4 Hose, hose connections, and flexible connectors shall comply with 2-4.6 and 6-3.3.1. Flexible connectors used in the piping system to compensate for stresses and vibration shall be limited to 3 ft (1 m) in overall length. Flexible connectors on existing LP-Gas cargo units replaced after December 1, 1967, shall comply with 2-4.6.

(a) Flexible connectors assembled from rubber hose and couplings installed after December 31, 1974, shall be permanently marked to indicate the date of assembly of the flexible connector and the flexible portion of the connector shall be replaced within six years of the indicated date of assembly of the connector.

(b) The rubber hose portion of flexible connectors shall be replaced whenever a cargo unit is remounted on a different chassis, or whenever the cargo unit is repiped, if such repiping encompasses that portion of piping in which the connector is located, unless the remounting and/or repiping is performed within one year of the date of assembly of the connector.

6-3.3.5 All threaded primary valves and fittings used in liquid filling or vapor equalization directly on the cargo container of transportation equipment shall be of steel, malleable or ductile iron construction. All existing equipment shall be so equipped not later than the scheduled requalification date of the container.

> **Experience with brass valves in liquid filling or vapor equalization lines (which are permitted on stationary containers) where the threads were stripped and the valves pulled out in driveaways with hose connected necessitated this provision prohibiting their further use on cargo vehicles.**

6-3.4 Equipment.

> **These provisions are also more detailed than DOT regulations.**

6-3.4.1 LP-Gas equipment, such as pumps, compressors, meters, dispensers, regulators, and strainers, shall comply with Section 2-5 as to design and construction and shall be installed in accordance with the applicable provisions of 3-2.10. Equipment on vehicles shall be securely mounted in place and connected into the piping system in accordance with the manufacturer's instructions, taking into account the greater (than for stationary service) jarring and vibration problems incident to vehicular use.

6-3.4.2 Pumps or compressors used for LP-Gas transfer may be mounted on tank trucks, trailers, semitrailers, or tractors, and may be driven by the truck or tractor motor power takeoff, by a separate internal combustion engine, or by hand, mechanical, hydraulic, or electrical means. If an electric drive is used, obtaining energy from the electrical installation at the delivery point, the installation on the vehicle (and at the delivery point) shall comply with 3-8.2.

6-3.4.3 The installation of compressors shall comply with the applicable provisions of 3-2.10.1 and 6-3.4.1.

6-3.4.4 The installation of liquid meters shall be in accordance with 3-2.10.5(a). If venting of LP-Gas to the air is necessary, provision shall be made to vent it at a safe location.

6-3.4.5 When wet hose is carried connected to the truck liquid pump discharge piping, an automatic device, such as a differential regulator, shall be installed between the pump discharge and the hose connection to prevent liquid discharge when the pump is not operating. When a meter or dispenser is used, this device shall be installed between the meter outlet and the hose connection. An excess-flow valve may also be used but shall not be the exclusive means of complying with this provision.

Generally, transport trucks which are large volume cargo vehicles carry "dry" hose which is disconnected completely following a liquid transfer operation and carried in protective containers during travel. The smaller retail bulk delivery tank trucks (bobtails) carry "wet" hose which contains liquid LP-Gas at all times. This latter type hose is protected by a differential regulator which is open only when the pump is running. (*See commentary on 4-2.3.4.*)

6-3.5 Protection of Container Appurtenances, Piping System and Equipment.

6-3.5.1 Container appurtenances, piping, and equipment comprising the complete LP-Gas system on the cargo vehicle shall be securely mounted in position (*see 6-3.2.1 for container mounting*), shall be protected against damage to the extent it is practical, and in accordance with DOT regulations.

6-3.6 Painting and Marking Liquid Cargo Vehicles.

6-3.6.1 Painting of cargo vehicles shall comply with Code of Federal Regulations, Title 49, Part 195. Placarding and marking shall comply with CFR 49.

6-3.7 Fire Extinguishers.

6-3.7.1 Each tank truck or tractor shall be provided with at least one approved portable fire extinguisher having a minimum capacity of 20 lb dry chemical with a B:C rating. (*Also see NFPA 10, Standard for Portable Fire Extinguishers.*)

See commentary on 5-5.1.1.

6-3.8 Chock Blocks for Liquid Cargo Vehicles.

6-3.8.1 Each tank truck and trailer shall carry chock blocks which shall be used to prevent rolling of the vehicle whenever it is being loaded or unloaded, or is parked.

6-3.9 Exhaust Systems.

6-3.9.1   The truck engine exhaust system shall comply with Federal Motor Carrier Safety Regulations.

6-3.10 Smoking Prohibition.

6-3.10.1   No person may smoke or carry lighted smoking material on or within 25 ft (7.6 m) of a vehicle required to be placarded per DOT regulations containing LP-Gas liquid or vapor. This shall also apply at points of liquid transfer and while delivering or connecting to containers.

This requirement was changed in the 1989 Edition of the standard to be consistent with federal regulations (49 CFR 397.13). Note that the separation distance for smoking of 25 ft (7.6 m) is greater than the separation distance for internal combustion engines during transfer operations specified in 3-8.4.1 (15 ft) (5 m). The Technical Committee agrees that 25 ft (7.6 m) is the correct separation distance for smoking and the separation distance of other sources of ignition may be changed in a future edition of the standard.

6-4 Trailers, Semitrailers, Movable Fuel Storage Tenders, or Farm Carts.

6-4.1 Application.

6-4.1.1   This section applies to all cargo vehicles, other than trucks, which may be parked at locations away from distributing points.

6-4.2 Trailers or Semitrailers Comprising Parts of Section 6-3 Vehicles.

Trailers or semitrailers are used where large volumes are needed on a temporary basis, such as at a temporary asphalt mixing plant for road construction or repair. They are not to be used as permanent storage at a small bulk plant for filling cylinders or as retail delivery tank trucks. If this were done, the system would have to meet the provisions of Sections 3-2 and 3-3 which would, in effect, be impractical.

6-4.2.1   When parked, cargo tank trailers or semitrailers covered by Section 6-3 shall be positioned so that the pressure relief valves shall communicate with the vapor space of the container.

6-4.3 Trailers, Including Movable Storage Tenders or Farm Carts.

6-4.3.1   Trailers, including fuel storage tenders or farm carts, shall comply with 6-4.3.2 through 6-4.3.6. If normally used over public ways they shall comply with applicable state regulations.

Figure 6.7 Farm Cart.

6-4.3.2   Cargo containers mounted on such vehicles shall be constructed in accordance with Section 2-2, and equipped with appurtenances as provided in Section 2-3. Container mounting shall be adequate for the service involved.

6-4.3.3   Threaded piping shall not be less than Schedule 80 and fittings shall be designed for not less than 250 psig (1.7 MPa gauge).

6-4.3.4   Piping, hoses and equipment, including valves, fittings, pressure relief valves and container accessories, shall be adequately protected against collision or upset.

6-4.3.5   Parked vehicles shall be so positioned that container safety relief valves communicate with the vapor space.

6-4.3.6   Such cargo units shall not be filled on a public way.

Movable fuel storage tenders and farm carts are limited in size to 1,200 gal (4 543 L) water capacity. Basically, they are non-highway vehicles but occasionally are moved over roads for short distances to be used as a temporary fuel supply for farm tractors, construction equipment, etc.

6-5 Transportation of Stationary Containers to and from Point of Installation.

6-5.1 Application.

6-5.1.1 This section applies to the transportation of containers designed for stationary service at the point of use and secured to the vehicle only for transportation. Such containers may be transported partially filled with LP-Gas.

These storage containers are moved to the consumer site with a small amount of LP-Gas (5 percent maximum) for initial startup. Section 173.315(j) of Title 49, CFR also sets forth rules consonant with NFPA 58 provisions for the shipment of storage containers to consumers' premises by private motor carrier. Occasionally, it may be necessary to move the container with more than 5 percent LP-Gas, such as for use in tobacco curing at one time of the year and home heating nearby at another time. This may be done subject to the limitations specified by the authority having jurisdiction.

6-5.2 Transportation of Containers.

6-5.2.1 Except as provided in 6-5.2.1(a), containers of 125 gal (0.5 m³) or more water capacity shall contain no more than 5 percent of their water capacity in liquid form during transportation.

(a) Containers containing more LP-Gas than 5 percent of their water capacity may be transported subject to such limitations as may be specified by the authority having jurisdiction.

Containers smaller than 125 gal (0.5 m³) are not required to be drained for two reasons. First, they are not required to be equipped with a connection for liquid evacuation (*see 2-2.3.3*) so they can be difficult to drain and their weight, even when full, is not so great that they cannot be safely handled.

The limit of 5 percent water capacity for transportation of containers is a practical one in that it represents a low weight, and recognizes that the gauges in use on containers cannot accurately measure contents below 5 percent.

6-5.2.2 Containers shall be safely secured to minimize movement relative to each other or to the carrying vehicle while in transit, giving consideration to the sudden stops, starts and changes of direction normal to vehicular operation.

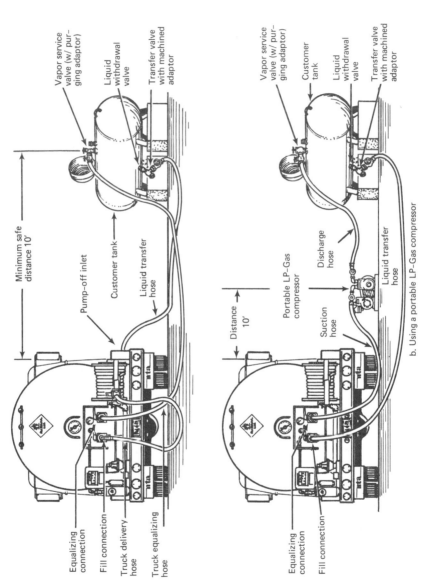

Vapor service valve (w/ purging adaptor)

Liquid withdrawal valve

Transfer valve with machined adaptor

Minimum safe distance 10'

Pump-off inlet

Customer tank

Liquid transfer hose

Equalizing connection

Fill connection

Truck delivery hose

Truck equalizing hose

Vapor service valve (w/ purging adaptor)

Customer tank

Liquid withdrawal valve

Transfer valve with machined adaptor

Distance 10'

Portable LP-Gas compressor

Discharge hose

Suction hose

Liquid transfer hose

Equalizing connection

Fill connection

b. Using a portable LP-Gas compressor

Figure 6.8 Removing Liquid from Container to Comply with 5 Percent Provision.

Figure 6.9 Transporting ASME Stationary Storage Containers. (Top) Flatbed truck with crane. (Bottom) Saddle trailer.

6-5.2.3 Valves, regulators and other container appurtenances shall be adequately protected against physical damage during transportation.

6-5.2.4 Pressure relief valves shall be in direct communication with the vapor space of the container.

6-5.2.5 Lifting lugs in good repair on containers filled to no more than five percent of their water capacity may be used for lifting and lowering.

Figure 6.10 Lifting ASME Stationary Storage Container.

(a) Additional means for securing and supporting the container shall be provided for transporting or when lifting or lowering with more than 5 percent of its water capacity [see 6-5.2.1(a)].

Lifting lugs on storage containers are designed to lift or lower the container with it being filled to 5 percent of its water capacity. If the container is filled to a higher level, the lugs cannot be used as the sole support. The additional support adds an extra measure of safety and thoroughly secures the container while in transit.

6-6 Parking and Garaging Vehicles Used to Carry LP-Gas Cargo.

6-6.1 Application.

6-6.1.1 This section applies to the parking (except parking associated with a liquid transfer operation) and garaging of vehicles used for the transportation of LP-Gas. Such vehicles include those used to carry portable containers (*see Section 6-2*) and those used to carry LP-Gas in cargo tanks (*cargo vehicles, see Section 6-3*).

This section sets out provisions for the parking of tank trucks and cylinder delivery trucks outdoors in congested and uncongested areas and indoors within public buildings or at the owner's premises for temporary parking or servicing and repairing.

6-6.2 Parking.

6-6.2.1 Vehicles carrying or containing LP-Gas parked out-of-doors shall comply with the following:

(a) Vehicles, except in an emergency and except as provided in 6-6.2.1(b), shall not be left unattended on any street, highway, avenue or alley, provided that this shall not prevent a driver from the necessary absence from the vehicle in connection with his normal duties, nor shall it prevent stops for meals or rest stops during the day or at night.

(b) Vehicles shall not be parked in congested areas. Such vehicles may be parked off the street in uncongested areas if at least 50 ft (15 m) from any building used for assembly, institutional or multiple residential occupancy. This shall not prohibit the parking of vehicles carrying portable containers or cargo vehicles of 3500 gal (13 m³) water capacity or less on streets adjacent to the driver's residence in uncongested residential areas, provided such points of parking are at least 50 ft (15 m) from a building used for assembly, institutional, or multiple residential occupancy.

In remote areas where the driver's residence may be a considerable distance from bulk plant facilities and consumers are in the area of the driver's home, it is permissible to park the retail bulk delivery tank truck or cylinder delivery truck near the driver's home under the conditions stipulated.

6-6.2.2 Vehicles parked indoors shall comply with the following:

(a) Cargo vehicles parked in any public garage or building shall have LP-Gas liquid removed from the cargo container, piping, pump, meter, hoses and related equipment and the pressure in the delivery hose and related equipment reduced to approximately atmospheric, and all valves closed before being moved inside. Delivery hose or valve outlets shall be plugged or capped before the vehicle is moved inside.

(b) Vehicles used to carry portable containers shall not be moved into any public garage or building for parking until all portable containers have been removed from the vehicle.

(c) Vehicles carrying or containing LP-Gas are permitted to be parked in buildings complying with Chapter 7 and located on premises owned or under the control of the operator of such vehicles, provided:

(1) The public is excluded from such buildings.

(2) There is adequate floor level ventilation in all parts of the building where these vehicles are parked.

(3) Leaks in the vehicle LP-Gas systems are repaired before the vehicle is moved inside.

(4) Primary shutoff valves on cargo tanks and other LP-Gas containers on the vehicle (except propulsion engine fuel containers) are closed and delivery hose outlets plugged or capped to contain system pressure before the vehicle is moved inside. Primary shutoff valves on LP-Gas propulsion engine fuel containers shall be closed when the vehicle is parked.

(5) No LP-Gas container is located near a source of heat or within the direct path of hot air being blown from a blower-type heater.

(6) LP-Gas containers are gauged or weighed to determine that they are not filled beyond the maximum filling density according to 4-5.1.

If it is necessary to park cargo vehicles in public buildings, all LP-Gas must be removed from the cargo tank and equipment on tank trucks and cylinders removed from cylinder trucks. In very cold climates it may be necessary to park these vehicles in garages on the owner's premises. The type of building construction required by 6-6.2.2(c) should not be overlooked.

6-6.2.3 Vehicles are permitted to be serviced or repaired indoors as follows:

(a) When it is necessary to take a vehicle into any building located on premises owned and/or operated by the operator of such vehicle for service on engine or chassis, the provisions of 6-6.2.2(a) or (c) shall be followed.

(b) When it is necessary to take a vehicle carrying or containing LP-Gas into any public garage or repair facility for service on the engine or chassis, the provisions of 6-6.2.2(a) or (b) shall be followed, unless the driver or qualified representative of an LP-Gas operator is in attendance at all times when the vehicle is inside. In that case, the following provisions shall be followed under the supervision of such qualified persons:

(1) Leaks in the vehicle LP-Gas systems shall be repaired before the vehicle is moved inside.

(2) Primary shutoff valves on cargo tanks, portable containers and other LP-Gas containers installed on the vehicle (except propulsion engine fuel containers) are closed. LP-Gas liquid shall be removed from the piping, pump, meter, delivery hose, and related equipment and the pressure therein reduced to approximately atmospheric before the vehicle is moved inside. Delivery hose or valve outlets shall be plugged or capped before the vehicle is moved inside.

(3) No container shall be located near a source of heat or within the direct path of hot air blown from a blower or from a blower-type heater.

(4) LP-Gas containers shall be gauged or weighed to determine that they are not filled beyond the maximum filling capacity according to 4-5.1.

(c) If repair work or servicing is to be performed on a cargo tank system, all LP-Gas shall be removed from the cargo tank and piping and the system thoroughly purged before the vehicle is moved inside.

All LP-Gas is to be removed when service or repairs are done on the vehicle in public garages, unless a qualified person is present at all times and certain precautions are taken. At the owner's premises certain measures are to be taken for servicing, which would not make it necessary to remove LP-Gas from the vehicle. However, if work is to be done on the cargo tank system, whether at a public garage or owner's premises or not, all LP-Gas must be removed and the system purged beforehand.

REFERENCES CITED IN COMMENTARY

The following publication is available from The U.S. Government Printing Office, Washington, DC.

Code of Federal Regulations, **Title 49, Parts 100-199, and 393, 397.**

7

BUILDINGS OR STRUCTURES HOUSING LP-GAS DISTRIBUTION FACILITIES

7-1 Scope.

7-1.1 Application.

7-1.1.1 This chapter includes the construction, ventilation and heating of structures housing certain types of LP-Gas systems as referenced in this standard. Such structures may be separate buildings used exclusively for the purpose (or for other purposes having similar hazards), or they may be rooms attached to, or located within, buildings used for other purposes.

> Chapter 7 was introduced in the 1972 Edition. It recognizes that certain systems or operations present combustion explosion hazards when they are located or conducted inside of structures; that is, that accidental escape of liquid or vapor could result in the formation of a quantity of flammable vapor/air mixture large enough that its ignition could produce destructive pressure in the structure. In such instances, the construction can limit the damage to the structure and its surroundings. For an in-depth discussion of this concept (commonly called "explosion-venting"), see NFPA 68, *Guide for Venting of Deflagrations.*
>
> Generally, Chapter 7 is applicable to structures where liquid LP-Gas is used in volume or high pressure gas is present. However, the application of Chapter 7 is stipulated in specific provisions throughout the standard—e.g., 4-3.1.1(a).
>
> The aerosol packaging industry using LP-Gas propellants has used Chapter 7 extensively for the design of filling rooms. It is noted, however, that NFPA 58 was not developed with such facilities in mind [they are considered chemical plants and excluded by 1-2.3.1(d)]. Therefore, Chapter 7 should be used with considerable judgment in such applications and may need to be augmented by other safeguards not cited in NFPA 58—e.g., higher ventilation rates, and explosion-suppression systems.

7-1.1.2 The provisions of this chapter apply only to buildings constructed or converted after December 31, 1972, except for those previously constructed under the provisions of 5-3.3. Also, see 1-2.4.1.

This applies Chapter 7 retroactively to storage within special buildings or rooms covered by 5-3.3, which includes all such structures not in a distributing plant. This provision recognizes that most of these locations are in occupancies having considerable public exposure, such as mercantile occupancies.

7-2 Separate Structures or Buildings.

Chapter 7 recognizes three basic locations for structures: separate (no physical connection with another structure); attached (connected to another structure with the common walls having a perimeter not exceeding 50 percent of the perimeter of the attached structure); and within (wholly or partly inside another structure). These are presented in an order of decreasing degree of overall safety to the facility; this order should be considered in the facility design. There should be good reasons, for example, to use a room within a structure if one of the other options will do the job.

7-2.1 Construction of Structures or Buildings.

7-2.1.1   Separate buildings or structures shall be one story in height and shall have walls, floors, ceilings, and roofs constructed of noncombustible materials. Exterior walls, ceilings, and roofs shall be constructed as follows:

(a) Of lightweight material designed for explosion venting, or

(b) If of heavy construction, such as solid brick masonry, concrete block or reinforced concrete construction, explosion venting windows or panels in walls or roofs shall be provided having an explosion venting area of at least 1 sq ft (0.1 m^2) for each 50 cu ft (1.4 m^3) of the enclosed volume.

From functional and economic considerations, the construction in 7-2.1.1(a) has definite advantages over that in 7-2.2.1(b). A simple steel framing to which roof and wall panels are affixed by fastenings just strong enough to withstand wind loadings provides for the maximum possible explosion venting area. If these panels are of nonfrangible material, such as steel or aluminum, they stay more or less in one piece when they blow off, are lightweight, and do not travel very far. While also reasonably light in weight, cement-asbestos panels tend to fragment and present more of a flying missile hazard. When such a structure explodes, the steel framing usually survives with little damage. The roof and walls can be replaced quickly and the structure can be returned to use promptly.

Aside from being much more costly, the heavy construction described in 7-2.2.1(b) does not have as much explosion venting area, and the masonry types present flying missile hazards and take much longer to rebuild. They do, however, have advantages where external security is a problem and aesthetics need to be considered.

7-2.1.2 The floor of such structures shall not be below ground level. Any space beneath the floor shall preferably be of solid fill. If not so filled, the perimeter of the space shall be left entirely unenclosed.

This provision recognizes that LP-Gas is heavier than air.

7-2.2 Structure or Building Ventilation.

7-2.2.1 The structure shall be ventilated utilizing air inlets and outlets arranged to provide air movement across the floor as uniformly as practical and in accordance with 7-2.2.1(a) or (b). The bottom of such openings shall not be more than 6 in. (152 mm) above the floor.

(a) When mechanical ventilation is used, air circulation shall be at least at the rate of one cu ft per minute per sq ft (0.4 m^3/s/m^2) of floor area. Outlets shall discharge at least five ft (1.5 m) away from any opening into the structure or any other structure.

(b) When natural ventilation is used, each exterior wall [up to 20 ft (6.1 m) in length] shall be provided with at least one opening, with an additional opening for each 20 ft (6.1 m) of length or fraction thereof. Each opening shall have a minimum size of 50 sq in. (12 900 mm^2) and the total of all openings shall be at least 1 sq in. (645 mm^2) per ft^2 (0.1 m^2) of floor area.

Mechanical ventilation is more predictable than natural ventilation and is preferred, especially where liquid transfer is done. However, where the occupancy is strictly for storage only, natural ventilation has proven to be generally adequate.

The ventilation provisions reflect the fact that LP-Gas is heavier than air. The criteria for mechanical ventilation are based upon the concept of continuously removing a stratum of air 1 ft (0.3 m) deep along the floor. It is important that the air discharged be made up by incoming air. The most common ventilation defect is blockage of inlets for mechanical systems and blockage of both inlets and outlets for natural arrangements in cold weather for comfort reasons. The inlet air should be heated to solve this problem. It is considerably easier to do this with a mechanical system because the inlets can be manifolded and a single heater installed in the common duct.

The requirements for natural ventilation openings were modified in the 1989 Edition. The change recognizes that in large buildings many smaller ventilation openings are preferable.

Previously the standard could be met with 2 openings, each opening having an area of 1 sq in. (645 mm²) per ft² (0.1 m²) of floor area, and with larger buildings being constructed the openings were very large, creating heating problems in colder regions.

7-2.3 Structure or Building Heating.

7-2.3.1   Heating shall be by steam or hot water radiation or other heating transfer medium with the heat source located outside of the building or structure (*see Section 3-8, Ignition Source Control*), or by electrical appliances installed in the building, if they are listed for Class I, Group D, Division 2 locations, in accordance with NFPA 70, *National Electrical Code* (*see Table 3-8.2.2*).

7-3 Attached Structures or Rooms within Structures.

7-3.1 Construction of Attached Structures.

The basic principle in 7-3.1 is to maintain the explosion venting function of the attached structure while minimizing the possibility of damage (explosion and fire) to the structure to which it is attached. The common wall(s) is built to be strong and with a degree of fire resistance, and the presence and character of openings is controlled.

While a wall designed for a static pressure of 100 psf (7 MPa m²) is a substantial wall [even a wall of minimal height would be at least 12 in. (0.3 m) of masonry or steel reinforced concrete block], such a wall should be as strong as is feasible to build.

7-3.1.1   Attached structures shall comply with 7-2.1 (attachment shall be limited to 50 percent of the perimeter of the space enclosed; otherwise such space shall be considered as a room within a structure — *see 7-3.2*), and with the following:

(a) Common walls at points at which structures are to be attached shall:

(1)   Have, as erected, a fire resistance rating of at least one hour, as determined by NFPA 251, *Standard Methods of Fire Tests of Building Construction and Materials*.

(2)   Have no openings. Common walls for attached structures used only for storage of LP-Gas are permitted to have doorways which shall be equipped with 1 ½ hour (B) fire doors. See NFPA 80, *Standard for Fire Doors and Windows*.

(3)   Be designed to withstand a static pressure of at least 100 lb (0.7 MPa) per sq ft (0.1 m²).

(b) The provisions of 7-3.1.1(a) may be waived if the building to which the structure is attached is occupied by operations or processes having a similar hazard.

(c) Ventilation and heating shall comply with 7-2.2.1 and 7-2.3.1.

7-3.2 Construction of Rooms within Structures.

The basic principle here is similar to that of 7-3.1, and the provisions are also similar. A first-story location is required to remove the floor as an area of exposure to the remainder of the building and to facilitate fire control activities by fire departments.

7-3.2.1 Rooms within structures shall be located in the first story and shall have at least one exterior wall with sufficient exposed area to permit explosion venting as provided in 7-3.2.1(a). The building in which the room is located shall not have a basement or unventilated crawl space and the room shall comply with the following:

(a) Walls, floors, ceilings, or roofs of such rooms shall be contructed of noncombustible materials. Exterior walls and ceilings shall either be of lightweight material designed for explosion venting, or, if of heavy construction (such as solid brick masonry, concrete block, or reinforced concrete construction), shall be provided with explosion venting windows or panels in the walls or roofs having an explosion venting area of at least 1 sq ft (0.1 m²) for each 50 cu ft (1.4 m³) of the enclosed volume.

(b) Walls and ceilings common to the room and to the building within which it is located shall:

(1) Have, as erected, a fire resistance rating of at least one hour as determined by NFPA 251, *Standard Methods of Fire Tests of Building Construction and Materials*.

(2) Not have openings. Common walls for rooms used only for storage of LP-Gas are permitted to have doorways which shall be equipped with 1 ½-hour (B) fire doors. See NFPA 80, *Standard for Fire Doors and Windows*.

(3) Be designed to withstand a static pressure of at least 100 lb (0.7 MPa) per sq ft (0.1 m²).

(c) The provisions of 7-3.2.1(b) may be waived if the building within which the room is located is occupied by operations or processes having a similar hazard.

(d) Ventilation and heating shall comply with 7-2.2.1 and 7-2.3.1.

REFERENCES CITED IN COMMENTARY

The following publication is available from the National Fire Protection Association, Batterymarch Park, Quincy, MA 02269.

NFPA 68, *Guide for Venting of Deflagrations.*

8

REFRIGERATED STORAGE

8-1 Refrigerated Containers.

This chapter was added in the 1989 Edition in conjunction with the addition of coverage of LP-Gas refrigerated storage to the standard.

Refrigerated LP-Gas storage systems are those in which the liquefied gas is stored at or near its boiling point at atmospheric pressure. The temperature of the liquid is maintained by a refrigeration process. The storage containers are characterized by their large size [over one million gallons (3 785 m³) is not unusual] and design pressure [½ psig (3 kPa) is common]. They are thermally insulated as a matter of operational necessity and they can withstand only very low internal pressures.

Such containers are not only very different from non-refrigerated LP-Gas containers, they are few in number and found usually in specialized applications in LP-Gas production facilities, marine and pipeline terminals, and in natural gas peak-shaving plants.

8-1.1 Refrigerated containers shall be built in accordance with applicable provisions of one of the following codes as appropriate for conditions of maximum allowable working pressure, design temperature, and hydrostatic testing.

8-1.1.1 For pressures of 15 psig (103 kPa gauge) or more, use the ASME Code, Section VIII, except that construction using joint efficiencies in Table UW 12, Column C, Division 1 is not permitted. Material shall be selected from those recognized by ASME which meet the requirements of Appendix R of ANSI/API 620.

Section VIII of the ASME Code would be used to design and fabricate shop built vessels or semi-refrigerated containers operating above 15 psig. While few of these vessels are anticipated their coverage is included to make the standard complete.

The exception of Table UW-12 of the ASME Code strengthens inspection requirements.

The material selection reference to Appendix R of API 620 reflects the somewhat unique problems associated with the design and construction of low temperature vessels.

353

8-1.1.2 For pressures below 15 psig (103 kPa gauge) use ANSI/API 620, *Recommended Rules for the Design and Construction of Large, Welded, Low Pressure Storage Tanks,* including Appendix R.

API 620 is recognized worldwide as a standard for the construction of large welded storage tanks. Appendix R of that standard specifically addresses the design and material requirements of such tanks operating at refrigerated LP-Gas temperatures.

8-1.2 Wind loading on containers shall be in accordance with paragraph 2-2.2.3(c).

Because of the larger size of refrigerated storage vessels, wind loading can be a significant factor in designing the vessel.

8-1.3 Seismic loading on containers shall be in accordance with paragraph 2-2.2.3(d).

Because of the larger size of refrigerated storage vessels, a detailed seismic analysis may be required in the more seismically active zones. A competent geotechnical engineer and structural engineer should perform this analysis.

8-1.4 Field-erected containers for refrigerated storage shall be designed as an integral part of the storage system including tank insulation, compressors, condensors, controls, and piping. Proper allowance shall be made for the service temperature limits of the particular process and the products to be stored when determining material specifications and the design pressure. Welded construction shall be used.

8-1.5 When austenitic steels or nonferrous materials are used, ANSI/API 620, Appendix Q shall be used as a guide in the selection of materials for use at the design temperature.

Appendix Q of API 620 contains guidance for the use of austenitic steels and nonferrous materials that are not included in Appendix R.

8-1.6 Prior to initial operation, containers shall be inspected to the extent necessary to assure compliance with the engineering design and material, fabrication, assembly, and test provisions of this standard. The operator shall be responsible for this inspection.

Both the ASME and API Codes contain requirement for test and inspection. This requirement reinforces the need to comply with these requirements and reiterates that they are the responsibility of the operator.

8-1.7 The operator may delegate performance of any part of the inspection to inspectors who may be employees of his own organization, an engineering or scientific organization, or of a recognized insurance or inspection company. Inspectors shall be qualified in accordance with the code or standard applicable to the container and as specified in this standard.

This is an extension of the previous paragraph and stresses the importance of inspector qualification.

8-1.8 The operator shall specify the maximum allowable working pressure, which includes a suitable margin above the operating pressure, and the maximum allowable vacuum.

Because of the large size of LP-Gas refrigerated storage containers small increments of design pressure and vacuum can carry large costs. Container pressure is significantly affected by small variations in filling rates, product composition, barometric pressure changes, and process upsets. The design pressure of these containers is usually quite low [½-2 psig (3-14 kPa gauge)] and their ability to withstand vacuum is minimal [usually ¼ psig (2 kPa gauge)]. The operator must take all these factors into account when specifying the container design pressure, and insure that the plant is designed and operated so that this design pressure is not exceeded.

8-1.9 All piping that is a part of an LPG container shall be in accordance with ANSI B31.3. This container piping shall include all piping internal to the container, within the insulation spaces, and external piping attached or connected to the container up to the first circumferential external joint of the piping. Inert gas purge systems wholly within the insulation spaces are exempt from this provision.

This requirement is an exception to API 620 because the requirements of ANSI B 31.3 are somewhat more stringent than the piping requirements of API 620.

8-1.10 LPG containers shall be installed on suitable foundations designed by a qualified engineer and constructed in accordance with recognized structural engineering practices. Prior to the start of design and construction of the foundation, a sub-surface investigation shall be conducted by a qualified soils engineer to determine the stratigraphy and physical properties of the soils underlying the site.

NOTE: See ASCE 56, *Sub-Surface Investigation for Design and Construction of Foundation for Buildings,* and Appendix C, API Standard 620, for further information.

8-1.11   The bottom of the outer tank shall be above the ground water table or otherwise protected from contact with ground water at all times, and the material in contact with the bottom of the outer tank shall be selected to minimize corrosion.

8-1.12   When the bottom of an outer tank is in contact with the soil, a heating system shall be provided to prevent the 32°F (0°C) isotherm from penetrating the soil. The heating system shall be designed so as to permit functional and performance monitoring, which shall be done, at a minimum, on a weekly basis. Where there is a discontinuity in the foundation, such as for bottom piping, careful attention and separate treatment shall be given to the heating system in this zone. Heating systems shall be installed so that any heating elements or temperature sensor used for control can be replaced. Provisions shall be incorporated to protect against the detrimental effects of moisture accumulation in the conduit which could result in galvanic corrosion or other forms of deterioration within the conduit or heating element.

The intent of this provision is to prevent frost heaving as a result of the formation of ice, a very real possibility.

8-1.13   If the foundation is installed to provide adequate air circulation in lieu of a heating system, then the bottom of the outer tank shall be of a material suitable for temperatures to which it will be exposed.

8-2 Marking on Refrigerated Containers.

8-2.1   Each refrigerated container shall be identified by the attachment of a nameplate on the outer covering in an accessible place marked as specified in the following:

(a) Manufacturers name and date built.

(b) Liquid volume of the container in gal (US Standard) or barrels.

(c) Maximum allowable working pressure in lbs per sq in.

(d) Minimum temperature in degrees Fahrenheit for which the container was designed.

(e) Maximum allowable water level to which the container may be filled for test purposes.

(f) Density of the product to be stored in lbs per cu ft for which the container was designed.

(g) Maximum level to which the container is permitted to be filled with the liquefied petroleum gas for which it was designed.

8-3 Refrigerated Container Impoundment.

8-3.1 Refrigerated containers shall be located within an impoundment area that complies with 8-3.2 through 8-3.8.

8-3.2 The following provisions shall be made to minimize the possibility of accidental discharge of LPG from containers from endangering adjoining property or important process equipment and structures, or reaching waterways.

8-3.3 Enclosed drainage channels for LP-Gas are prohibited.

Enclosed drainage channels are prohibited to prevent accumulation of vapor in enclosed spaces which can impede the flow of liquid or promote detonation if ignition should occur.

Exception: Container downcomers used to rapidly conduct spilled LP-Gas away from critical areas may be enclosed provided that an adequate drainage rate is achieved.

8-3.4 Dikes, impounding walls, and drainage systems for LP-Gas and flammable refrigerant containment shall be of compacted earth, concrete, metal, and/or other suitable materials. They may be independent of the container or they may be mounted, integral to, or constructed against the container. They, and any penetrations thereof, shall be designed to withstand the full hydrostatic head of impounded LP-Gas or flammable refrigerant, the effect of rapid cooling to the temperature of the liquid to be confined, any anticipated fire exposure, and natural forces such as earthquake, wind, and rain.

8-3.5 Dikes, impounding walls, and drainage channels for flammable liquid containment shall conform to NFPA 30, *Flammable and Combustible Liquids Code.*

8-3.6 To assure that any accidentally discharged liquid stays within an area enclosed by a dike or impounding wall and yet to provide a reasonably wide margin for area configuration design, the dike or impounding wall, height, and distance shall be determined in accordance with Figure 8-3.6.

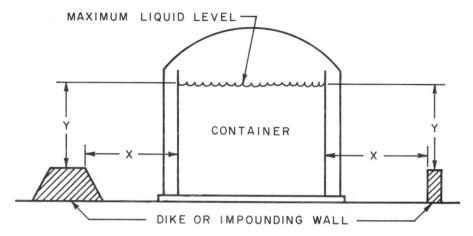

Figure 8-3.6 Dike or Impounding Wall Proximity to Containers.

Notes to Figure 8-3.6:

Dimension "X" must equal or exceed the sum of dimension "Y" plus the equivalent head in LP-Gas of the pressure in the vapor space above the liquid.

Exception: When the height of the dike or impounding wall is equal to, or greater than the maximum liquid level, "X" may have any value.

Dimension "X" is the distance from the inner wall of the container to the closest face of the dike or impounding wall.

Dimension "Y" is the distance from the maximum liquid level in the container to the top of the dike or impounding wall.

8-3.7   Provision shall be made to clear rain or other water from the impounding area. Automatically controlled sump pumps are permitted if equipped with an automatic cutoff device which shall prevent their operation when exposed to LP-Gas temperatures. Piping, valves, and fittings whose failure could permit liquid to escape from the impounding area shall be suitable for continuous exposure to LP-Gas temperatures. If gravity drainage is employed for water removal, provision shall be made to prevent the escape of LP-Gas by way of the drainage system.

8-3.8   Insulation systems used for impounding surfaces shall be, in the installed condition, noncombustible and suitable for the intended service considering the anticipated thermal and mechanical stresses and loadings. If flotation is a problem, mitigation measures shall be provided. Such insulation systems shall be inspected as appropriate for their intended service.

8-4   Refrigerated Aboveground Containers.

8-4.1   Containers shall be located outside of buildings.

8-4.2   A container or containers with aggregate water capacity in excess of 120,000 gal (454 m³) shall be located 100 ft (31 m) or more from buildings associated with the LP-Gas plant which are occupied for generation, compression, or purification of manufactured gas, from natural gas stationary refrigerated containers, or from natural gas compressor buildings or from outdoor installations essential to the maintenance of operation in such buildings. Such a container or containers shall be 100 ft (31 m) or more from aboveground storage of flammable liquids and from any buildings of such construction or occupancy which constitute a material hazard of exposure to the containers in the event of fire or explosion in said buildings. If the container or containers are located closer to any such buildings or installations, then the latter shall be protected by walls adjacent to such storage containers or by other appropriate means against the entry of escaped liquefied petroleum gas, or of drainage from the storage container area and its loading points — all in such a manner as may be required and approved by the authority having jurisdiction.

8-4.3   Refrigerated liquefied petroleum gas containers shall not be located within dikes enclosing flammable liquid tanks or within dikes enclosing nonrefrigerated liquefied petroleum gas containers.

8-4.4   Refrigerated containers shall not be installed one above the other.

This requirement is consistent with requirements for non-refrigerated containers, which may not be stacked.

8-4.5   The ground within 25 ft (8 m) of any aboveground refrigerated container and all ground within a diked area shall be kept clear of readily ignitible material such as weeds and long dry grass.

This requirement is similar to that for non-refrigerated containers except that the distance required to be kept clear is greater than the 10 ft (0.3 m) required for non-refrigerated containers. This recognizes the larger size of refrigerated containers. Most refrigerated container installations design their facilities to prevent the growth of vegetation near containers by the use of gravel or crushed stone surfacing.

Plant operators must take steps to insure that any brush clearing operations do not create hazards. At least one incident has occurred where a portable gasoline powered weed cutter cut a pipe, resulting in liquid release and ignition.

9 REFERENCED PUBLICATIONS

9-1 The following documents or portions thereof are referenced within this standard and shall be considered part of the requirements of this document. The edition indicated for each reference is the current edition as of the date of the NFPA issuance of this document.

9-1.1 NFPA Publications. National Fire Protection Association, Batterymarch Park, Quincy, MA 02269.

NFPA 10-1988, *Standard for Portable Fire Extinguishers*

NFPA 15-1985, *Standard for Water Spray Fixed Systems for Fire Protection*

NFPA 30-1987, *Flammable and Combustible Liquids Code*

NFPA 37-1984, *Standard for the Installation and Use of Stationary Combustion Engines and Gas Turbines*

NFPA 50-1985, *Standard for Bulk Oxygen Systems at Consumer Sites*

NFPA 50A-1989, *Standard for Gaseous Hydrogen Systems at Consumer Sites*

NFPA 50B-1989, *Standard for Liquefied Hydrogen Systems at Consumer Sites*

NFPA 51-1987, *Standard for the Design and Installation of Oxygen-Fuel Gas Systems for Welding, Cutting, and Allied Processes*

NFPA 54 (ANSI Z223.1)-1988, *National Fuel Gas Code*

NFPA 59-1989, *Standard for the Storage and Handling of Liquefied Petroleum Gases at Utility Gas Plants*

NFPA 61B-1989, *Standard for the Prevention of Fires and Explosions in Grain Elevators and Facilities Handling Bulk Raw Agricultural Commodities*

NFPA 70-1987, *National Electrical Code*

NFPA 80-1986, *Standard for Fire Doors and Windows*

NFPA 82-1983, *Standard on Incinerators, Waste and Linen Handling Systems and Equipment*

NFPA 86-1985, *Standard for Ovens and Furnaces*

NFPA 96-1987, *Standard for the Installation of Equipment for the Removal of Smoke and Grease-Laden Vapors from Commercial Cooking Equipment*

NFPA 220-1985, *Standard on Types of Building Construction*

NFPA 251-1985, *Standard Methods of Fire Tests of Building Construction and Materials*

NFPA 302-1989, *Fire Protection Standard for Pleasure and Commerial Motor Craft*

NFPA 321-1987, *Standard on Basic Classification of Flammable and Combustible Liquids*

NFPA 501A-1987, *Standard for Firesafety Criteria for Manufactured Home Installations, Sites, and Communities*

NFPA 501C-1986, *Standard on Firesafety Criteria for Recreational Vehicles*

NFPA 505-1987, *Firesafety Standard for Powered Industrial Trucks Including Type Designations, Areas of Use, Maintenance, and Operation*

9-1.2 ANSI Publications. American National Standards Institute, 1430 Broadway, New York, NY 10018.

ANSI A58.1-1972, *Design Loads for Buildings and Other Structures*

ANSI B36.10-1979, *Welded and Seamless Wrought Steel Pipe*

ANSI B95.1-1977, *Standard Terminology for Pressure Relief Devices*

9-1.3 API Publications. American Petroleum Institute, 2101 L ST., NW, Washington, DC 20037.

API-ASME *Code for Unfired Pressure Vessels for Petroleum Liquids and Gases*, Pre - July 1, 1961.

ANSI/API 620, *Recommended Rules for Design and Construction of Large, Welded, Low-Pressure Storage Tanks*, 1982.

9-1.4 ASME Publications. American Society for Mechanical Engineers, 345 East 47th St., New York, NY 10017.

"Rules for the Construction of Unfired Pressure Vessels," Section VIII, Division 1, *ASME Boiler and Pressure Vessel Code*, 1986, and all addenda and errata thru 1988.

ANSI/ASME B31.3-1988, *Chemical Plant and Refinery Piping*

9-1.5 ASTM Publications. American Society for Testing and Materials, 1916 Race St., Philadelphia, PA 19103.

ASTM A 47-1984, *Standard Specification for Ferritic Malleable Iron Castings*

ASTM A 48-1983, *Standard SpeciTube for Air Conditioning and Refrigeration Field Service*

ASTM D 2513-1987, *Standard Specification for Thermoplastic Gas Pressure Pipe, Tubing and Fittings*

ASTM D 2683-1987, *Standard Specification for Socket-Type Polyethylene (PE) Fittings for Outside Diameter—Controlled Polyethylene Pipe*

ASTM D 3261-1987, *Standard Specification for Butt Heat Fusion Polyethylene (PE) Plastic Fittings for Polyethylene (PE) Plastic Pipe and Tubing*

9-1.6 AWS Publication. American Welding Society, 2501 NW 7th St., Miami, FL 33125.

ANSI Z49.1-1988, *Safety in Welding and Cutting.*

9-1.7 CGA Publications. Compressed Gas Association, Inc., 1235 Jefferson Davis Highway, Arlington, VA 22202.

ANSI/CGA C-4-1978, *Method of Marking Portable Compressed Gas Containers to Identify the Material Contained*

Pressure-Relief Device Standards,

S-1.1-1979, *Cylinders for Compressed Gases* (Errata, 1982)

S-1.2-1980, *Cargo and Portable Tanks for Compressed Gases*

S-1.3-1980, *Compressed Gas Storage Containers*

9-1.8 Federal Regulations. U.S. Government Printing Office, Washington, DC.

Code of Federal Regulations, Title 49, Parts 171-192 and Parts 393 and 397. (Also available from the Association of American Railroads, American

Railroads Bldg., 1920 L St. NW, Washington, DC 20036 and American Trucking Assns., Inc., 2201 Mill Rd., Alexandria, VA 22314.)

9-1.9 ICBO Publication.   International Conference of Building Officials, 5360 S. Workman Mill Rd., Whittier, CA 90601.

Uniform Building Code, 1988.

9-1.10 UL Publication.   Underwriters Laboratories, Inc., 333 Pfingsten Rd., Northbrook, IL 60062.

UL 132-1984, *Safety Relief Valves for Anhydrous Ammonia and LP-Gas.*

APPENDIX A

This Appendix is not a part of the requirements of this NFPA document, but is included for information purposes only.

The material contained in Appendix A of NFPA 58, *Standard for the Storage and Handling of Liquefied Petroleum Gases,* is included in the text within this Handbook, and therefore Appendix A is not repeated here.

APPENDIX B
PROPERTIES OF LP-GASES

This Appendix is not a part of the requirements of this NFPA document, but is included for information purposes only.

B-1 Approximate Properties of LP-Gases.

B-1.1 Source of Property Values.

B-1.1.1 The property values for the LP-Gases are based on average industry values and include values for LP-Gases coming from natural gas liquids plants as well as those coming from petroleum refineries. Thus, any particular commerical propane or butane might have properties varying slightly from the values shown. Similarly, any propane-butane mixture might have properties varying from those obtained by computation from these average values (*see B-1.2.1 for computation method used*). Since these are average values, the interrelationships between them (i.e., lb per gal, specific gravity, etc.) will not cross-check perfectly in all cases.

B-1.1.2 Such variations are not sufficient to prevent the use of these average values for most engineering and design purposes. They stem from minor variations in composition. The commerical grades are not pure (CP-Chemically Pure) propane or butane, or mixtures of the two, but may also contain small and varying percentages of ethane, ethylene, propylene, isobutane, or butylene which can cause slight variations in property values. There are limits to the accuracy of even the most advanced testing methods used to determine the percentages of these minor components in any LP-Gas.

B-1.2 Approximate Properties of LP-Gases.

B-1.2.1 The principal properties of commercial propane and commercial butane are shown in Table B-1.2.1. Reasonably accurate property values for propane-butane mixtures may be obtained by computation, applying the percentages by weight of each in the mixture to the values for the property it is desired to obtain. Slightly more accurate results for vapor pressure are obtained by using the percentages by volume. Very accurate results can be obtained using data and methods explained in petroleum and chemical engineering data books.

B-1.3 Specifications of LP-Gases. Specifications of LP-Gases covered by this standard are listed in Gas Processors Association, *Liquefied Petroleum Gas Specifications for Test Methods Standard 2140* and/or *Specification for Liquefied Petroleum (LP) Gases*, ASTM D 1835.

Table B-1.2.1 (English)   Approximate Properties of LP-Gases

	Commercial Propane	Commercial Butane
Vapor Pressure in psig at:		
70°F	127	17
100°F	196	37
105°F	210	41
130°F	287	69
Specific Gravity of Liquid at 60°F	0.504	0.582
Initial Boiling Point at 14.7 psia, °F	−44	15
Weight per Gallon of Liquid at 60°F, lb	4.20	4.81
Specific Heat of Liquid, Btu/lb at 60°F	0.630	0.549
Cu. ft of Vapor per Gallon at 60°F	36.38	31.26
Cu. ft of Vapor per Pound at 60°F	8.66	6.51
Specific Gravity of Vapor (Air = 1) at 60°F	1.50	2.01
Ignition Temperature in Air, °F	920–1120	900–1000
Maximum Flame Temperature in Air, °F	3,595	3,615
Limits of Flammability in Air, Percent of Vapor in Air-Gas Mixture:		
(a) Lower	2.15	1.55
(b) Upper	9.60	8.60
Latent Heat of Vaporization at Boiling Point:		
(a) Btu per Pound	184	167
(b) Btu per Gallon	773	808
Total Heating Values after Vaporization:		
(a) Btu per Cubic Foot	2,488	3,280
(b) Btu per Pound	21,548	21,221
(c) Btu per Gallon	91,502	102,032

Table B-1.2.1 (Metric) Approximate Properties of LP-Gases

	Commercial Propane	Commercial Butane
Vapor Pressure in kPa gauge at:		
20°C	895	103
40°C	1 482	285
45°C	1 672	345
55°C	1 980	462
Specific Gravity	0.504	0.582
Initial Boiling Point at atm, Pressure, °C	−42	−9
Weight per Cubic Metre of Liquid at 15.56°C, kg	504	582
Specific Heat of Liquid, Kilojoule per Kilogram, at 15.56°C	1.464	1.276
Cubic Metre of Vapor per Litre of Liquid at 15.56°C	0.271	0.235
Cubic Metre of Vapor per Kilogram of Liquid at 15.56°C	0.539	0.410
Specific Gravity of Vapor (Air = 1) at 15.56°C	1.50	2.01
Ignition Temperature in Air, °C	493–549	482–538
Maximum Flame Temperature in Air, °C	1 980	2 008
Limits of Flammability in Air, % of Vapor in Air-Gas Mixture:		
(a) Lower	2.15	1.55
(b) Upper	9.60	8.60
Latent Heat of Vaporization at Boiling Point:		
(a) Kilojoule per kilogram	428	388
(b) Kilojoule per litre	216	226
Total Heating Value after Vaporization:		
(a) Kilojoule per Cubic Metre	92 430	121 280
(b) Kilojoule per Kilogram	49 920	49 140
(c) Kilojoule per Litre	25 140	28 100

APPENDIX C
DESIGN, CONSTRUCTION, AND REQUALIFICATION
OF DOT (ICC) CYLINDER SPECIFICATION
CONTAINERS

This Appendix is not a part of the requirements of this NFPA document, but is included for information purposes only.

C-1 Scope.

C-1.1 Application.

C-1.1.1 This appendix provides general information on DOT cylinder specification containers referred to in this standard. For complete information consult the applicable specification (*see C-2.1.1*). The water capacity of such cylinders may not be more than 1,000 lb (454 kg).

C-1.1.2 This appendix is not applicable to DOT tank car portable tank container or cargo tank specifications. Portable and cargo tanks are basically ASME containers and are covered in Appendix D.

C-1.1.3 Prior to April 1, 1967, these specifications were promulgated by the Interstate Commerce Commission (ICC). On this date, certain functions of the ICC, including the promulgation of specifications and regulations dealing with LP-Gas cylinders, were transferred to the Department of Transportation (DOT). Throughout this appendix both ICC and DOT are used; ICC applying to dates prior to April 1, 1967, and DOT to subsequent dates.

C-2 LP-Gas Cylinder Specifications.

C-2.1 Publishing of DOT Cylinder Specifications.

C-2.1.1 DOT cylinder specifications are published under Title 49, *Code of Federal Regulations*, Parts 171-190, available from U.S. Government Printing Office, Washington, D.C. The information in this publication is also issued as a Tariff at approximately three-year intervals by the Bureau of Explosives, American Railroads Building, 1920 L Street, NW, Washington, DC 20036.

C-2.2 DOT Specification Nomenclature.

C-2.2.1 The specificaion designation consists of a one-digit number, sometimes followed by one or more capital letters, then by a dash and a three-digit number. The one-digit number alone, or in combination with one or more capital letters, designates the specification number. The three-digit number following the dash shows the service pressure for which the container is designed. Thus, "4B-240" indicates a cylinder built to Specification 4B for a 240 psig service pressure. (*See C-2.2.3.*)

C-2.2.2 The specification gives the details of cylinder construction, such as material used, method of fabrication, tests required and inspection method, and prescribes the service pressure, or range of service pressures for which that specification may be used.

C-2.2.3 The term "service pressure" is analagous to, and serves the same purpose as, the ASME "design pressure." However, it is not identical, representing instead the highest pressure to which the container will normally be subjected in transit or in use but not necessarily the maximum pressure to which it may be subjected under emergency conditions in transportation. The service pressure stipulated for the LP-Gases is based on the vapor pressures exerted by the product in the container at two different temperatures, the higher pressure of the two becoming the service pressure, as follows:

(a) The pressure in the container at 70°F must be less than the service pressure for which the container is marked, and

(b) The pressure in the container at 130°F must not exceed $\frac{5}{4}$ times the pressure for which the container is marked.

EXAMPLE: Commercial propane has a vapor pressure at 70°F of 132 psig. However, its vapor pressure at 130°F is 300 psig, so service pressure ($\frac{5}{4}$ times which must not exceed 300 psig) is 300 divided by $\frac{5}{4}$, or 240 psig. Thus commercial propane requires at least 240 psig service pressure cylinder.

C-2.3 DOT Cylinder Specifications Used for LP-Gas.

C-2.3.1 A number of different specifications were approved by ICC, and since 1967 by DOT, for use with LP-Gases. Some of these are no longer published or used for new construction. However, containers built under these old specifications, if properly maintained and requalified, are still acceptable for LP-Gas transportation.

C-2.3.2 DOT specifications cover primarily safety in transportation. However, in order for the product to be used, it is necessary for it to come to rest at the point of use and serve as LP-Gas storage during the period of use. Containers adequate for transportation are also deemed to be adequate for use as provided in NFPA 58. As small size ASME containers were not available at the time tank truck delivery was started, ICC (now DOT) cylinders have been equipped for tank truck deliveries and permanently installed.

C-2.3.3 The DOT cylinder specifications most widely used for the LP-Gases are shown in Table C-2.3.3. The differing materials of construction, method of fabrication and the date of the specification reflect the progress made in knowledge of the products to be contained and improvement in metallurgy and methods of fabrication.

C-3 Requalification, Retesting and Repair of DOT Cylinder Specification Containers.

C-3.1 Application.

C-3.1.1 This section outlines the requalification, retesting and repair requirements for DOT cylinder specification containers but should be used only as a guide. For official information, the applicable DOT regulations should be consulted.

Table C-2.3.3

Specification No. & Marking	Material of Construction	Method of Fabrication
26–150*	Steel	Welded and Brazed
3B–300	Steel	Seamless
4–300	Steel	Welded
4B–300	Steel	2 piece Welded & Brazed
4B–240	Steel	2 piece Welded & Brazed
4BA–240	Alloy Steel	2 piece Welded & Brazed
4E–240	Aluminum	Welded and Brazed
4BW–240	Steel	3 piece Welded

* The term "service pressure" had a different connotation at the time the specification was adopted.

C-3.2 Requalification (Including Retesting) of DOT Cylinders.

C-3.2.1 DOT cylinders may not be refilled, continued in service or transported unless they are properly qualified or requalified for LP-Gas service in accordance with DOT regulations.

C-3.2.2 A careful examination must be made of every container each time it is to be filled and it must be rejected if there is evidence of exposure to fire, bad gouges or dents, seriously corroded areas, leaks or other conditions indicating possible weaknesses which might render it unfit for service. The following disposition is to be made of rejected cylinders:

(a) Containers subjected to fire must be requalified, reconditioned or repaired in accordance with C-3.3.1, or permanently removed from service except that DOT 4E (aluminum) cylinders must be permanently removed from service.

(b) Containers showing serious physical damage, leaks or with a reduction in the marked tare weight of 5 percent or more must be retested in accordance with C-3.2.4(a) or (b) and, if necessary, repaired in accordance with C-3.3.1.

C-3.2.3 All containers, including those apparently undamaged, must be periodically requalified for continued service. The first requalification for a new cylinder is required within 12 years after the date of manufacture. Subsequent requalifications are required within the periods specified under the requalification method used.

C-3.2.4 DOT regulations permit three alternative methods of requalification for most commonly used LP-Gas specification containers (*see DOT regulations for permissible requalification methods for specific cylinder specifications*). Two use hydrostatic testing, and the third uses a carefully made and duly recorded visual examination by a competent person. In the case of the two hydrostatic test methods, only test results are recorded but a careful visual examination of each container is also required. DOT regulations cite in detail the data to be recorded for the hydrostatic test methods, the observations to be made during the recorded visual examination method, and the marking of containers to indicate the requalification date and the method used. The three methods are outlined as follows:

(a) The water jacket type hydrostatic test may be used to requalify containers for 12 years before the next requalification is due. A pressure of twice the marked service pressure is applied, using a water jacket (or the equivalent) so that the total expansion of the container during the application of the test pressure can be observed and recorded for comparison with the permanent expansion of the container after depressurization. The following disposition is made of containers tested in this manner:

(1) Containers which pass the retest, and the visual examination required with it (*see C-3.2.4*), are marked with the date and year of the test (Example: "6-70," indicating requalification by the water jacket test method in June 1970) and may be placed back in service.

(2) Containers which leak, or for which the permanent expansion exceeds 10 percent of the total expansion (12 percent for Specification 4E aluminum cylinders) must be rejected. If rejected for leakage, containers may be repaired in accordance with C-3.3.1.

(b) The simple hydrostatic test may be used to requalify containers for seven years before the next requalification is due. A pressure of twice the marked service pressure is applied but no provision is made for measuring total and permanent expansion during the test outlined in C-3.2.4(a) above. The container is carefully observed while under the test pressure for leaks, undue swelling or bulging indicating weaknesses. The following disposition is made of containers tested in this matter:

(1) Containers which pass the test, and the visual examination required with it (*see C-3-2.4*), are marked with the date and year of the retest followed by an "S" (Example: "8-71S," indicating requalification by the simple hydrostatic test method in August 1971), and may be placed back in service.

(2) Containers developing leaks or showing undue swelling or bulging must be rejected. If rejected for leaks, containers may be repaired in accordance with C-3.3.1.

(c) The recorded visual examination may be used to requalify containers for five years before the next qualification is due provided the container has been used exclusively for LP-Gas commerically free from corroding components. Inspection is to be made by a competent person, using as a guide Compressed Gas Association Standard for Visual Inspection of Steel Compressed Gas Cylinders (CGA Pamphlet C-6), and recording the inspection results as required by DOT regulations. [Note: reference to NLPGA Safety Bulletin, *Recommended Procedures for Visual Inspection and Requalification of DOT (ICC) Cylinders in LP-Gas Service*, is also recommended.] The following disposition is to be made of containers inspected in this manner:

The Compressed Gas Association's *Standard for Visual Inspection of Steel Compressed Gas Cylinders*, pamphlet C-6, is a guide which can be used to requalify cylinders as required by DOT regulations. It contains examples of various types of pitting, corrosion, and failed cylinders and is very useful for persons not completely familiar with the subject.

The National Propane Gas Association publishes two safety pamphlets which provide useful information on requalifying cylinders, NPGA 117, *Visual Cylinder Inspection*, which is a guide for visual inspection of cylinders prior to refilling, and NPGA 118, *Recommended Procedures for Visual Inspection and Requalification of DOT (ICC) Cylinders in LP-Gas Service*, which provides the complete set of DOT requirements along with illustrations demonstrating the use of the gauges needed to complete the inspection and a "visual inspection report form". (*See Figure C.1.*)

(1) Containers which pass the visual examination are marked with the date and year of the examination followed by an "E" (Example: "7-70E," indicating requalification by the recorded visual examination method in July 1970), and may be placed back in service.

(2) Containers which leak, show serious denting or gouging, or excessive corrosion must either be scrapped or repaired in accordance with C-3.3.1.

C-3.3 Repair of DOT Cylinder Specification Containers.

C-3.3.1 Repair of DOT cylinders must be performed by a manufacturer of the type of cylinder to be repaired or by a repair facility authorized by DOT.

Repairs normally made are for fire damage, leaks, denting, gouges and for broken or detached valve protecting collars or foot rings.

FIVE YEAR VISUAL INSPECTION REPORT

_____ Co.   Date _____

Cylinder Identification					Protective Coating		Cylinders Checked For							Disposition		
Serial Number	Date Mf'd.	ICC Spec. Number	Symbol	Name of Manufacturer	Type	Condition	Fire Damage	Dents or Digs	Footring	Leaks	Corrosion	Collar and/or Opening	Bulges	Disposition (ser.) (code)	Date Inspect.	Inspector's Initials
90165	6-58	4BA240	GAS INC.	ABCyl	PAINT	6000	√	√	√	√	√	√	√	OK	1-6-72	JHD
220196	10-70	4BW240	ABCCyl	ABCyl	GALVANIZED	FAIR	√	SC	√	√	√	√	√	SC	1-6-72	JHD
109640	5-60	4BA240	XYZCoy	ABCyl	PAINT	GOOD	√	√	√	√	√	R	√	R	1-6-72	JHD

Disposition Code
OK—Return To Service
SC—Scrap
RM—Return To Manufacturer For Repair

Figure C-1   Visual Inspection Report Form

APPENDIX D
DESIGN OF ASME AND API-ASME CONTAINERS

This Appendix is not a part of the requirements of this NFPA document, but is included for information purposes only.

D-1 General.

D-1.1 Application.

D-1.1.1 This appendix provides general information on containers designed and constructed in accordance with ASME or API-ASME Codes, usually referred to as ASME containers. For complete information on either ASME or API-ASME containers the applicable code should be consulted. Construction of containers to the API-ASME Code has not been authorized since July 1, 1961.

D-1.1.2 DOT (ICC) specification portable tank containers and cargo tanks are basically either ASME or API-ASME containers. In writing these specifications, which should be consulted for complete information, additions were made to these pressure vessel codes to cover the following:

(a) Protection of container valves and appurtenances against physical damage in transportation.

(b) Holddown devices for securing cargo containers to conventional vehicles.

(c) Attachments to relatively large [6,000 gal (22.7 m³) or more water capacity] cargo containers in which the container serves as a stress member in lieu of a frame.

D-1.2 Development of ASME and API-ASME Codes.

D-1.2.1 ASME type containers of approximately 12,000 gal (45.4 m³) water capacity or more were initially used for bulk storage in processing, distribution and industrial plants. As the industry expanded and residential and commercial usage increased, the need for small ASME containers with capacities greater than the upper limit for DOT cylinders grew. This ultimately resulted in the development of cargo containers for tank trucks and the wide use of ASME containers ranging in size from less than 25 gal (0.1 m³) to 120,000 gal (454 m³) water capacity.

D-1.2.2 The American Society of Mechanical Engineers (ASME) in 1911 set up the Boiler and Pressure Vessel Committee to formulate "standard rules for the construction of steam boilers and other pressure vessels." The

ASME *Boiler and Pressure Vessel Code*, first published in 1925, has been revised and republished in 22 separate editions including the 1980 edition. During this period there have been changes in the code as materials of construction improved and more was known about them, and as fabrication methods changed and inspection procedures were refined.

D-1.2.3 One major change involved the so-called "factor of safety" (the ratio of the ultimate strength of the metal to the design stress used). Prior to 1946, a 5:1 safety factor was used. Fabrication changed from the riveting widely used when the code was first written (some forge welding was used), to fusion welding. This latter method was incorporated into the code as welding techniques were perfected, and now predominates.

D-1.2.4 The safety factor change in the ASME Code was based on the technical progress made since 1925 and on experience with the use of the API-ASME Code. This offshoot of the ASME Code, initiated in 1931, was formulated and published by the American Petroleum Institute (API) in cooperation with the ASME. It justified the 4:1 safety factor on the basis of certain quality and inspection controls not at that time incorporated in the ASME Code editions.

D-1.2.5 ASME Code Case Interpretations and Addenda are published between Code editions and normally become part of the Code in the new edition. Adherence to these is considered compliance with the Code. [*See 2-2.1.3(a)*.]

D-2 Design of Containers for LP-Gas.

D-2.1 ASME Container Design.

D-2.1.1 When ASME containers were first used to store LP-Gas, the properties of the CP grades of the principal constituents were available, but the average properties for the commercial grades of propane and butane were not. Also there was no experience as to what temperatures and pressures to expect for product stored in areas with high atmospheric temperatures. A 200 psi (1.4 MPa) design pressure was deemed appropriate for propane [the CP grade of which has a vapor pressure of 176 psi (1.2 MPa) at 100°F (38°C) and 80 psi (0.6 MPa) for butane (CP grade has vapor pressure of 37 psi (0.6 MPa) at 100°F (38°C).] These containers were built with a 5:1 safety factor (*see D-1.2.3*).

D-2.1.2 Pressure vessel codes, following boiler pressure relief valve practice, require that the pressure relief valve start-to-leak setting be the design pressure of the container. In specifying pressure relief valve capacity, however, they stipulate that this relieving capacity be adequate to prevent the internal pressure from rising above 120 percent of the design pressure under fire exposure conditions.

D-2.1.3 Containers built in accordance with D-2.1.1 were entirely adequate for the commercial grades of the LP-Gases [the vapor pressure of propane at 100°F (38°C) is 205 psi (1.43 MPa); the vapor pressure of butane at 100°F (38°C) is 37 psi (0.26 MPa)]. However, as they were equipped with pressure relief valves set to start-to-leak at the design pressure of the container, these relief valves occasionally opened on an unusually warm day. Since any unnecessary release of a flammable gas is potentially dangerous, and giving weight to recommendations of fire prevention and insurance groups as well as to the favorable experience with API-ASME containers (*see D-2.2.1*), relief valve settings above the design pressure [up to 250 psi (1.7 MPa) for propane and 100 psi (0.7 MPa) for butane] were widely used.

D-2.1.4 In determining safe filling densities for compressed liquefied gases, DOT (ICC) uses the criterion that the container shall not become liquid full at the highest temperature the liquid may be expected to reach due to the normal atmospheric conditions to which the container may be exposed. For containers of more than 1,200 gal (4.5 m³) water capacity, the liquid temperature selected is 115°F (46°C). The vapor pressure of the gas to be contained at 115°F (46°C) is specified by DOT as the minimum design pressure for the container. The vapor pressure of CP propane at 115°F (46°C) is 211 psig (1.4 MPa gauge), and of commercial propane, 243 psig (1.5 MPa gauge). The vapor pressure of both normal butane and commercial butane at 115°F (46°C) is 51 psig (0.4 MPa gauge).

D-2.1.5 The ASME *Boiler and Pressure Vessel Code* editions generally applicable to LP-Gas containers, and the design pressures, safety factors and exceptions to these editions for LP-Gas use, are shown in Table D-2.1.5. These reflect the use of the information in D-2.1.1 through D-2.1.4.

Table D-2.1.5

| Year ASME Code Edition Published | Design Pressure, psi (Pascals) | | Safety Factor |
	Butane	Propane	
1931 through 46[2]	100[1] (0.7)	200 (1.4)	5:1
1949 Par. U-68 & U-69[2]	100 (0.7)	200 (1.4)	5:1
1949 Par. U-200 & U-201[2]	125 (0.9)	250 (1.7)	4:1
1952 through 80	125 (0.9)	250[3] (1.7)	4:1

[1]Until December 31, 1947, containers designed for 80 psi (0.6 MPa) under prior (5:1 safety factor) codes were authorized for butane. Since that time, either 100 psi (0.7 MPa) (under prior codes) or 125 psi (0.9 MPa) (under present codes) is required.

[2]Containers constructed in accordance with 1949 and prior editions of the ASME Code were not required to be in compliance with paragraphs U-2 to U-10 inclusive, or with paragraph U-19. Construction in accordance with paragraph U-70 of these editions was not authorized.

[3]Higher design pressure [312.5 psi (2.2 MPa)] is required for small ASME containers used for vehicular installations (such as forklift trucks used in buildings or those installed in enclosed spaces) because they may be exposed to higher temperatures and consequently develop higher internal pressure.

D-2.2 API-ASME Container Design.

D-2.2.1   The API-ASME Code was first published in 1931 (*see D-2.1.4*). Based on petroleum industry experience using certain material quality and inspection controls not at that time incorporated in the ASME Code, the 4:1 safety factor was first used. Many LP-Gas containers were built under this code with design pressures of 125 psi (0.9 MPa) [100 psi (0.7 MPa) until December 31, 1947] for butane and 250 psi (1.7 MPa) for propane. Containers constructed in accordance with the API-ASME Code were not required to comply with Section 1, or the appendix to Section 1. Paragraphs W-601 through W-606 of the 1943 and earlier editions were not applicable to LP-Gas containers.

D-2.2.2   The ASME Code, by changing from the 5:1 to the 4:1 safety factor through consideration of the factors described in D-2.1.1 through D-2.1.4, became nearly identical in effect to the API-ASME Code by the 1950's. Thus, the API-ASME Code was phased out and construction was not authorized after July 1, 1961.

D-2.3 Design Criteria for LP-Gas Containers.

D-2.3.1   To prevent confusion in earlier editions of NFPA 58, the nomenclature "container type" was used to designate the design pressure of the container to be used for various types of LP-Gases. With the adoption of the 4:1 safety factor in the ASME Code and the phasing out of the API-ASME Code, the need for "container type" ceased to exist. Table D-2.3.1 makes it possible to compare older containers which may have carried this designation with the new containers complying with 2-2.2.2 and Table 2-2.2.2 in this standard.

Table D-2.3.1   Vapor Pressure, Design Pressures and Container Type

As Shown in Table 2-2.2.2		Design Press. Earlier Codes		
Maximum Vapor Press. at 100°F (37.8°C)	Design Press. Present ASME Code[1]	API-ASME	ASME[2]	Container Type
80 (.6)	100 (.7)	100 (.7)	80 (.6)	80 (.6)
100 (.7)	125 (.9)	125 (.9)	100 (.7)	100 (.7)
125 (.9)	156 (1.1)	156 (1.1)	125 (.9)	125 (.9)
150 (1.0)	187 (1.3)	187 (1.3)	150 (1.0)	150 (1.0)
175 (1.2)	219 (1.5)	219 (1.5)	175 (1.2)	175 (1.2)
215 (1.5)	250 (1.7)	250 (1.7)	200 (1.4)	200 (1.4)
215 (1.5)	312.5 (2.2)	312.5 (2.2)	—	250 (1.7)

[1]ASME Code edition for 1949, Par. U-200 and U-201 and all later editions (*See D-2.1.5*).

[2]All ASME Codes up to the 1946 edition and paragraphs U-68 and U-69 of the 1949 edition (*See D-2.1.5*).

D-2.4 DOT (ICC) Specifications Utilizing ASME or API-ASME Containers.

D-2.4.1 DOT (ICC) Specifications for portable tank containers and cargo tanks require ASME or API-ASME construction for the container proper (*see D-1.1.2*). Several such specifications were written by the ICC prior to 1967 and DOT has continued this practice.

D-2.4.2 ICC Specifications written prior to 1946, and to some extent through 1952, used ASME containers with a 200 psig (1.4 MPa gauge) design pressure for propane and 80 psig (0.6 MPa gauge) for butane [100 psig (0.7 MPa gauge) after 1947] with a 5:1 safety factor. During this period and until 1961, ICC Specifications also permitted API-ASME containers with a 250 psig (1.7 MPa gauge) design pressure for propane and 100 psig (0.7 MPa gauge) for butane [125 psig (0.9 MPa gauge) after 1947].

D-2.4.3 To prevent any unnecessary release of flammable vapor during transportation (*see D-2.1.3*), the use of safety relief valve settings 25 percent above the design pressure was common for ASME 5:1 safety factor containers. To eliminate confusion, and in line with the good experience with API-ASME containers, the ICC permitted the rerating of these particular ASME containers used under its specifications to 125 percent of the originally marked design pressure.

D-2.4.4 DOT (ICC) Specifications applicable to portable tank containers and cargo tanks currently in use are listed in Table D-2.4.4. New construction is not permitted under the older specifications. However, these older containers may continue to be used provided they have been maintained in accordance with DOT (ICC) regulations.

Table D-2.4.4

Spec. Number	ASME Construction Design Pressure, psig		Safety Factor	API-ASME Construction Design Pressure, psig		Safety Factor
	Propane	Butane		Propane	Butane	
ICC-50[1]	200[3]	100[3]	5:1	250	125	4:1
ICC-51[1]	250	125	4:1	250	125	4:1
MC-320[2,4]	200[3]	100[3]	5:1	250	125	4:1
MC-330[2]	250	125	4:1	250	125	4:1
MC-331[2]	250	125	4:1	250	125	4:1

[1]Portable Tank Container.
[2]Cargo Tank.
[3]Permitted to be rerated to 125 percent of original ASME Design Pressure.
[4]Require DOT Exemption.

For SI Units

100 psig = 0.7 MPa gauge; 125 psig = 0.9 MPa gauge; 200 psig = 1.4 MPa gauge; 250 psig = 1.7 MPa gauge

D-3 Underground ASME or API-ASME Containers.

D-3.1 Use of Containers Underground.

D-3.1.1 ASME or API-ASME containers are used for underground or partially underground installation in accordance with 3-2.3.8 or 3-2.3.9. The temperature of the soil is normally low so that the average liquid temperature and vapor pressure of product stored in underground containers will be lower than in aboveground containers. This lower operating pressure provides a substantial corrosion allowance for underground containers.

D-3.1.2 Containers listed to be used interchangeably for either installation aboveground or underground must comply as to pressure relief valve rated relieving capacity and filling density with aboveground provisions when installed aboveground [*see 2-3.2.4(a)*]. When installed underground the pressure relief valve rated relieving capacity and filling density may be in accordance with underground provisions (*see E-2.3.1*), provided all other underground installation provisions are met. Partially underground containers are considered as aboveground insofar as loading density and pressure relief valve rated relieving capacity are concerned.

APPENDIX E
PRESSURE RELIEF DEVICES

This Appendix is not a part of the requirements of this NFPA document, but is included for information purposes only.

(This Appendix contains non-NFPA mandated provisions.)

E-1 Pressure Relief Devices for DOT (ICC) Cylinders.

E-1.1 Source of Provisions for Relief Devices.

E-1.1.1 The requirements for relief devices on DOT cylinders are established by the Bureau of Explosives with DOT approval. Complete technical information as to these requirements will be found in the Compressed Gas Association (CGA) Pamphlet S-1.1, Pressure-Relief Device Standards, Part 1 — *Cylinders for Compressed Gases.*

E-1.2 Essential Requirements of LP-Gas Cylinder Relief Devices.

E-1.2.1 CGA Pamphlet S-1.1 provides that LP-Gas cylinders shall be equipped with fusible plugs, spring-loaded pressure relief valves, or a combination of the two. Fusible plugs are not permitted on cylinders used in certain vehicular installations [*see 3-6.2.3(a)(4)*]. The provisions of E-1.2.2 through E-1.2.4 outline the generally accepted industry practice in the use of fusible plugs and safety pressure devices on LP-Gas cylinders.

E-1.2.2 If fusible plugs constitute the only relief devices the plugs used shall comply with the flow capacity requirements of CGA S-1.1 with a nominal melting or yield point of 165°F (74°C) [not less than 157°F (69°C) nor more than 170°F (77°C)]. For cylinders over 30 in. (0.8 m) long (exclusive of neck), a plug is required in each end of the cylinder.

E-1.2.3 If a spring-loaded pressure relief valve(s) constitutes the only relief device, the valves used shall comply with the flow capacity requirements of CGA S-1.1, with the set pressure not less than 75 percent nor more than 100 percent of the minimum required test pressure of the cylinder. For example, the test pressure for a 240 psig (1.6 MPa gauge) service pressure is 480 psig (3.2 MPa gauge); 75 percent of this is 360 psig (2.5 MPa gauge). In practice, such valves are set at 375 psig (2.6 MPa gauge).

E-1.2.4 If fusible plugs and spring-loaded pressure relief valves are used in combination, this combined use shall be in accordance with CGA S-1.1, or in substance as follows:

(a) If 100 percent of the relief device capacity is provided by the pressure relief valve, the supplementary fuse plug may be of any convenient size provided the total plug area does not exceed 0.25 sq in. (1.6 cm²), and may have a melting point of more than 170°F (77°C) [usually a nominal yield point of 212°F (100°C), with the upper limit not to exceed 220°F (104°C)]. Combination devices are required at one end of a cylinder only, or may be separated and installed at opposite ends.

(b) If at least 70 percent of the relief device capacity is provided by the pressure relief valve, the balance of the capacity requirement shall be supplied by a 165°F (74°C) nominal yield point fusible plug in accordance with E-1.2.2. Combined devices are required at one end of a cylinder only or may be separated and installed at opposite ends.

E-2 Pressure Relief Devices for ASME Containers.

E-2.1 Source of Provisions for Relief Devices.

E-2.1.1 Capacity requirements for relief devices are in accordance with the applicable provisions of Compressed Gas Association (CGA) Pamphlet S-1.2, Pressure-Relief Device Standards, Part 2 — *Cargo and Portable Tanks for Compressed Gases*; or with CGA Pamphlet S-1.3, Safety Relief-Device Standards, Part 3 — *Compressed Gas Storage Containers*.

E-2.2 Spring-Loaded Pressure Relief Valves for Aboveground and Cargo Containers.

E-2.2.1 The minimum rate of discharge for spring-loaded pressure relief valves is based on the outside surface of the containers on which the valves are installed. Paragraph 2-2.6.5(g) provides that new containers shall be marked with the surface area in sq ft. The surface area of containers not so marked (or not legibly marked) may be computed by use of the applicable formula:

(a) Cylindrical container with hemispherical heads:

Surface area = overall length × outside diameter × 3.1416.

(b) Cylindrical container with other than hemispherical heads:

Surface area = (overall length + 0.3 outside diameter) × outside diameter × 3.1416.

NOTE: This formula is not precise, but will give results with limits of practical accuracy in sizing relief valves.

(c) Spherical containers:

Surface area = outside diameter squared × 3.1416.

E-2.2.2 The minimum required relieving capacity in cu ft per minute of air at 120 percent of the maximum permitted start-to-leak pressure (or Flow Rate CFM Air), under standard conditions of 60°F (16°C) and atmospheric pressure [14.7 psia (0.1 MPa absolute)], shall be as shown in Table E-2.2.2 for the surface area in sq ft of the container on which the pressure relief valve is to be installed. The flow rate may be interpolated for intermediate values of surface area. For containers with a total outside surface area exceeding 2,000 sq ft, the required flow rate shall be calculated, using the formula: Flow Rate CFM Air $= 53.632 \times A^{0.82}$ where A = total outside surface area of container in sq ft.

Table E-2.2.2

Surface Area Sq. Ft.	Flow Rate CFM Air	Surface Area Sq. Ft.	Flow Rate CFM Air	Surface Area Sq. Ft.	Flow Rate CFM Air
20 or less	626	170	3620	600	10170
25	751	175	3700	650	10860
30	872	180	3790	700	11550
35	990	185	3880	750	12220
40	1100	190	3960	800	12880
45	1220	195	4050	850	13540
50	1330	200	4130	900	14190
55	1430	210	4300	950	14830
60	1540	220	4470	1000	15470
65	1640	230	4630	1050	16100
70	1750	240	4800	1100	16720
75	1850	250	4960	1150	17350
80	1950	260	5130	1200	17960
85	2050	270	5290	1250	18570
90	2150	280	5450	1300	19180
95	2240	290	5610	1350	19780
100	2340	300	5760	1400	20380
105	2440	310	5920	1450	20980
110	2530	320	6080	1500	21570
115	2630	330	6230	1550	22160
120	2720	340	6390	1600	22740
125	2810	350	6540	1650	23320
130	2900	360	6690	1700	23900
135	2990	370	6840	1750	24470
140	3080	380	7000	1800	25050
145	3170	390	7150	1850	25620
150	3260	400	7300	1900	26180
155	3350	450	8040	1950	26750
160	3440	500	8760	2000	27310
165	3530	550	9470		

E-2.3 Spring-Loaded Pressure Relief Valves for Underground or Mounded Containers.

E-2.3.1   In the case of containers installed underground or mounded, the pressure relief valve relieving capacities may be as small as 30 percent of those specified in Table D-2.2.2 provided the container is empty of liquid when installed, that no liquid is placed in it until it is completely covered with earth, and that it is not uncovered for removal until all liquid has been removed.

E-2.3.2 Containers partially underground must have pressure relief valve relieving capacitites in accordance with 2-3.2.4.

E-2.4 Provisions for Fusible Plugs.

E-2.4.1 Fusible plugs, supplementing spring-loaded pressure relief valves, and complying with 2-3.2.4(e), are permitted only with aboveground stationary containers of 1,200 gal (4.5 m³) or less water capacity. They shall not be used on larger containers nor on portable or cargo containers of ASME construction. The total fusible plug discharge area is limited to 0.25 sq in. (1.6 cm²) per container.

E-2.5 Pressure Relief Valve Testing.

E-2.5.1 Frequent testing of pressure relief valves on LP-Gas containers is not considered necessary for the following reasons:

(a) The LP-Gases are so-called "sweet gases" having no corrosive or other deleterious effect on the metal of the containers or relief valves.

(b) The relief valves are constructed of corrosion-resistant materials, and are installed so as to be protected against the weather. The variations of temperature and pressure due to atmospheric conditions are not sufficient to cause any permanent set in the valve springs.

(c) The required odorization of the LP-Gases makes escape almost instantly evident.

(d) Experience over the years with the storage of LP-Gases has shown a good safety record on the functioning of pressure relief valves.

E-2.5.2 Since no mechanical device can be expected to remain in operative condition indefinitely, it is suggested that the pressure relief valves on containers of more than 2,000 gal (7.6 m³) water capacity be tested at approximately 10-year intervals. Some types of valves may be tested by the use of an external lifting device having an indicator to show the pressure equivalent at which the valve may be expected to open. Others must be removed from the container for testing, requiring that the container first be emptied.

APPENDIX F
LIQUID VOLUME TABLES, COMPUTATIONS, AND GRAPHS

This Appendix is not a part of the requirements of this NFPA document, but is included for information purposes only.

F-1 Scope.

F-1.1 Application.

F-1.1.1 This appendix explains the basis for Table 4-5.2.1, includes the LP-Gas liquid volume temperature correction table, Table E-3.1.3, and describes its use. It also explains the methods of making liquid volume computations to determine the maximum permissible LP-Gas content of containers in accordance with Tables 4-5.2.3(a), (b), and (c).

F-2 Basis for Determination of LP-Gas Container Capacity.

F-2.1 The basis for determination of the maximum permitted filling densities shown in Table 4-5.2.1 is the maximum safe quantity which will assure that the container will not become liquid full when the liquid is at the highest anticipated temperature.

(a) For portable containers built to DOT specifications and other above-ground containers with water capacities of 1,200 gal (4.5 m³) or less, this temperature is assumed to be 130°F (54°C).

(b) For other aboveground uninsulated containers with water capacities in excess of 1,200 gal (4.5 m³), including those built to DOT portable or cargo tank specifications, this temperature is assumed to be 115°F (46°C).

(c) For all containers installed underground, this temperature is assumed to be 105°F (41°C).

F-3 Liquid Volume Correction Table.

F-3.1 Correction of Observed Volume to Standard Temperature Condition (60°F and Equilibrium Pressure).

F-3.1.1 The volume of a given quantity of LP-Gas liquid in a container is directly related to its temperature, expanding as temperature increases and contracting as temperature decreases. Standard conditions, often used for weights and measures purposes and, in some cases, to comply with safety regulations, specify correction of the observed volume to what it would be at 60°F.

F-3.1.2 To correct the observed volume to 60°F (16°C), the specific gravity of LP-Gas at 60°F (16°C) in relation to water at 60°F (16°C) (usually referred to as "60°F/60°F"), and its average temperature must be known. The specific gravity normally appears on the shipping papers. The average liquid temperature may be obtained as follows:

(a) Insert a thermometer in a thermometer well in the container into which the liquid has been transferred and read the temperature after the completion of the transfer [*see F-3.1.2(c) as to proper use of a thermometer*].

(b) If the container is not equipped with a well, but is essentially empty of liquid prior to loading, the temperature of the liquid in the container from which liquid is being withdrawn may be used. Otherwise, a thermometer may be inserted in a thermometer well or other temperature sensing device installed in the loading line at a point close to the container being loaded, reading temperatures at intervals during transfer and averaging. [*See F-3.1.2(c)*].

(c) A suitable liquid should be used in thermometer wells to obtain an efficient heat transfer from the LP-Gas liquid in the container to the thermometer bulb. The liquid used should be noncorrosive and should not freeze at the temperatures to which it will be subjected. Water should not be used.

F-3.1.3 The volume observed or measured is corrected to 60°F (16°C) by use of Table F-3.1.3. The column headings, across the top of the tabulation, list the range of specific gravities for the LP-Gases complying with 1-2.1.1. Specific gravities are shown from 0.500 to 0.590 by 0.010 increments, except that special columns are inserted for chemically pure propane, isobutane and normal butane. To obtain a correction factor, follow down the column for the specific gravity of the particular LP-Gas to the factor corresponding with the liquid temperature. Interpolation between the specific gravities and temperatures shown may be used if necessary.

F-3.2 Use of Liquid Volume Correction Factors, Table F-3.1.3.

F-3.2.1 To correct the observed volume in gal for any LP-Gas (the specific gravity and temperature of which is known) to gal at 60°F (16°C), Table F-3.1.3 is used as follows:

(a) Obtain the correction factor for the specific gravity and temperature as described in F-3.1.3.

(b) *Multiply* the gal observed by this correction factor to obtain the gal at 60°F (16°C).

EXAMPLE: A container has in it 4,055 gal of LP-Gas with a specific gravity of 0.560 at a liquid temperature of 75°F. The correction factors in the 0.560

Table F-3.1.3 Liquid Volume Correction Factors.

Observed Temperature Degrees Fahrenheit	0.500	Propane 0.5079	0.510	0.520	0.530	0.540	0.550	0.560	iso-Butane 0.5631	0.570	0.580	n-Butane 0.5844	0.590
SPECIFIC GRAVITIES AT 60°F./60°F. — VOLUME CORRECTION FACTORS													
−50	1.160	1.155	1.153	1.146	1.140	1.133	1.127	1.122	1.120	1.116	1.111	1.108	1.106
−45	1.153	1.148	1.146	1.140	1.134	1.128	1.122	1.117	1.115	1.111	1.106	1.103	1.101
−40	1.147	1.142	1.140	1.134	1.128	1.122	1.117	1.111	1.110	1.106	1.101	1.099	1.097
−35	1.140	1.135	1.134	1.128	1.122	1.116	1.112	1.106	1.105	1.101	1.096	1.094	1.092
−30	1.134	1.129	1.128	1.122	1.116	1.111	1.106	1.101	1.100	1.096	1.092	1.090	1.088
−25	1.127	1.122	1.121	1.115	1.110	1.105	1.100	1.095	1.094	1.091	1.087	1.085	1.083
−20	1.120	1.115	1.114	1.109	1.104	1.099	1.095	1.090	1.089	1.086	1.082	1.080	1.079
−15	1.112	1.109	1.107	1.102	1.097	1.093	1.089	1.084	1.083	1.080	1.077	1.075	1.074
−10	1.105	1.102	1.100	1.095	1.091	1.087	1.083	1.079	1.078	1.075	1.072	1.071	1.069
− 5	1.098	1.094	1.094	1.089	1.085	1.081	1.077	1.074	1.073	1.070	1.067	1.066	1.065
0	1.092	1.088	1.088	1.084	1.080	1.076	1.073	1.069	1.068	1.066	1.063	1.062	1.061
2	1.089	1.086	1.085	1.081	1.077	1.074	1.070	1.067	1.066	1.064	1.061	1.060	1.059
4	1.086	1.083	1.082	1.079	1.075	1.071	1.068	1.065	1.064	1.062	1.059	1.058	1.057
6	1.084	1.080	1.080	1.076	1.072	1.069	1.065	1.062	1.061	1.059	1.057	1.055	1.054
8	1.081	1.078	1.077	1.074	1.070	1.066	1.063	1.060	1.059	1.057	1.055	1.053	1.052
10	1.078	1.075	1.074	1.071	1.067	1.064	1.061	1.058	1.057	1.055	1.053	1.051	1.050
12	1.075	1.072	1.071	1.068	1.064	1.061	1.059	1.056	1.055	1.053	1.051	1.049	1.048
14	1.072	1.070	1.069	1.066	1.062	1.059	1.056	1.053	1.053	1.051	1.049	1.047	1.046
16	1.070	1.067	1.066	1.063	1.060	1.056	1.054	1.051	1.050	1.048	1.046	1.045	1.044
18	1.067	1.065	1.064	1.061	1.057	1.054	1.051	1.049	1.048	1.046	1.044	1.043	1.042
20	1.064	1.062	1.061	1.058	1.054	1.051	1.049	1.046	1.046	1.044	1.042	1.041	1.040
22	1.061	1.059	1.058	1.055	1.052	1.049	1.046	1.044	1.044	1.042	1.040	1.039	1.038
24	1.058	1.056	1.055	1.052	1.049	1.046	1.044	1.042	1.042	1.040	1.038	1.037	1.036
26	1.055	1.053	1.052	1.049	1.047	1.044	1.042	1.039	1.039	1.037	1.036	1.036	1.034
28	1.052	1.050	1.049	1.047	1.044	1.041	1.039	1.037	1.037	1.035	1.034	1.034	1.032
30	1.049	1.047	1.046	1.044	1.041	1.039	1.037	1.035	1.035	1.033	1.032	1.032	1.030
32	1.046	1.044	1.043	1.041	1.038	1.036	1.035	1.033	1.033	1.031	1.030	1.030	1.028
34	1.043	1.041	1.040	1.038	1.036	1.034	1.032	1.031	1.030	1.029	1.028	1.028	1.026
36	1.039	1.038	1.037	1.035	1.033	1.031	1.030	1.028	1.028	1.027	1.025	1.025	1.024
38	1.036	1.035	1.034	1.032	1.031	1.029	1.027	1.026	1.025	1.025	1.023	1.023	1.022
40	1.033	1.032	1.031	1.029	1.028	1.026	1.025	1.024	1.023	1.023	1.021	1.021	1.020
42	1.030	1.029	1.028	1.027	1.025	1.024	1.023	1.022	1.021	1.021	1.019	1.019	1.018
44	1.027	1.026	1.025	1.023	1.022	1.021	1.020	1.019	1.019	1.018	1.017	1.017	1.016
46	1.023	1.022	1.022	1.021	1.020	1.018	1.018	1.017	1.016	1.016	1.015	1.015	1.014
48	1.020	1.019	1.019	1.018	1.017	1.016	1.015	1.014	1.014	1.013	1.013	1.013	1.012
50	1.017	1.016	1.016	1.015	1.014	1.013	1.013	1.012	1.012	1.011	1.011	1.011	1.010
52	1.014	1.013	1.012	1.012	1.011	1.010	1.010	1.009	1.009	1.009	1.009	1.009	1.008
54	1.010	1.010	1.009	1.009	1.008	1.008	1.007	1.007	1.007	1.007	1.006	1.006	1.006
56	1.007	1.007	1.006	1.006	1.005	1.005	1.005	1.005	1.005	1.005	1.004	1.004	1.004
58	1.003	1.003	1.003	1.003	1.003	1.003	1.002	1.002	1.002	1.002	1.002	1.002	1.002
60	1.000	1.000	1.000	1.000	1.000	1.000	1.000	1.000	1.000	1.000	1.000	1.000	1.000
62	0.997	0.997	0.997	0.997	0.997	0.997	0.997	0.998	0.998	0.998	0.998	0.998	0.998
64	0.993	0.993	0.994	0.994	0.994	0.994	0.995	0.995	0.995	0.995	0.996	0.996	0.996
66	0.990	0.990	0.990	0.990	0.991	0.992	0.992	0.993	0.993	0.993	0.993	0.993	0.993
68	0.986	0.986	0.987	0.987	0.988	0.989	0.990	0.990	0.990	0.990	0.991	0.991	0.991
70	0.983	0.983	0.984	0.984	0.985	0.986	0.987	0.988	0.988	0.988	0.989	0.989	0.989
72	0.979	0.980	0.981	0.981	0.982	0.983	0.984	0.985	0.986	0.986	0.987	0.987	0.987
74	0.976	0.976	0.977	0.978	0.980	0.980	0.982	0.983	0.983	0.984	0.985	0.985	0.985
76	0.972	0.973	0.974	0.975	0.977	0.978	0.979	0.980	0.981	0.981	0.982	0.982	0.983
78	0.969	0.970	0.970	0.972	0.974	0.975	0.977	0.978	0.978	0.979	0.980	0.980	0.981
80	0.965	0.967	0.967	0.969	0.971	0.972	0.974	0.975	0.976	0.977	0.978	0.978	0.979
82	0.961	0.963	0.963	0.966	0.968	0.969	0.971	0.972	0.973	0.974	0.976	0.976	0.977
84	0.957	0.959	0.960	0.962	0.965	0.966	0.968	0.970	0.971	0.972	0.974	0.974	0.975
86	0.954	0.956	0.956	0.959	0.961	0.964	0.966	0.967	0.968	0.969	0.971	0.971	0.972
88	0.950	0.952	0.953	0.955	0.958	0.961	0.963	0.965	0.966	0.967	0.969	0.969	0.970
90	0.946	0.949	0.949	0.952	0.955	0.958	0.960	0.962	0.963	0.964	0.967	0.967	0.968
92	0.942	0.945	0.946	0.949	0.952	0.955	0.957	0.959	0.960	0.962	0.964	0.965	0.966
94	0.938	0.941	0.942	0.946	0.949	0.952	0.954	0.957	0.958	0.959	0.962	0.962	0.964
96	0.935	0.938	0.939	0.942	0.946	0.949	0.952	0.954	0.955	0.957	0.959	0.960	0.961
98	0.931	0.934	0.935	0.939	0.943	0.946	0.949	0.952	0.953	0.954	0.957	0.957	0.959
100	0.927	0.930	0.932	0.936	0.940	0.943	0.946	0.949	0.950	0.952	0.954	0.955	0.957
105	0.917	0.920	0.923	0.927	0.931	0.935	0.939	0.943	0.943	0.946	0.949	0.949	0.951
110	0.907	0.911	0.913	0.918	0.923	0.927	0.932	0.936	0.937	0.939	0.943	0.944	0.946
115	0.897	0.902	0.904	0.909	0.915	0.920	0.925	0.930	0.930	0.933	0.937	0.938	0.940
120	0.887	0.892	0.894	0.900	0.907	0.912	0.918	0.923	0.924	0.927	0.931	0.932	0.934
125	0.876	0.881	0.884	0.890	0.898	0.903	0.909	0.916	0.916	0.920	0.925	0.927	0.928
130	0.865	0.871	0.873	0.880	0.888	0.895	0.901	0.908	0.909	0.913	0.918	0.921	0.923
135	0.854	0.861	0.863	0.871	0.879	0.887	0.894	0.901	0.902	0.907	0.912	0.914	0.916
140	0.842	0.850	0.852	0.861	0.870	0.879	0.886	0.893	0.895	0.900	0.905	0.907	0.910

column are 0.980 at 76°F and 0.983 at 74°F, or, interpolating, 0.9815 for 75°F. The volume of liquid at 60°F is 4,055 × 0.9815, or 3980 gal.

F-3.2.2 To determine the volume in gal of a particular LP-Gas at temperature "t" to correspond with a given number of gal at 60°F (16°C), Table F-3.1.3 is used as follows:

(a) Obtain the correction factor for the LP-Gas, using the column for its specific gravity and reading the factor for temperature "t."

(b) *Divide* the number of gal at 60°F (16°C) by this correction factor to obtain the volume at temperature "t."

EXAMPLE: It is desired to pump 800 gal at 60°F into a container. The LP-Gas has a specific gravity of 0.510 and the liquid temperature is 44°F. The correction factor in the 0.510 column for 44°F is 1.025. Volume to be pumped at 44°F is 800 ÷ 1.025 = 780 gal.

F-4 Maximum Liquid Volume Computations.

F-4.1 Maximum Liquid LP-Gas Content of a Container at Any Given Temperature.

F-4.1.1 The maximum liquid LP-Gas content of any container depends upon the size of the container, whether it is installed aboveground or underground, the maximum permitted filling density and the temperature of the liquid [*see Tables 4-5.2.3(a), (b), and (c).*]

F-4.1.2 The maximum volume "V_t" (in percent of container capacity) of an LP-Gas at temperature "t," having a specific gravity "G" and a filling density of "L," is computed by use of the formula:

$$V_t = \frac{L}{G} \div F, \text{ or } V_t = \frac{L}{G \times F} \text{ where:}$$

V_t = percent of container capacity which may be filled with liquid
L = filling density
G = specific gravity of particular LP-Gas
F = correction factor to correct volume at temperature "t" to 60°F (16°C).

EXAMPLE 1: The maximum liquid content, in percent of container capacity, for an aboveground 500 gal water capacity container of an LP-Gas having a specific gravity of 0.550 and at a liquid temperature of 45°F is computed as follows:

From Table 4-5.2.1, L = 0.47, and from Table F-3.1.3, °F = 1.019.

$$\text{Thus } V_{45} = \frac{0.47}{0.550 \times 1.019} = 0.838 \ (83\%), \text{ or } 415 \text{ gallons.}$$

EXAMPLE 2: The maximum liquid content, in percent of container capacity, for an aboveground 30,000 gal water capacity container of LP-Gas having a specific gravity of 0.508 and at a liquid temperature of 80°F is computed as follows:

From Table 4-5.2.1, L = 0.45, and from Table F-3.1.3, °F = 0.967.

$$\text{Thus } V_{80} = \frac{0.45}{0.508 \times 0.967} = 0.915 \text{ (91\%), or 27,300 gallons.}$$

F-4.2 Alternate Method of Filling Containers.

F-4.2.1 Containers equipped only with fixed maximum level gauges or only with variable liquid level gauges, when temperature determinations are not practical, may be filled with either gauge provided the fixed maximum liquid level gauge is installed, or the variable gauge is set, to indicate the volume equal to the maximum permitted filling density as provided in 4-5.3.3(a). This level is computed on the basis that the liquid temperature will be 40°F (4°C) for aboveground containers, or 50°F (10°C) for underground containers.

F-4.2.2 The percentage of container capacity which may be filled with liquid is computed by use of the formula shown in F-4.1.2, substituting the appropriate values as follows:

$$V_t = \frac{L}{G \times F}, \text{ where:}$$

t = the liquid temperature. Assumed to be 40°F (4°C) for aboveground containers or 50°F (10°C) for underground containers.
L = the loading density obtained from Table 4-5.2.1 for:

(a) the specific gravity of the LP-Gas to be contained.

(b) the method of installation, aboveground or underground, and if aboveground, then:

(1) for containers of 1,200 gal (4.5 m³) water capacity or less.

(2) for containers of more than 1,200 gal (4.5 m³) water capacity.

G = the specific gravity of the LP-Gas to be contained.
F = the correction factor. Obtained from Table F-3.1.3, using G and 40°F (4°C) for aboveground containers or 50°F (10°C) for underground containers.

EXAMPLE: The maximum volume of LP-Gas with a specific gravity of 0.550 which may be in a 1,000 gal water capacity aboveground container which is filled by use of a fixed maximum liquid level gauge is computed as follows:

t is 40°F for an aboveground container.
L for 0.550 specific gravity, and an aboveground container of less than 1,200 gal water capacity, from Table 4-5.2.1, is 47 percent.
G is 0.550.
F for 0.550 specific gravity at 40°F from Table F-3.1.3 is 1.025.

$$\text{Thus } V_{40} = \frac{.47}{0.550 \times 1.025} = 0.834 \ (83\%), \text{ or } 830 \text{ gallons.}$$

F-4.2.3   Percentage values, such as in the example in F-4.2.2, are rounded off to the next lower full percentage point, or to 83 percent in this example.

F-4.3 Location of Fixed Maximum Liquid Level Gauges in Containers.

F-4.3.1   Due to the diversity of fixed liquid gauges, and the many sizes [from DOT cylinders to 120,000 gal (454 m^3) ASME vessels] and types (vertical, horizontal, cylindrical and spherical) of containers in which gauges are installed, it is not possible to tabulate the liquid levels such gauges should indicate for the maximum permitted filling densities [*see Table 4-5.2.1 and 4-5.3.3(a)*].

F-4.3.2   The percentage of container capacity which these gauges should indicate is computed by use of the formula in F-4.1.2. The liquid level this gauge should indicate is obtained by applying this percentage to the water capacity of the container in gal [water at 60°F (16°C)], then using the strapping table for the container (obtained from its manufacturer) to determine the liquid level for this gallonage. If such a table is not available, this liquid level is computed from the internal dimensions of the container, using data from engineering handbooks.

F-4.3.3   The formula of F-4.1.2 is used to determine the maximum LP-Gas liquid content of a container to comply with Table 4-5.2.1 and 4-5.3.3(a), as follows:

$$\text{Volumetric percentage, or } V_t = \frac{L}{G \times F}, \text{ and}$$

Volume in Gallons = V_t × Container Gallons Water Capacity, or

Vol. in Gal. at t =

$$\frac{L \text{ (Table 4-5.2.1)} \times \text{Container Gallons Water Capacity}}{G \text{ (Spec. Grav.)} \times F \text{ (For G and at temperature t)}}$$

EXAMPLE 1:   Assume a 100 gal water capacity container for underground storage of propane with a specific gravity of 0.510. From Table 4-5.2.1, L = 46 percent; from 4-5.3.3(a), t = 50°F; and from Table F-3.1.3, °F for 0.510 specific gravity and a temperature of 50°F is 1.016; or

$$\text{Vol. in Gal. at } 50°F = \frac{.46 \times 100}{0.510 \times 1.016} = 88.7 \text{ gallons.}$$

EXAMPLE 2: Assume an 18,000 gal water capacity container for above-ground storage of a mixture with a specific gravity of 0.550. From Table 4-5.2.1, L = 50 percent; from 4-5.3.3(a), t = 40°F; and from Table F-3.1.3, °F for 0.550 specific gravity and 40°F temperature is 1.025; or

$$\text{Vol. in Gal. at } 40°F = \frac{.50 \times 18,000}{0.550 \times 1.025} = 15,950 \text{ gallons.}$$

APPENDIX G
WALL THICKNESS OF COPPER TUBING

This Appendix is not a part of the requirements of this NFPA document, but is included for information purposes only.

Table G-1 Wall Thickness of Copper Tubing
(Specification for Copper Water Tube, ASTM B 88)

Standard Size Inches	Nominal OD Inches	Nominal Wall Thickness Inches	
		Type K	Type L
¼	0.375	0.035	0.030
⅜	0.500	0.049	0.035
½	0.625	0.049	0.040
⅝	0.750	0.049	0.042
¾	0.875	0.065	0.045

Table G-2 Wall Thickness of Copper Tubing
(Specification for Seamless Copper Tube for
Air Conditioning and Refrigeration Field Service, ASTM B 280)

Standard Size Inches	Outside Diam. Inches	Wall Thickness Inches
¼	0.250	0.030
⁵⁄₁₆	0.312	0.032
⅜	0.375	0.032
½	0.500	0.032
⅝	0.625	0.035
¾	0.750	0.042
⅞	0.875	0.045

APPENDIX H
PROCEDURE FOR TORCH FIRE AND HOSE STREAM TESTING OF THERMAL INSULATING SYSTEMS FOR LP-GAS CONTAINERS

This Appendix is not a part of the requirements of this NFPA document, but is included for information purposes only.

This Appendix was added in the 1989 Edition. It was developed by the NFPA Committee on Fire Tests at the request of the Technical Committee on Liquefied Petroleum Gases.

Paragraph 3-10.3.1 provides a performance requirement for container insulation, and Appendix A gives additional information. The Committee recognized that no test methods were in existence for insulation where the critical condition was flame impingement with hose streams, which would be expected to prevent a BLEVE. The Technical Committee has placed this new insulation test procedure in the Appendix so that experience can be gained with it prior to making it a mandatory test procedure.

A. Performance Standard. Thermal protection insulating systems, proposed for use on LP-Gas containers as a means of "Special Protection" under paragraph 3-10.3.1, are required to undergo thermal performance testing as a precondition for acceptance. The intent of this testing procedure is to identify insulation systems which retard or prevent the release of the container's contents in a fire environment of a 50 minute duration; and which will resist a concurrent hose stream of a 10 minute duration.

B. Reference Test Standards. The testing procedure described herein was taken with some modification from segments of the two following test standards:

1. *Code of Federal Regulations* - Title 49, Part 179.105-4, "Thermal Protection."

2. National Fire Code - NFPA 252, *Standard Methods of Fire Tests of Door Assemblies*, Chapter 4, Part 4-3 "Hose Stream Test."

C. Thermal Insulation Test.

1. A torch fire environment shall be created in the following manner:

(i) The source of the simulated torch shall be a hydrocarbon fuel. The flame temperature from the simulated torch shall be 2,200°F ±100°F

(1200°C ± 56°C) throughout the duration of the test. Torch velocities shall be 40 miles per hour (64 km/h) ± 10 miles per hour (16 km/h) throughout the duration of the test.

(ii)   An uninsulated square steel plate with thermal properties equivalent to ASME pressure vessel steel shall be used. The plate dimensions shall be not less than 4 feet by 4 feet (1.2 m x 1.2 m) by nominal ⅝ in. (16 mm) thick. The plate shall be instrumented with not less than 9 thermocouples to record the thermal response of the plate. The thermocouples shall be attached to the surface not exposed to the simulated torch, and shall be divided into 9 equal squares with a thermocouple placed in the center of each square.

(iii)   The steel-plate holder shall be constructed in such a manner that the only heat transfer to the back side of the plate is by heat conduction through the plate and not by other heat paths. The apex of the flame shall be directed at the center of the plate.

(iv)   Before exposure to the torch fire, none of the temperature recording devices shall indicate a plate temperature in excess of 100°F (38°C) or less than 32°F (0°C).

(v)   A minimum of two thermocouples shall indicate 800°F (427°C) in a time of 4.0 ± 0.5 minutes of torch fire exposure.

2. A thermal insulation system shall be tested in the torch fire environment described in paragraph (1) of this section in the following manner:

(i)   The thermal insulation system shall cover one side of a steel plate identical to that used under paragraph C.1.(ii) of this section.

(ii)   The back of the steel plate shall be instrumented with not less than 9 thermocouples placed as described in paragraph C.1.(ii) of this section to record the thermal response of the steel.

(iii)   Before exposure to the torch fire, none of the thermocouples on the thermal insulation system steel plate configuration shall indicate a plate temperature in excess of 100°F (38°C) or less than 32°F (0°C).

(iv)   The entire outside surface of the thermal insulation system shall be exposed to the torch fire environment.

(v)   A torch fire test shall be run for a minimum of 50 minutes. The thermal insulation system shall retard the heat flow to the steel plates so that none of the thermocouples on the uninsulated side of the steel plate indicates a plate temperature in excess of 800°F (427°C).

D. Hose Stream Resistance Test.

1. After 20 minutes exposure to the torch test, the test sample shall be hit with a hose stream concurrently with the torch for a period of 10 minutes. The hose stream test shall be conducted in the following manner:

(i) The stream shall be directed first at the middle and then at all parts of the exposed surface, making changes in direction slowly.

(ii) The hose stream shall be delivered through a 2 ½ inch (64 mm) hose discharging through a National Standard Playpipe of corresponding size equipped with 1 ⅛ in. (29 mm) discharge tip of the standard-taper smooth-bore pattern without shoulder at the orifice. The water pressure at the base of the nozzle and for the duration of the test shall be 30 psi (207 kPa). [Estimated delivery rate is 205 gallons per minute (776 l/min).]

(iii) The tip of the nozzle shall be located 20 ft (6 m) from and on a line normal to the center of the test specimen. If impossible to be so located, the nozzle may be on a line deviating not to exceed 30 degrees from the line normal to the center of the test specimen. When so located the distance from the center shall be less than 20 ft (6 m) by an amount equal to 1 ft (0.3 m) for each 10 degrees of deviation from the normal.

(iv) Subsequent to the application of the hose stream, the torching shall continue until any thermocouple on the uninsulated side of the steel plate indicates a plate temperature in excess of 800°F (427°C).

(v) The thermal insulation system shall be judged to be resistant to the action of the hose stream if the time from initiation of torching for any thermocouple on the uninsulated side of the steel plate to reach in excess of 800°F (427°C) is 50 minutes or greater.

(vi) One (1) successful combination torch fire and hose stream test shall be required for certification.

APPENDIX I
CONTAINER SPACING

This Appendix is not a part of the requirements of this NFPA document, but is included for information purposes only.

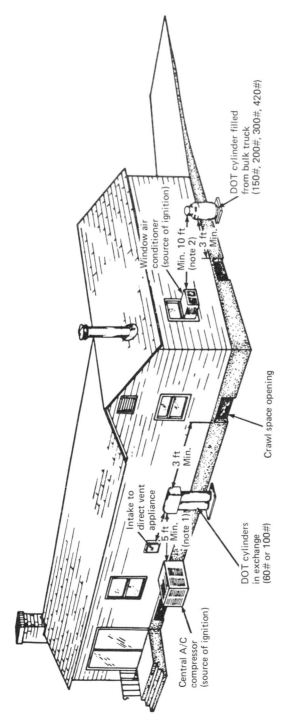

Central A/C compressor (source of ignition)

Intake to direct vent appliance

5 ft Min. (note 1)

DOT cylinders in exchange (60# or 100#)

3 ft Min.

Crawl space opening

Window air conditioner (source of ignition)

Min. 10 ft (note 2)

3 ft Min.

DOT cylinder filled from bulk truck (150#, 200#, 300#, 420#)

NOTE 1 : 5 ft minimum between relief valve discharge and external source of ignition (air conditioner), direct vent, or mechanical ventilation system (attic fan).

NOTE 2 : If the DOT cylinder is filled on-site from a bulk truck, the filling connection and vent valve must be at least 10 ft from any external source of ignition, direct vent, or mechanical ventilation system.

(For SI Units: 1 ft = 0.3048 m)

Figure I-1 DOT Cylinders.

(This figure for illustrative purposes only; text shall govern.)

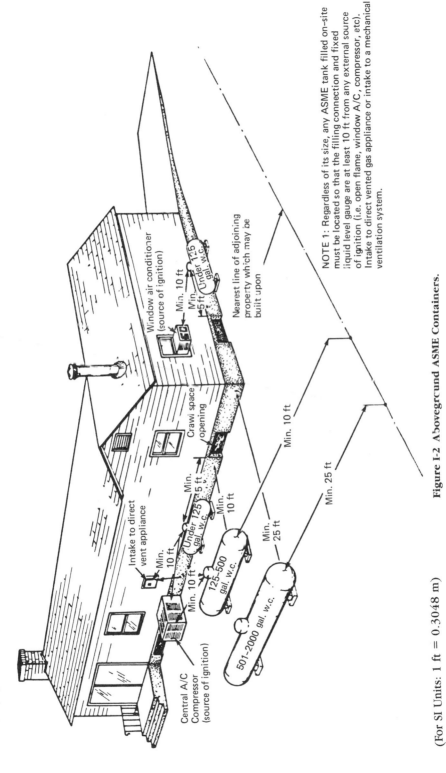

Window air conditioner
(source of ignition)

Min. 10 ft

Min. 5 ft   Under 125 gal. w.c.

Nearest line of adjoining property which may be built upon

Crawl space opening

Intake to direct vent appliance

Min. 5 ft   Under 125 gal. w.c.

Min. 10 ft

Min. 10 ft

Min. 10 ft   Under 125 gal. w.c.

Min. 10 ft

125-500 gal. w.c.

Min. 25 ft

Min. 25 ft

501-2000 gal. w.c.

Central A/C Compressor (source of ignition)

NOTE 1: Regardless of its size, any ASME tank filled on-site must be located so that the filling connection and fixed liquid level gauge are at least 10 ft from any external source of ignition (i.e. open flame, window A/C, compressor, etc). Intake to direct vented gas appliance or intake to a mechanical ventilation system.

Figure I-2 Aboveground ASME Containers.

(This figure for illustrative purposes only; text shall govern.)

(For SI Units: 1 ft = 0.3048 m)

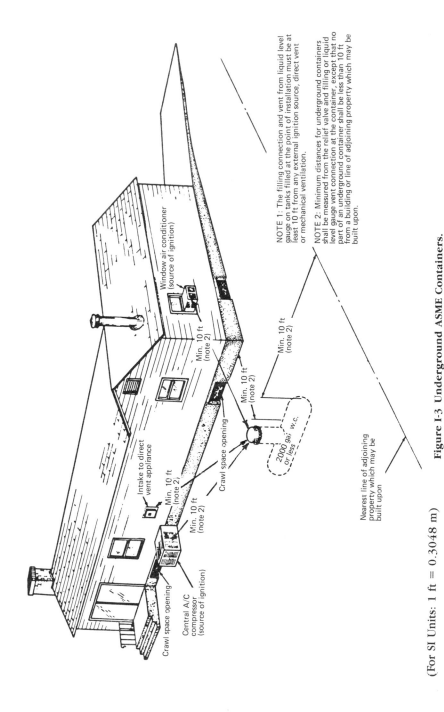

Window air conditioner (source of ignition)

Min. 10 ft (note 2)

Intake to direct vent appliance

Min. 10 ft (note 2)

Min. 10 ft (note 2)

Crawl space opening

Crawl space opening

Central A/C compressor (source of ignition)

Min. 10 ft (note 2)

Min. 10 ft (note 2)

2000 gal. or less w.c.

Nearest line of adjoining property which may be built upon

NOTE 1: The filling connection and vent from liquid level gauge on tanks filled at the point of installation must be at least 10 ft from any external ignition source, direct vent or mechanical ventilation.

NOTE 2: Minimum distances for underground containers shall be measured from the relief valve and filling or liquid level gauge vent connection at the container, except that no part of an underground container shall be less than 10 ft from a building or line of adjoining property which may be built upon.

(For SI Units: 1 ft = 0.3048 m)

Figure I-3 Underground ASME Containers.

(This figure for illustrative purposes only; text shall govern.)

APPENDIX J
REFERENCED PUBLICATIONS

J-1 The following documents or portions thereof are referenced within this standard for informational purposes only and thus are not considered part of the requirements of this document. The edition indicated for each reference is the current edition as of the date of the NFPA issuance of this document.

J-1.1 NFPA Publications. National Fire Protection Association, Batterymarch Park, Quincy, MA 02269.

NFPA 77-1988, *Recommended Practice on Static Electricity*

NFPA 78-1986, *Lightning Protection Code*

J-1.2 ASCE Publication. American Society of Civil Engineers, United Engineering Center, 345 East 47th St., New York, NY 10017.

ASCE *Manual and Report on Engineering Practice No. 56.*

J-1.3 ASTM Publication. American Society for Testing and Materials, 1916 Race St., Philadelphia, PA 19103.

ASTM D 1835-1982, *Specification for Liquefied Petroleum (LP) Gases.*

J-1.4 CGA Publication. Compressed Gas Association, Inc., 1235 Jefferson Davis Highway, Arlington, VA 22202.

CGA Pamphlet C-6-1984, *Standards for Visual Inspection of Steel Compressed Gas Cylinders.*

J-1.5 Federal Publications. National Technical Information Service, U.S. Dept. of Commerce, Springfield, VA 22161.

BERC/RI-77/1, September 1977, *New Look at Odorization Levels for Propane Gas*, United States Energy Research and Development Administration, Technical Information Center.

Code of Federal Regulations, Title 49, Part 195, Transportation of Hazardous Liquids by Pipeline.

J-1.6 GPA Publication. Gas Processors Association, 1812 First National Bank Bldg., Tulsa, OK 74103.

Liquefied Petroleum Gas Specifications for Test Methods Standard 2140.

J-1.7 NPGA Publications. National Propane Gas Association, 1301 W. 22nd St., Oak Brook, IL 60521.

NPGA 122-1970, *Recommendations for Prevention of Ammonia Contamination of LP-Gas.*

NPGA Safety Bulletin 118-1979, *Recommended Procedures for Visual Inspection and Requalification of DOT (ICC) Cylinders in LP-Gas Service.*

SUPPLEMENT—
GUIDELINES FOR CONDUCTING A FIRESAFETY ANALYSIS

This supplement is not a part of the Code text of this Handbook, but is included for information purposes only. It is printed in black for ease of reading.

The following material was prepared as a Safety Bulletin by the Safety Committee of the National Propane Gas Association to assist in conducting the firesafety analysis that is required in the section on fire protection (3-10) of NFPA 58. The section on fire protection was first introduced in the 1976 Edition of NFPA 58. There was confusion in the minds of both enforcement officials and the LP-Gas industry on how to conduct the safety analysis. The bulletin provides the information needed to conduct the analysis in a clear, usable form.

Special thanks are extended to the National Propane Gas Association for granting permission to reprint this material.

Introduction

NFPA 58 includes provisions in paragraph 3-10.2.3 for fire protection to augment the leak control and ignition source control provisions of the standard. A firesafety analysis may be required for installations having storage containers with an aggregate water capacity of more than 4000 gallons subject to exposure from a single fire. These guidelines are presented to aid you in conducting such a firesafety analysis.

In spite of efforts to eliminate hazards within an LP-Gas facility, prudent management must consider that accidents may occur and take reasonable steps to control them. The safety of employees and customers within the facility, its neighbors, and emergency personnel are of paramount concern. Conservation of the physical plant is also important - especially as it relates to the safety of neighbors and protection of their property.

Hazard Characterization

LP-Gas is stored, transported and handled in closed containers and piping until ultimately used in consuming equipment - usually by burning. An emergency is created whenever LP-Gas is released in other than the intended manner.

LP-Gas is present in containers and equipment in both liquid and gaseous (vapor) form. It can escape in either form. If it escapes in liquid form, it will vaporize rapidly to the gaseous form.

When escaping gas mixes with air, a flammable mixture will form. If ignited, it will result in a fire. It could also result in an explosion - depending upon how large the flammable mixture is or whether the mixture is confined inside a structure.

If fire is impinging on a container, the possibility of an explosion (a Boiling Liquid-Expanding Vapor Explosion, or BLEVE) exists. A BLEVE will produce a fireball and shock wave, the degree depending upon how much liquid LP-Gas is inside the container when it ruptures. It may also produce flying fragments in the form of container pieces.

Because ignition sources are controlled to a substantial degree in LP-Gas facilities, there is often a time interval between LP-Gas escape and ignition. There is also a time lag between flame contact with the unwetted portion of a container and container failure. This time can be used to advantage in controlling the emergency given adequate resources, planning, and training.

Emergency Control Objectives

The emergency control objectives consist of three operational phases:

1. Stopping or slowing down the rate of LP-Gas release.

2. Dissipating LP-Gas vapors and/or presenting flammable gas-air mixtures from reaching ignition sources and entering structures.

3. Keeping fire exposed containers and equipment cool.

Attaining these objectives requires the application of both equipment arrangement and human performance.

The Firesafety Analysis

An adequate degree of emergency control is subject to widely varying circumstances specific to each facility. Because of the large number of variables each cannot be specifically addressed in NFPA 58. Furthermore, while it is possible that a facility could be made entirely self-sufficient in fire protection, it is seldom necessary or cost-efficient to do so. These factors mean that judgement must be brought to bear and agreements reached with local authorities. While parties involved share a common goal, there may be differences in views as to how best to achieve the goal and these views must be resolved. These guidelines are offered to aid in such decision-making.

A. No Fire Protection Needed

There may be situations where no fire protection is needed beyond those provided in the facility in compliance with specific provisions in NFPA 58. A candidate for such a decision would be a well isolated facility.

If an adequate water supply is not available, agreement should be reached with public safety authorities to limit their emergency forces to control of onlookers.

B. Fire Protection Needed

Unless the circumstances described in item A exist, it must be recognized that a degree of exposure exists to the facility, its employees, the persons and property of neighbors, and that fire protection is needed beyond that specifically required by NFPA 58. This fire protection could be provided by a plant fire brigade and/or fire department.

Such a facility will undoubtedly be within an area served by one or more emergency services — e.g., public fire and police departments and ambulance services. Facility management has a right to expect assistance from these sources. Management has a duty to obtain this assistance without requiring the emergency personnel to accept undue risks. The resolution of this risk factor is a key element in the firesafety analysis process.

If we study the previously cited Emergency Control Objectives, it will be evident that a fire department is suited to accomplish Objectives 2 and 3. Water in the proper form can disperse gas clouds and cool containers and equipment. Aside from its rescue function, a fire department is essentially a man-machine system designed to apply water.

These are the reasons why NFPA 58 states that the first consideration in the analysis should be directed at the fire department's capabilities.

C. Fire Department Capability

The most important step in evaluating the fire department's adequacy is to recognize that the facility management and fire department officers must communicate with each other at all stages of the analysis. Neither should make any assumptions about the other's capability. Often the initial tendency of the facility management is to overestimate the fire department's capability. Conversely, often the initial tendency of the fire department is to underestimate its own capability.

All to often, the basic reason why a fire department will underestimate its capability is due to lack of knowledge. Most are accustomed to fighting fires in buildings and Class A and B combustibles in which the tactic consists of "attack-and-extinguish". The "control-and-not-extinguish" tactic for LP-Gas (or any other flammable gas) incidents are alien to their basic training. Furthermore, LP-Gas incidents are infrequent and fire departments don't get much on-the-job training. When the need for fast action and the potential severity of an incident are added to these factors their position is understandable (accounts of severe incidents are readily available — accounts of those readily handled are not).

The answer to this problem is adequate education and training. It will be necessary during this training to consider the particular character of the facility and the general knowledge of the fire department.

In those instances where the general knowledge is not present, the facility management should do all that it can to encourage the department to obtain it. If feasible, key fire department personnel should be directed to attend a school that offers "hands-on" field exercises in its curriculums. This may require financial assistance. Facility management should also be prepared to provide instructional material which includes visual aids. Good detailed accounts of both successfully handled, and not so successfully handled, incidents involving facilities similar to the one being analyzed have proven to be useful tools.

In any event, it must be recognized that it will be difficult, if not impossible, to reach a sound agreement on the fire department's capability to handle an incident at a specific facility until this education and training has been achieved.

D. Analysis Factors

Once a basic appreciation for the problems has been obtained, the analysis can proceed on a sound basis. As noted in NFPA 58, the analysis should include the following factors:

1. Local conditions of hazards within the container site.

2. Exposure to and from other properties.

3. The available water supply.

4. The probable effectiveness of plant fire brigades.

5. The time of response and probable effectiveness of the fire department(s).

Aspects relating to fire department capability include the following (remember, we are considering a fire department as a man-machine water distribution system in the analysis):

1. Local Conditions of Hazards Within the Facility

Experience has shown that by far the most frequent cause of accidents leading to leaks and fires in an LP-gas facility are associated with liquid transfer operations. The character and arrangement of these are important factors.

The number of simultaneous operations involving small containers, transports, delivery vehicles, and tank cars should be identified and located with respect to each other and to the storage containers and facility structures. The fewer such operations and the greater the distance between them, the greater the fire department's chances for successful containment.

Other features pertinent to access by fire apparatus and personnel — e.g., fence gates, roadways — should be noted. The position of the long axis of storage containers and cargo vehicles is important in this context because apparatus and personnel should be positioned to the sides of such containers as much as possible.

The presence of any overhead power lines should be noted.

If tank cars are involved, it is important to note that they are all insulated and do not present immediate BLEVE hazards.

Features for leak shutdown should be noted. Consider the (automatic) leak control provisions of the recently available Emergency Shut-Off Valves (ESVs).

2. Exposure to and from Other Properties

If the facility is located in what NFPA 58 refers to as a "heavily populated or congested area," the fire department will tend to place less reliance on their capability. This is so, even though their tactics would remain the same. The consequences of an unsuccessful operation could be more serious.

Facility management should recognize that vulnerable locations for property damage and/or loss of life are prime candidates for Special Protection (*see definition in the standard*). It should be recognized that pieces of large storage and cargo containers that BLEVE, may travel 1000-2000 feet and occasionally 2500-3000 feet.

3. Available Water Supply

One of the major points of consideration in this analysis will undoubtedly be the water supply. The time required to make it available, quantity, and reliability are the essential factors.

The first arriving fire company may find either unignited escaping gas or burning LP-Gas. If ignition has not occurred, its job is to try to prevent ignition and the migration of gas into an enclosed structure. A water fog is generally used to dissipate LP-gas vapors. While the quantity of water needed will vary greatly with the size of the leak and its location, a leak that can be controlled with manpower available at this stage cannot be very large. If a large leak has occurred, or is in progress, the gas may have spread to outside the facility by the time the fire department arrives and could be beyond currently available control measures. Water would have to be supplied from a source not in the vicinity of the facility. Fortunately, experience shows that the likelihood of a leak large enough to escape beyond the facility boundaries is of a lesser degree at facilities complying with the leak control provisions in NFPA 58. Ignited escaping gas that is feasible to control does not pose a severe hazard to the facility, its neighbors, or the emergency personnel, unless the fire impinges on containers which do not have Special Protection.

Obviously, LP-Gas facilities may have conventional fires in offices, shops, garages, and warehouses. Because these present lesser control problems, a water supply designed for the hazards in this analysis will more than suffice for the ordinary hazards.

If a fire is impinging on containers not provided with Special Protection, a mot severe hazard is presented. The larger the container, the more severe the BLEVE hazard. Therefore, the water supply needed is determined by the magnitude of the BLEVE hazard.

The quantity of water needed for adequate container cooling can be determined by the container area directly contacted by flames. Considerable test work indicates that around 0.20 to 0.25 gpm per sq ft of container area directly contacted by flames is needed. Based upon reports of fire exposing 30,000 gallon containers, exposure to about ⅓ of the area (700 sq ft) is not unusual. On this basis, the water needed would be about 140 to 175 gpm per each container exposed to a single fire.

A more important consideration exists in the initial stages of fire department control activity. Experience reveals rather conclusively that a container is in danger of a BLEVE after about 10 minutes of intense flame impingment on the unwetted portion of the shell. This time span will often coincide rather closely with the time it takes the fire department to get streams into operation. The instant the initial stream of water contacts the container is usually the most dangerous moment in the entire operation. Therefore, those manning the nozzle must do so from a reasonably safe position. Unless protective structures (i.e., buildings, unexposed storage tanks, tank cars, cargo vehicles, strongwalls, etc.) are available, then maintaining suitable distance becomes most important. Firefighters should be able to position a nozzle at least 50 ft away, preferably more.

The hydraulics of fire streams are such that a discharge of 250-500 gpm is needed to carry an effective stream 50-100 ft. Therefore, strictly from a cooling standpoint, this consideration will result in roughly twice the quantity needed.

More than 1000 gallons cannot be carried to the facility by the first arriving company. This 1000 gallons of water is only 2-4 minutes of operation and usually there is little prospect of more arriving before the first unit runs out. Therefore, it is necessary for a water supply to be located at or very near the facility. This can be obtained from a body of water or a reliable public or private water system and used during the time necessary to permit the transportation of water to the site.

Fire departments have demonstrated their ability to transport very large quantities of water to LP-Gas facilities. However, it takes time to set up such a system. As a minimum, a one-half to one hour supply should be available at or very near the facility.

In the most severe exposure locations, of course, it is likely that a public water system is in place. This is also the case in the larger industrial locations — often from a private water system. In such circumstances, the problem is one of being able to provide a suitable rate of flow.

Note any features present that relate to water supply in your analysis. The fire department and water department are in a much better position than facility management to evaluate this factor.

4. Plant Fire Brigade

This term is used in NFPA 58 in the context of what any plant employee can do to accomplish the Emergency Control Objectives. In any facility in which the storage and distribution of LP-Gas is the primary purpose, the employee's role reflects primarily Obejective 1 and 2. Usually the number of employees in these facilities is small and totally inadequate to accomplish Objective 3. This must be accomplished through use of fire department equipment and techniques. (In this respect, the facility management should avoid providing connected hoses, nozzles, etc. at the facility. The fire department may not trust their condition and probably will not use them.)

With respect to Objective 1 (stopping or slowing down the rate of LP-Gas release), facility employees are in a far better position to accomplish this than the fire department or anyone else. This should be their primary activity.

With respect to Objective 2 (preventing flammable gas-air mixtures from reaching ignition sources and entering structures), employees can shut down equipment which constitute ignition sources, close doors and windows, etc.

Every facility should have a written emergency plan which includes activities to accomplish Objectives 1 and 2. An evaluation of this emergency plan should be part of the fire analysis. (*See NPGA Bulletin "Guidelines for Developing Plant Emergency Procedures."*)

If Special Protection of a w.ter system type is present, facility employees could be involved with Objective 3 (keeping fire exposed containers and equipment cool).

Some large industrial facilities do maintain highly organized fire brigades large enough to address Objective 3. They are essentially fire departments.

5. Time of Response and Probable Effectiveness of Fire Department

Several aspects of fire department effectiveness have been addressed in the preceding four factors. There are others that must also be considered in the analysis.

Response time is very important and the shorter the response, the better. The importance of 10 minutes has been discussed. If this is exceeded, this alone can result in a decision that the fire department cannot effectively handle the emergency.

Response time from the receipt of an alarm to the first control activity can be predetermined by the fire department. A key component, which cannot be determined by the fire department alone, is the time between start of the emergency and receipt of the alarm. Delayed alarms are a major problem and a common occurrence in the experience of all fire departments. They will carefully evaluate this with respect to the facility.

LP-Gas facility management should emphasize to their employees the importance of prompt notification of the fire department. The written emergency plan should require prompt notification of any fire except the most minor unignited leak situations. Late notification could result in the fire department not effectively handling the emergency.

It must also be recognized that delayed alarms can occur in spite of the best efforts of facility employees. As has been noted earlier, distribution facilities do not have many employees. A leak may occur where no one is present or an employee who is in the vicinity may be immobilized as a consequence of the leak. Also, facilities do not usually operate 24 hours a day; however, accidents rarely occur during non-operating periods.

The telephone may not be a reliable alarm communicator. Telephone lines may not operate because of overload, physical damage, or they may be slow. In some cases, radios or special alarm circuits may be needed. The exact alarm procedure should be written in the emergency plan.

The nature of the fire department equipment and the manning of it is very important, especially for the first alarm response. Except in heavily populated areas, one pumper and 2 or 3 men is all that can be expected initially. Considering the 250-500 gpm needed and the 50-100 ft range discussed earlier (*see "Available Water Supply"*) it could be very difficult for the first crew to accomplish much using only the equipment they have hauled to the scene.

Conclusion

It must be concluded that considerable study is needed to evaluate the capability of a fire department. In many instances the capability may be marginal at best. Nevertheless, the fire department remains the most flexible man-machine system possible to accomplish Objective 3 of the Emergency Control Objectives.

If the analysis initially reveals fire department control inadequacies, the next most frequent step is for the fire officer to recommend Special Protection. NFPA 58 is used to support such a recommendation. However, NFPA 58 stipulates that two conditions must be met — namely, that a serious hazard must exist and that the fire department is incapable — before Special Protection is required.

Even if the hazard can be shown to be serious (here an analysis of "Exposure to and from Other Properties" is a key), the next step should be to see if measures can be taken to upgrade the fire department.

Measures can include: provision of one or two fixed monitors arranged for the fire department to hook up a pumper to (as a substitute for having to get hose streams into service[1]); upgrading of the public water supply in the vicinity (often this is long overdue anyway as a result of construction of other facilities in the area); and special alarm systems to assure prompt response.

Often, partial Special Protection can change the picture drastically. For example, if a storage container is insulated and the fire department must only address a cargo vehicle, the initial response problem is greatly simplified. This can also be accomplished by providing more space between storage containers and cargo vehicles.

The most important single step is for the facility management and the fire department to work closely together in an atmosphere of cooperation. While vital to a sound analysis, it should be an ongoing practice. The personnel in both organizations are subject to frequent change and this relationship must be sustained for the good of the facility, its neighbors, the fire department, and the community at large.

Brief Outline for Conducting A Firesafety Analysis

The standard requires fire protection for an installation having storage containers with an aggregate water capacity of more than 4000 gallons subject to exposure from a single fire, unless a firesafety analysis indicates a serious hazard does not exist. It also provides for alternate ways of protecting the installation.

I.   A firesafety analysis considers the following:

1.   Local conditions of hazards within the container site.

---

[1]This could also reduce water demand by reducing the stream range needed.

2. Exposure to and from other properties.

3. The available water supply.

4. The probable effectiveness of plant fire brigades.

5. The time of response and probable effectiveness of the fire department.

II. Prepare a plot plan to scale showing and locating:

1. The property lines.

2. Buildings, storage containers, loading and unloading sites, cylinder filling and storage areas, etc.

3. Fence, gates, rail sidings, vehicle parking areas and spacing, and roadways.

4. Buildings and other exposures on adjacent properties.

5. Fire hydrants or nearest water supply.

6. Existing protection required by building or other codes.

III. Determine water supply available, the flow quantity, the location and accessibility.

IV. Determine the effectiveness of the plant fire brigade or plant emergency procedures.

1. Emergency shut-off valves at loading and unloading locations.

2. Emergency shut-off valves for cylinder filling operations.

3. Product venting procedures and facilities.

4. Reliability and speed of communications within the plant area and with Public Emergency Services (fire, police, and medical, etc.)

5. Shut-offs for electric, gas pilots, and other ignition sources.

V. Determine the time of response and probable effectiveness of the local fire department.

VI. Take the information you have gathered, analyze it, and make a tentative judgement of your own. Then take it to the fire department and initiate dialogue.

INDEX